THE CHEMISTRY OF NONAQUEOUS SOLVENTS

Volume IV

SOLUTION PHENOMENA AND APROTIC SOLVENTS

Contributors

BARBARA J. BARKER

E. C. BAUGHAN

JOSEPH A. CARUSO

W. H. LEE

ANN T. LEMLEY

JUKKA MARTINMAA

JOHN H. ROBERTS

MICHEL RUMEAU

THE CHEMISTRY OF NONAQUEOUS SOLVENTS

Edited by J. J. LAGOWSKI

DEPARTMENT OF CHEMISTRY
THE UNIVERSITY OF TEXAS AT AUSTIN
AUSTIN, TEXAS

Volume IV

SOLUTION PHENOMENA AND APROTIC SOLVENTS

1976

ACADEMIC PRESS New York San Francisco London
A Subsidiary of Harcourt Brace Jovanovich, Publishers

ACADEMIC PRESS, INC.
111 Fifth Avenue, New York, New York 10003

United Kingdom Edition published by
ACADEMIC PRESS, INC. (LONDON) LTD.
24/28 Oval Road, London NW1

Library of Congress Cataloging in Publication Data

Lagowski, J J ed.
The chemistry of non-aqueous solvents.

Includes bibliographies.
CONTENTS.—v. 1. Principles and techniques.—v. 2. Acidic and basic solvents.—v. 3. Inert, aprotic, and acidic solvents.—v. 4. Solution phenomena and aprotic solvents.
1. Nonaqueous solvents. I. Title.
TP247.5.L3 660.2′9′482 66-16441
ISBN 0–12–433804–6

PRINTED IN THE UNITED STATES OF AMERICA

Contents

4. Tetramethylurea

BARBARA J. BARKER and JOSEPH A. CARUSO

5. Inorganic Acid Chlorides of High Dielectric Constant (with Special Reference to Antimony Trichloride)

E. C. BAUGHAN

6. Cyclic Carbonates

W. H. LEE

List of Contributors

Numbers in parentheses indicate the pages on which the authors' contributions begin.

BARBARA J. BARKER, Department of Chemistry, Hope College, Holland, Michigan (109)

E. C. BAUGHAN, Department of Chemistry and Metallurgy, Royal Military College of Science, Shrivenham, near Swindon, Wilts., England (129)

JOSEPH A. CARUSO, Department of Chemistry, University of Cincinnati, Cincinnati, Ohio (109)

W. H. LEE, Department of Chemistry, University of Surrey, Guildford, Surrey, England (167)

ANN T. LEMLEY,* Department of Chemistry, Cornell University, Ithaca, New York (19)

JUKKA MARTINMAA, Department of Wood and Polymer Chemistry, University of Helsinki, Helsinki, Finland (247)

JOHN H. ROBERTS, Department of Chemistry, The University of Texas, Austin, Texas (1)

MICHEL RUMEAU, Faculté des Sciences et des Techniques, Centre Universitaire de Savoie, Chambery, France (75)

* Present address: Department of Applied and Engineering Physics, Cornell University, Ithaca, New York.

Preface

The contributions to Volume IV of this treatise complement different parts of the first three volumes. The first three chapters—Conductivity in Nonaqueous Solvents; Hydrogen Bonding Phenomena; and Redox Systems in Nonaqueous Solvents—are a continuation of the themes developed in Volume I in which the discussion of phenomena or techniques stands apart from the nature of the solvent although solvent effects are important and are discussed. The remaining chapters are critical reviews of specific aprotic solvents and, hence, can be considered as an extension of a part of Volume III, e.g., aprotic solvents.

The cooperation of the staff of Academic Press in many diverse areas is gratefully acknowledged as is the effort expended by the authors in meeting the necessary deadlines. I should also like to acknowledge the help of Ms. R. Schall who assisted in numerous ways in the preparation of this volume.

J. J. LAGOWSKI

Contents of Previous Volumes

VOLUME I PRINCIPLES AND TECHNIQUES

VOLUME II ACIDIC AND BASIC SOLVENTS

VOLUME III INERT, APROTIC, AND ACIDIC SOLVENTS

~ 1 ~

Conductivity in Nonaqueous Solvents

JOHN H. ROBERTS

Department of Chemistry
The University of Texas, Austin, Texas

I. INTRODUCTION

Interest in the nature of electrolytic solutions has been of great importance historically in the development of presently held concepts of the nature of ionic compounds and their physical chemistry and electrochemistry. Observation, understanding, and description of electrolytic conductivity were particularly significant for the early development of solution theory and today electrolytic conductivity remains one of the primary investigatory tools for the study of electrolytic solutions. Numerous experimental techniques have been developed to determine the mobilities of ions in solution and the fraction free to conduct. This in turn allows calculation of thermodynamic quantities such as association constants.

Parallel development of the theory of conductivity has resulted in a hydrodynamic model for solutions which is widely used and accepted in many fields of science. The ease of mathematical calculation brought about by the development of large computers has allowed an increasingly better fit of precise experimental data to theoretical expectations. The latest forms of the most widely used conductivity equations now contain many higher terms. New developments appear in the literature frequently and on many points of interest there is still controversy.

Conductometric studies have also been of great importance in elucidating the nature of phenomena in nonaqueous solvents. Unanticipated behavior, in terms of what one would expect for aqueous solutions, is more often the rule than the exception. After discussing the theory of conductivity and elementary experimental considerations some of the interesting recent research in nonaqueous solvents will be discussed.

II. Theory of Electrical Conductivity

A. Definition of Terms

According to Ohm's Law the current I, in amperes, flowing through a conductor is proportional to the electromotive force E, in volts, and is inversely proportional to the resistance of the conductor R, in ohms.

$$I = \frac{E}{R} \tag{1}$$

The resistance R depends on the quantity and shape of the material. For a material of uniform cross section and of area a cm^2 and length l cm we have

$$R = \frac{E}{I} = \frac{rl}{a} \tag{2}$$

where r is the specific resistance. The specific conductivity L is defined as the reciprocal of r.

$$L = \frac{K}{R} \tag{3}$$

The cell constant K depends on the size, shape, and surface of the electrodes of the conductivity cell and on the distance between them.

For a solution of an electrolyte the specific conductivity depends on the ions present, and therefore it is useful to consider the conductivity per unit of concentration Λ, the equivalent conductivity

$$\Lambda = \frac{10^3 L}{C} \tag{4}$$

where C is the concentration in gram equivalents per liter.

B. Fundamental Conductivity Equation

The equivalent conductivity is proportional to the current which is carried through the solution in the conductivity cell. Since the current is carried only by the ions of the dissolved electrolyte it is also necessary to consider factors which affect ion transport. Thus,

$$\Lambda = (\text{current carried by positive ions}) + (\text{current carried by negative ions}) \tag{5}$$

Since Faraday's law states that one gram equivalent weight of a substance is discharged at each electrode by 96,487 ($\mathscr{F}$) coulombs of electricity passed through an electrolytic solution and current is defined as coulombs per seconds, Eq. 5 may be stated in terms of equivalents as

$$\Lambda = \mathscr{F} c^+ m^+ + \mathscr{F} c^- m^- \tag{6}$$

where c^+ and c^- are the numbers of positive and negative ions per equivalent of solute in the conductivity cell and m^+ and m^- are the mobilities of the respective ions. For a 1:1 electrolyte in solution $c^+ = c^- = \alpha$, the number of equivalents of either ion per equivalent of solute, and so

$$\Lambda = \mathscr{F} \alpha (m^+ + m^-) \tag{7}$$

This fundamental conductivity equation is a concise statement of the source of conductivity of electrolytic solutions, namely, that the conductivity is a function of the number of ions and their mobilities. Considerations of the factors which affect these two variables have led to the development of a number of conductivity equations which will subsequently be discussed.

C. Factors Affecting the Mobility of Ions

In an infinitely dilute solution the ions are far apart so the only hindrance to their motion toward the electrodes is the friction of their passage through the solvent. Consequently the mobilities should remain constant, and

$$m^{\pm} = m^{\pm 0}$$

where $m^{\pm 0}$ is the mobility of the ion at infinite dilution.

The properties of electrolytic solutions result from the interaction of electrostatic forces, which impels the ions toward a definite arrangement, and thermal motion, which tends to produce random orientations of the ions and solvent molecules. At significant concentrations imbalances in these forces occur, so that

$$m^{\pm} \neq m^{\pm 0}$$

and a description of the velocity of an ion becomes quite complex.

If one views an individual ion in a time-averaged environment, the ion will be surrounded more by ions of the opposite charge than of like charge, simply owing to electrostatic attraction. This environment is referred to as the *ionic atmosphere*. Thermodynamically the result is reflected as a reduction in the activity coefficients of the ions. The ionic atmosphere has spherical symmetry about the central ion until it is perturbed by an external force such as an applied electric field. Then two concentration-dependent forces can be described, both of which decrease ionic mobilities with increasing concentration. These are the *electrophoretic effect* and the *relaxation effect*.

When an external electric field is applied, ions of opposite charge tend to move in opposite directions. An individual ion becomes displaced from the center of its ionic atmosphere and the previously spherical ionic atmosphere becomes egg-shaped (Fig. 1). More importantly the center of charge of the drifting ion and the center of charge of the now asymmetrical, oppositely charged ionic atmosphere no longer coincide. As the ion drifts, the ionic atmosphere behind it relaxes continuously and is continuously built up in front of the ion. The center of charge of the *relaxation field* of the ionic atmosphere and the ion lay on the path made by the moving ion, with the charge center of the relaxation field considerably behind that of the ion. Consequently the relaxation field acts in opposition to the externally applied field, which is

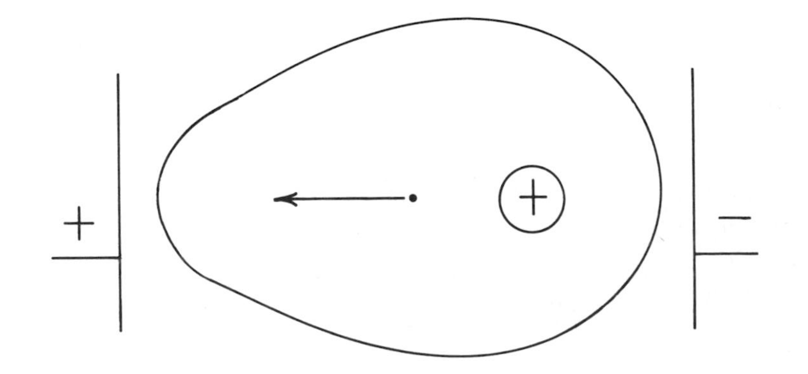

FIG. 1. Egg-shaped ionic atmosphere.

the driving force causing the ion to move. Thus, the velocity of the ion is retarded by the relaxation field of its own egg-shaped ionic atmosphere.

The second major effect results because of the action of the external field on the ionic atmosphere of an ion. This force moves the ionic atmosphere in the direction opposite to the motion of the central ion and so is also a braking effect on the central ion. Part of the effect is due to the solvent molecules associated with the entire ionic atmosphere, the size of which is about 100 Å diameter and allows a comparison with the description of motion of colloidal particles in an electric field. Therefore, the motion of the aggregate comprising the ionic atmosphere and its associated solvent molecules is referred to as the electrophoretic effect.

D. Conductivity Equations

Both the relaxation effect and the electrophoretic effect must be taken into account in the formulation of a conductivity equation. In the limit of an infinitely dilute solution the equation of Onsager[1,2] has been shown to be correct for strong electrolytes. Expressed in linear form it is

$$\Lambda = \Lambda_0 - (a\Lambda_0 + b)\, C^{1/2} \tag{8}$$

For 1 : 1 electrolytes

$$a = \frac{8.204 \times 10^5}{(DT)^{3/2}} \qquad b = \frac{82.43}{\eta(DT)^{1/2}}$$

where D is the dielectric constant, T is the absolute temperature, and η is the viscosity. This expression reflects the fact that the size of the ionic atmosphere increases as a function of the square root of ionic strength, for which concentration is an adequate measure in extremely dilute solutions. Note also that the electrophoretic correction (b) for movement of the ionic atmosphere and associated solvent is viscosity-dependent.

For many nonaqueous solutions, especially with solvents of low dielectric constant, the ionic strength cannot be adequately represented by the analytical concentration. With regard to liquid ammonia, for example, with $D = 23$–28 for the usual working range of temperature, this situation is summed up by the statement that in liquid ammonia there are no strong electrolytes. Departures from ideal behavior are observed in most solvents at significant concentrations, usually at lower concentrations for a lower D value. Not only does incomplete dissociation complicate matters, but even where dissociation is essentially 100% complete the reduced insulting power of the solvent results in ion-pair formation, and at higher concentrations higher aggregates form. Although the primary effect is to reduce the number of carriers, the nature of the ionic atmosphere is also changed. The interplay of these effects is such that for many solvents of low to medium dielectric constant plots of Λ versus $C^{1/2}$ decrease, pass through a minimum, and then increase at higher concentrations. On the microscopic level it is not seriously suggested that the applied field and the bulk dielectric constant are translated quantitatively into a specifically calculable environment for a given ion. At present there is no complete theory of electrolytic conductivity which allows calculation of ionic mobilities at infinite dilutions and the subsequent change as a function of concentration. There are numerous refinements of the Onsager equation, the merits of which are currently the subject of continuing debate. However, for many applications the differences that result from the use of different equations for data analysis are small enough that there is no ambiguity in interpreting the implication of the conductivity experiment.

Just as the Onsager equation was found to apply in the limit for strong electrolytes, the case for extremely weak electrolytes is also simple. The original Arrhenius equation for the conductivity of partially dissociate solutes is

$$\Lambda = \alpha\Lambda_0$$

where α is the degree of dissociation. These two cases are the extremes, and solutions of most electrolytes in most solvents exhibit behavior somewhere between these extremes. The first acceptable solution to the problem was devised by Fuoss and Kraus.[3] The conductivity equation for 1:1 electrolytes became

$$\Lambda = \alpha[\Lambda_0 - (a\Lambda_0 + b)(C\alpha)^{1/2}] \tag{9}$$

where a and b have the same values as in Eq. 8. In this formulation it is recognized that the conductivity of a solution depends on both the number of carrier ions and their mobilities.

Shedlovsky suggested a further modification to extend the applicability of the equation to solutions of higher concentration.[4] If α is 1, the equation is

$$\Lambda = \Lambda_0 - (a\Lambda_0 + b)\frac{\Lambda}{\Lambda_0}(C)^{1/2} \tag{10}$$

and, if α is not 1,

$$\Lambda = \alpha\Lambda_0 - (a\Lambda_0 + b)\frac{\Lambda}{\Lambda_0}(C\alpha)^{1/2} \tag{11}$$

When solved for α

$$\alpha = \frac{\Lambda}{\Lambda_0} + \frac{(a\Lambda_0 + b)}{\Lambda_0}\frac{\Lambda}{\Lambda_0}(C\alpha)^{1/2} \tag{12}$$

the equation can be arranged as

$$\alpha = \frac{\Lambda}{\Lambda_0}\left[1 + \frac{(a\Lambda_0 + b)}{\Lambda_0}(C\alpha)^{1/2}\right] \tag{13}$$

which may be written as

$$\alpha = \frac{\Lambda}{\Lambda_0} S \tag{14}$$

Now if we represent the ionization of a weak electrolyte as

$$XY \rightleftharpoons X^+ + Y^- \tag{15}$$

the ionization constant is

$$K = \frac{[X^+]f_+[Y^-]f_-}{[XY]} \tag{16}$$

where the terms in brackets are analytical concentrations and f_+ and f_- are ionic activity coefficients defined so that the ionic activity $a_\pm$ is given by

$$a_\pm = f_\pm C_\pm \tag{17}$$

where $C_\pm$ is the concentration of the ions. For a 1:1 electrolyte $[X^+] = C_+ = [Y^-] = C_-$ and the equation for K becomes

$$K = \frac{C\alpha f_+ \times C\alpha f_-}{C(1-\alpha)} \tag{18}$$

or

$$K = \frac{C\alpha^2 f^2}{1-\alpha} \tag{19}$$

where f is the mean ionic activity coefficient

$$f = [(f_+)(f_-)]^{1/2} \tag{20}$$

If Eq. 14 is substituted into Eq. 19 with rearrangement of terms and division by ΛS, the result is

$$\frac{1}{\Lambda S} = \frac{1}{K\Lambda_0{}^2} C\Lambda f^2 S + \frac{1}{\Lambda_0} \tag{21}$$

Thus, if $1/\Lambda S$ is plotted versus $C\Lambda f^2 S$ as in Fig. 2, the intercept is $1/\Lambda_0$ and the slope is $1/K\Lambda_0{}^2$. A straight-line fit of the data in the Shedlovsky analysis is the basic criterion for the appropriateness of this analysis. Also, experimental error is exaggerated in this type of plot so a fair degree of precision (1% or better) is required. In practice an initial value of Λ_0 is estimated from

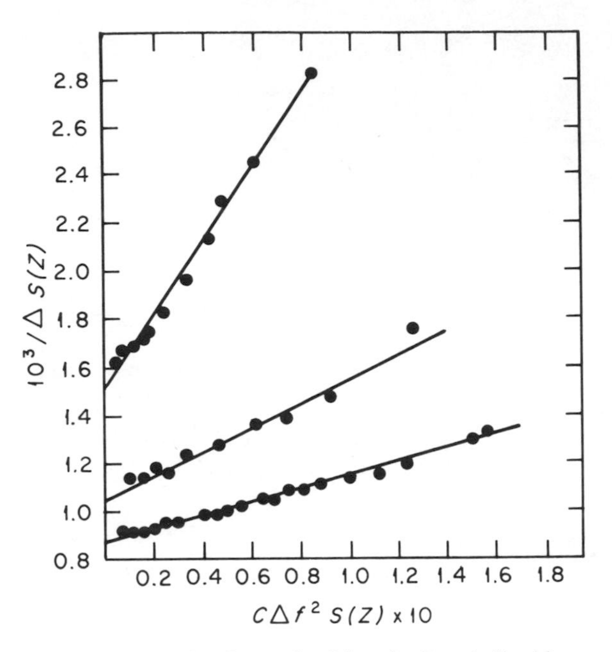

FIG. 2. Shedlovsky plots for the conductivity of sodium in liquid ammonia.[38]

a plot of Λ versus $C^{1/2}$. A computer program employing an iterative method is then used to calculate K and Λ_0. Usually only a few iterations are necessary to achieve a constant Λ_0.

E. Which Conductivity Equation to Use?

Three basic parameters are obtained from the analysis of conductivity data: Λ_0, the equivalent conductivity at infinite dilution; K_A, the association constant of the primary nonconducting species (or K_D the dissociation constant, which is the reciprocal of K_A); and a distance or ion size parameter. All the equations use an extrapolation function to obtain Λ_0 and the result is usually identical for all equations. The physical meaning of Λ_0 in the various hydrodynamic models is clear.

More divergence is encountered in the values for K_A calculated by the various equations, even though the physical significance of K_A clearly represents the process of conversion of conducting species into nonconducting species. However, the greatest single area of disagreement concerns the use and interpretation of the distance parameters. A contributing factor is that earlier forms of the various equations often omitted higher terms that added little to the total calculation in most cases. With increasing use of electronic

TABLE I

CONDUCTIVITY EQUATION AND COEFFICIENTS FOR 1:1 ELECTROLYTES

I. Fuoss–Onsager equation (1957)

$$\Lambda = \Lambda_0 - S(C\gamma)^{1/2} + EC\gamma \log C\gamma + JC\gamma - K_A C\gamma f^2 \Lambda$$
$$S = \alpha\Lambda_0 + \beta$$
$$\alpha = 8.204 \times 10^5/(DT)^{3/2}$$
$$\beta = 82.43/\eta(DT)^{1/2}$$
$$E = E_1\Lambda_0 - 2E_2$$
$$E_1 = 2.303\kappa^2a^2b^2/24C$$
$$E_2 = 2.303\kappa ab\beta/16C^{1/2}$$
$$J = \sigma_1\Lambda_0 + \sigma_2$$
$$\sigma_1 = (\kappa^2a^2b^2/12C)[2b^3+2b-1)/b^3 + 0.9074 + \ln(\kappa a/C^{1/2})]$$
$$\sigma_2 = \alpha\beta + 11\beta\kappa a/(12C^{1/2}) - (\kappa ab\beta/8C^{1/2})[1.0170+\ln(\kappa a/C^{1/2})]$$

II. Modified Fuoss–Onsager equation, Fuoss–Hsia equation

$$\Lambda = \Lambda_0 - S(C\gamma)^{1/2} + EC\gamma \log C\gamma + J_1 C\gamma - J_2(C\gamma)^{3/2} - K_A C\gamma f^2 \Lambda$$
$$J_1 = \sigma_1\Lambda_0 + \sigma_2$$
$$J_2 = \sigma_3\Lambda_0 + \sigma_4$$
$$\sigma_1 = [(\kappa ab)^2/24C][1.8147 + 2\ln(\kappa a/C^{1/2}) + (2/b^3)(2b^2+2b-1)]$$
$$\sigma_2 = \alpha\beta + \beta(\kappa a/C^{1/2}) - \beta[(\kappa ab)/16C^{1/2}][1.5337 + (4/3b) + 2\ln(\kappa a/C^{1/2})]$$
$$\sigma_3 = [b^2(\kappa a)^3/24C^{3/2}][0.6094 + (4.4748/b) + (3.8284/b^2)]$$
$$\sigma_4 = [\beta(\kappa ab)^2/24C][2/b^3)(2b^2+2b-1) - 1.9384] + \alpha\beta(\kappa a/C^{1/2}) + [\beta(\kappa a)^2/C] - [\beta b(\kappa a)^2/16C][1.5405 + (2.2761/b)] - (\beta^2\kappa ab/16\Lambda_0 C^{1/2})[(4/3b) - 2.2194)]$$

III. Pitts equation

$$\Lambda = \Lambda_0 - S(C\gamma)^{1/2} + EC\gamma \log C\gamma + J_1 C\gamma - J_2(C\gamma)^{3/2} - K_A C\gamma f^2 \Lambda$$
$$J_1 = \sigma_1\Lambda_0 + \sigma_2$$
$$J_2 = \sigma_3\Lambda_0 + \sigma_4$$
$$\sigma_1 = \frac{(b\kappa a)^2}{12C}\left[\ln\left(\frac{\kappa a}{C^{1/2}}\right) + \frac{2}{b} + 1.7718\right]$$
$$\sigma_2 = \frac{\beta\kappa a}{C^{1/2}} + \frac{\beta b\kappa a}{8C^{1/2}}\left[0.01387 - \ln\left(\frac{\kappa a}{C^{1/2}}\right)\right]$$
$$\sigma_3 = \frac{(b\kappa a)^3}{6C^{3/2}}\left[\frac{1.2929}{b^2} + \frac{1.5732}{b}\right]$$
$$\sigma_4 = \frac{\beta(\kappa a)^2}{C} + 0.23484\frac{\beta b(\kappa a)^2}{3C}$$
$$b = e^2/aDkT$$
$$\kappa = \left(\frac{8\pi Nz^2e^2c}{1000DkT}\right)^{1/2}$$

where a, the distance or ion-size parameter; C, concentration (equivalents/liter); T, absolute temperature; D, solvent dielectric constant; $\pm z$, valencies of the ions; N, Avogadro's number; k, Boltzmann constant; e, electronic charge; η, solvent viscosity; K_A, the association constant; and γ, the degree of association.

For unassociated electrolytes $\gamma = 1$ and $K_A = 0$; for associated electrolytes $\gamma < 1$ and $K_A > 0$.

computers and higher precision data it became obviously advantageous to include the higher terms. These and other minor corrections have been made recently, and the continuing resurgence of interest in the field suggests that other improvements remain to be made.

The Fuoss–Hsia equation is currently one of the most highly regarded and most widely used.[5–7] This equation has evolved from the older Fuoss–Onsager equation.[8,9] It has been recently corrected,[7,10] and is often used in the expanded form due to Fernández-Prini.[11]

The other most widely used equation is that of Pitts.[12–14] It has been pointed out that one of the differences between the Fuoss–Hsia (F–H) equation and the Pitts (P) equation is that the Pitts equation implies a smaller interionic interaction for ions of the same size.[11] This results in smaller association constants and smaller contact distances for P than for F–H.[15,16]

Another conductivity equation has been used by Justice which is a modification of the F–H theory.[17–20] This has recently been shown to be of limited utility and that the effect of the modification is to change the hydrodynamic model.[21] Presently, there is still no agreement upon this point.[22–24]

Both the Fuoss–Hsia[11] and the Pitts[25] equations have been expressed in the form

$$\Lambda = \Lambda_0 - SC^{1/2} + EC \ln C + J_1 C - J_2 C^{3/2} \tag{22}$$

Differences in the equations occur in the coefficients J_1 and J_2 which are given with S and E in Table I. Either of these equations will yield parameters that can be interpreted easily in terms of the physical properties of solutions. In practice, both equations are frequently used,[16] as is occasionally the Fuoss–Onsager equation. Comparisons of several sets of data with both equations almost invariably reveal the same trends in the parameters and, consequently, except for minor numerical differences, the resulting physical interpretation is the same.

III. Experimental Techniques*

The measurement of electrolytic conductivity can be one of the most precise physicochemical measurements, but in order to achieve this a high level of technical competence is required. When the prerequisites of the conductivity experiment are superimposed on the usual requirements for handling nonaqueous solvents the experimentalist is often truly challenged. After first discussing the technical aspects of the conductivity experiment and applications to nonaqueous solvents some interesting examples of solutions to typical problems will be given.

* For a detailed discussion of general experimental techniques useful for nonaqueous solution chemistry see Vol. I.

A. Measurement of Electrolytic Conductivity

When electrical current is passed through a solution of an electrolyte, a chemical reaction occurs at the electrodes which results in local changes in concentration and the formation of potentials at the electrodes. The dielectric nature of the solution between the electrodes gives rise to parallel capacitance, while the buildup of ions oppositely charged on the surface of the electrodes results in double-layer capacitance. Electron transfer through the electrode–solution interface causes faradic impedance. These and other polarization effects are serious experimental problems in the measurement of conductivity and have been discussed at length.[26–28] Unless the solution in contact with the electrodes is electrochemically reversible to one of the ions, the direct current method cannot be used. Most studies thus use an alternating current method employing a bridge designed for this purpose. Bridge designs have been discussed at length elsewhere.[26, 29–32] Polarization effects are minimized by the use of an a.c. bridge at low frequencies with an appropriately designed conductivity cell.

B. Conductivity Cells

Figure 3 shows the basic design of a typical conductivity cell. The cell consists of a suitable container with a tube for filling the cell and two electrodes. The metal electrodes are sealed by a glass-to-metal seal, which is often the source of considerable experimental difficulty. Techniques for construction of glass-to-metal seals have been discussed elsewhere.[27] Platinum is the usual electrode material because of its inertness to a wide variety of solutions. It is also relatively easy to clean. In practice the resistance of a solution is measured and its specific conductivity is computed from Eq. 3. As noted before, the cell constant K depends on the size, shape, and surface of the electrodes and on the

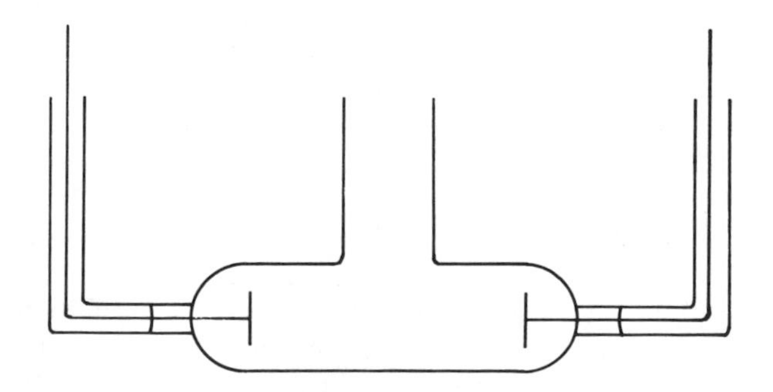

FIG. 3. Simple conductivity cell.

geometry of the conductivity cell. Jones and Bradshaw[33] have accurately determined the conductivity of potassium chloride solutions in cells of known geometry. Standard KCl solutions can therefore be used to determine the cell constants of new cells of unknown geometry.[34] When it is desirable to use solutions of concentration lower than 0.01 mole/liter KCl, the following equation[35] may be used.

$$\Lambda_c = 149.93 - 94.65C^{1/2} + 58.74C \log C + 698.46C$$

Cell constants should not vary over the concentration range studied. Commonly, several cells are used in a single study over a large concentration range and the measured resistances are kept between 1000 and 30,000 ohms. This avoids high current densities which lead to polarization problems at low resistances, and to insulation leakage at high resistances. These problems and other sources of experimental error have been discussed in detail by Shedlovsky.[26]

Another procedure used to minimize polarization effects is the platinizing of electrodes. In this process a spongy, black deposit of finely divided platinum with a large surface area is deposited on the shiny clean platinum electrodes. After cleaning the electrodes with aqua regia the cell is filled with a solution of 3% chloroplatinic acid containing a trace amount of lead acetate, which favors the adherence of a fine deposit. A small current is sufficient to produce platinum black on first one electrode and then the other by reversing the current.

The many catalytic and adsorptive properties of platinum are widely recognized, so it is not surprising that many solutions of interest may be incompatible with platinized platinum electrodes. Occasionally gold-plated electrodes can be used successfully,[36] but more often clean bright platinum electrodes are used. The effects of polarization are then noted as a small change in the measured resistance as a function of frequency. When resistance is plotted against the reciprocal of the square root of frequency, the intercept gives the resistance at infinite frequency. This procedure should be followed for measuring the cell constant as well as the solutions studied. The use of potassium iodide solutions containing iodine has been recommended for measuring the cell constants of bright platinum electrodes.[37]

C. Auxiliary Apparatus

In addition to measuring accurately the resistance of a solution it is necessary to know its concentration and the temperature at which the measurement is made. Temperature coefficients of conductivity are typically about 2% per degree Centigrade, so it can be seen that control to 0.01°C or even 0.001°C is necessary for precise measurements. Bath fluids with low dielectric constants should be used to prevent leakage of electrical energy from the cell. Oil or silicone oil is suitable. Silicone oil of low viscosity is particularly attractive when working at low temperatures.

Although many studies have varied the concentration of the solution by a dilution technique, that is the addition of successive increments of the solvent, this method is subject to large errors in the dilute region, especially if a solvent correction must be made, or if the solution is metastable. Barthel[38] has pointed out the advantages of stepwise increments of added electrolyte to a single sample of solvent. In this method the solvent correction is constant and other advantages are gained in simplifying the construction of the apparatus used. Several recent articles have discussed appropriate apparatus for solute addition to anaerobic or closed systems.[27,38] A third method of obtaining the concentration dependence of the conductivity is by independent preparation of each sample. Although this method requires considerably more working time in the laboratory, it allows the use of exceedingly simple apparatus. This method has been used successfully to determine the conductivity of metal–ammonia solutions.[39]

IV. Recent Research in Nonaqueous Solvents

A number of the most electropositive metals are known to dissolve in ammonia, amines, polyethers, and a few other solvents to give solutions of varying stability.* In dilute solutions it is thought that the dissolution process is accompanied by the formation of solvated cations and solvated electrons. Indeed the possibility of preparing relatively stable solutions containing solvated electrons has had the result that many physical studies have been made on these solutions and interest in their physical properties continues. Metal–ammonia solutions have the highest conductivities of any known electrolytic solutions, with a Λ_0 of 1142 for cesium–ammonia at −33.9°C[40] and of 1127 for sodium–ammonia at −33.9°C.[39] The electrolyte in these behaves as a weak electrolyte owing to the extensive ion pairing in solution. At high concentrations there is a phase separation, with the formation of a bronze-colored phase, more concentrated in metal and less dense than the dilute blue phase. The electrical conductivity of the bronze phase is extremely high, but its nature appears to be metallic rather than electrolytic.

In measuring the conductivity of metal–ammonia solutions a number of difficult experimental problems were encountered. The methods which provided solutions to these problems can be used to advantage in many other systems.

* See Vol. II for a discussion of the nature of solutions containing dissolved metals.

The vapor pressure over liquid ammonia at room temperature is adout 10 atm. Therefore, unless pressure vessels are used, all work must be done near $-33.9°C$, the boiling point, or lower. Ammonia is hygroscopic and metal–ammonia solutions are even more sensitive.* They react with oxygen and water and are catalytically decomposed by the reaction products, trace impurities, especially transition metal ions, and platinized platinum electrodes. To make meaningful measurements on these systems thus requires metals and solvents of high purity, working within completely closed systems, temperature control and measurement at low temperatures, and the use of electrodes which do not decompose the solution.

Since earlier workers using a dilution technique to study the conductivity of metal–ammonia solutions had experienced considerable decomposition,[41] it was decided that a separate sample should be made up for each data point.[39] This allowed the apparatus (Fig. 4) to be of extremely simple design, although considerable glass blowing was necessary before each conductivity run. The alkali metal was freshly distilled *in vacuo* from the side arm. Solution volumes could be measured directly in the calibrated bulb. The conductivity cell employed bright platinum or gold-plated electrodes for sodium. Gold-

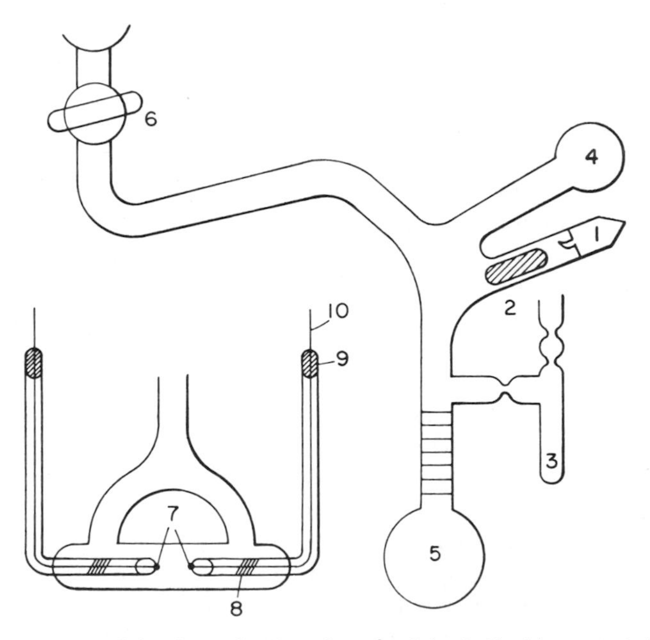

Fig. 4. Apparatus used for determination of conductivity in liquid ammonia.[38] (1) Ammonium bromide sample, (2) glass-encased magnet, (3) side arm, (4) conductivity cell, (5) calibrated bulb, (6) stopcock, (7) platinum electrodes, (8) graded seal, (9) Torr seal (Varian), (10) leads to bridge.

* See Vol. II for a discussion of the properties of ammonia solutions.

plated electrodes could not be used for cesium solutions, because the cesium–ammonia solution dissolved the gold plate.[42]

Conductivity measurements can be used to determine ionization constants. Shedlovsky and Kay[43] have used this method to determine the ionization constant of acetic acid in water–methanol mixtures as a function of concentration. Although the system was complicated by ion association as well as incomplete ionization, their results agree remarkably well with those of other workers. The method has also been used to estimate the ionization constant of water in liquid ammonia.

Conductivity measurements can also be used for analytical purposes, such as the determination of concentration, solubility, and the extent of hydrolysis.[26]

Recently there has been considerable interest in crown compounds which form cationic complexes with alkali metal and alkaline earth cations. Hogen-Esch and Smid[44] have used conductivity to study the ion-pairing properties and thermodynamics of association of a number of fluorenyl salts and their complexes with dimethyldibenzo-18-crown-6. As in numerous other studies conductometric data were used with supporting evidence from other physical measurements, in this case, characteristic spectra of the complexes. A number of equilibria have been established between contact ion-pairs, partial contact and solvent separated ion-pairs and crown complex ion-pairs. Accurate thermodynamic data were obtained by determining the conductivity of the system as a function of temperature. Use was made of the relation

$$R[d \ln \Lambda / d(1/T)] = E_{\text{vis}} - \tfrac{1}{2}\Delta H^0 d$$

where E_{vis} denotes the activation energy of viscous flow.[45]

Ion-pairing processes were found to be important in a conductometric study of hydrophobic interactions by double long-chain electrolytes.[46] Oakenfull and Fennick examined electrolytes such as decyltrimethylammonium decanoate in ethanol–water mixtures and were able to separate the electrostatic and hydrophobic contributions to the free energy of ion-pair formation by examining the effect of hydrocarbon chain length. The authors conclude that ΔG_{HI}, the hydrophobic contribution to the free energy of ion-pair formation, does not depend on the surface tension of the solvent as previously believed. Excellent agreement was obtained between the variation of $\Delta G_{\text{electrostatic}}$ and dielectric constant according to the theory of Bjerrum.[47,48] Also, a correlation was found between the isothermal compressibility of the solution and ΔG_{HI}. No correlation was found between ΔG_{HI} and the interfacial tension, which appears to contradict the assumption that interfacial tension largely determines hydrophobic interactions.[49,50]

Propylene carbonate (PC) has been suggested as a medium for the electrodeposition of alkali metals.[51] Although the alkali metal halides have a low solubility in pure PC, in the presence of $AlCl_3$ they dissolve owing to the formation of alkali metal cations and $AlCl_4^-$. It is thought that the dissolution of $AlCl_3$ in PC follows the scheme[52]

$$AlCl_3 + \tfrac{6}{4}PC \rightarrow \tfrac{1}{4}Al(PC)_6^{3+} + \tfrac{3}{4}AlCl_4^-$$

Then, when an alkali metal chloride is added, the following reaction occurs

$$xMCl + \tfrac{1}{4}Al(PC)_6^{3+} + \tfrac{3}{4}AlCl_4^- \rightarrow xM^+ + (1-x)/4Al(PC)_6^{3+} + (3+x)/4AlCl_4^- + 3x/2PC$$

Studies of the conductivity of solutions of alkali metal chlorides in $AlCl_3$–PC have shown that anions are poorly solvated with limiting ionic conductivities two to three times higher than most cations. Although minima in the conductance curves of alkali metal chlorides in $AlCl_3$–PC can be explained by ion association, a minimum in the curve for $AlCl_3$ in PC seems to be due to complex multiple ionic equilibria.[51]

References

1. L. Onsager, *Phys. Z.* **27**, 388 (1926).
2. L. Onsager, *Phys. Z.* **28**, 277 (1927).
3. R. M. Fuoss and C. A. Kraus, *J. Amer. Chem. Soc.* **55**, 476 (1933).
4. T. Shedlovsky, *J. Franklin Inst.* **225**, 739 (1938).
5. R. M. Fuoss and K.-L. Hsia, *Proc. Nat. Acad. Sci. U.S.* **57**, 1550 (1967).
6. R. M. Fuoss and K.-L. Hsia, *Proc. Nat. Acad. Sci. U.S.* **58**, 1818 (1967).
7. K.-L. Hsia and R. M. Fuoss, *J. Amer. Chem. Soc.* **90**, 3055 (1968).
8. R. M. Fuoss and L. Onsager, *J. Phys. Chem.* **61**, 668 (1957).
9. R. M. Fuoss and F. Accascina, *in* "Electrolytic Conductance," Chapters XV and XVII. Wiley (Interscience), New York, 1959.
10. P. C. Carman, *J. Phys. Chem.* **74**, 1653 (1970).
11. R. Fernández-Prini, *Trans. Faraday Soc.* **65**, 3311 (1969).
12. E. Pitts, *Proc. Roy. Soc., Ser. A* **217**, 43 (1953).
13. E. Pitts, B. E. Tabor, and J. Daly, *Trans. Faraday Soc.* **65**, 849 (1969).
14. E. Pitts, B. E. Tabor, and J. Daly, *Trans. Faraday Soc.* **66**, 693 (1970).
15. E. M. Hanna, A. D. Pethybridge, and J. E. Prue, *Electrochim. Acta* **16**, 677 (1971).
16. B. J. Barker, H. L. Huffman, Jr., and P. G. Sears, *J. Phys. Chem.* **78**, 2689 (1974).
17. J. C. Justice, *J. Chim. Phys.* **65**, 353 (1968).
18. J. C. Justice, R. Bury, and C. Treiner, *J. Chim. Phys.* **65**, 1708 (1968).
19. M. C. Justice, R. Bury, and J. C. Justice, *Electrochim. Acta* **16**, 687 (1971).
20. J. C. Justice, *Electrochim. Acta* **16**, 701 (1971).
21. R. M. Fuoss, *J. Phys. Chem.* **78**, 1383 (1974).
22. J. C. Justice, *J. Phys. Chem.* **79**, 454 (1975).
23. R. M. Fuoss, *J. Phys. Chem.* **79**, 1038 (1975).
24. J. C. Justice, *J. Phys. Chem.* **79**, 1039 (1975).
25. R. Fernández-Prini and J. E. Prue, *Z. Phys. Chem. (Leipzig)* **228**, 373 (1965).
26. T. Shedlovsky, *in* "Techniques of Organic Chemistry" (A. Weissberger, ed.), 3rd ed., Vol. 1, Part 4. Wiley (Interscience), New York, 1959.

27. D. F. Evans and M. A. Matesich, *in* "Techniques of Electrochemistry" (E. Yeager and A. J. Salkind, eds.) **2**, p. 1. Wiley, New York, 1973.
28. J. O'M. Bockris and A. K. N. Reddy, "Modern Electrochemistry." Plenum, New York, 1970.
29. G. Jones and R. C. Josephs, *J. Amer. Chem. Soc.* **50**, 1049 (1928).
30. T. Shedlovsky, *J. Amer. Chem. Soc.* **52**, 1793 (1930).
31. P. H. Dike, *Rev. Sci. Instrum.* **2**, 379 (1931).
32. J. Braunstein and G. D. Robbins, *J. Chem. Educ.* **48**, 52 (1971).
33. G. Jones and B. C. Bradshaw, *J. Amer. Chem. Soc.* **55**, 1780 (1933).
34. H. S. Harned and B. B. Owen, "The Physical Chemistry of Electrolytic Solutions," 3rd ed. Van Nostrand-Reinhold, Princeton, New Jersey, 1958.
35. J. E. Lind, Jr., J. J. Zwolenik, and R. M. Fuoss, *J. Amer. Chem. Soc.* **81**, 1557 (1959).
36. R. R. Dewald and R. V. Tsina, *J. Phys. Chem.* **72**, 4520 (1968).
37. C. A. Kraus, *J. Amer. Chem. Soc.* **43**, 749 (1921).
38. J. Barthel, *Angew. Chem., Int. Ed. Engl.* **7**, 260 (1968).
39. R. R. Dewald and J. H. Roberts, *J. Phys. Chem.* **72**, 4224 (1968).
40. R. R. Dewald, *J. Phys. Chem.* **73**, 2615 (1969).
41. C. A. Kraus, *J. Amer. Chem. Soc.* **43**, 749 (1921).
42. J. H. Roberts and R. R. Dewald, unpublished results.
43. T. Shedlovsky and R. L. Kay, *J. Phys. Chem.* **60**, 151 (1956).
44. T. E. Hogen-Esch and J. Smid, *J. Phys. Chem.* **79**, 233 (1975).
45. T. Ellingson and J. Smid, *J. Phys. Chem.* **73**, 2712 (1969).
46. D. G. Oakenfull and D. E. Fenwick, *J. Phys. Chem.* **78**, 1759 (1974).
47. N. Bjerrum, *Kgl. Dan. Vidensk. Selsk., Mat.-Fys. Medd.* **7**, 1 (1926).
48. R. A. Robinson and R. H. Stokes, "Electrolyte Solutions," 2nd ed. Butterworth, London, 1959.
49. S. Lewin, *Nature (London), New Biol.* **231**, 80 (1971).
50. O. Sinanoglu and S. Abdulnur, *Fed. Proc., Fed. Amer. Soc. Exp. Biol.* **24**, Suppl. 15, S-12 (1965).
51. J. Jorné and C. W. Tobias, *J. Phys. Chem.* **78**, 2521 (1974).
52. R. Keller, NASA Contract. Rep. 1968, NASA-CR-72407.

~ 2 ~

Hydrogen Bonding Phenomena

ANN T. LEMLEY

Department of Chemistry
Cornell University, Ithaca, New York

I. INTRODUCTION

Hydrogen bonding is a very broad subject which has been covered by many workers. The definitive work by Pimentel and McClellan[1] discussed many aspects of the work up to 1958. Their follow-up work in 1971,[2] while not as comprehensive, pointed out many other summaries on specialized topics. This chapter attempts to deal with hydrogen bonding phenomena in nonaqueous systems as probed by four spectroscopic techniques.

Infrared and Raman spectroscopy are uniquely adapted to the study of bonding. Both the fundamental vibrational modes of molecules involved in hydrogen bonding and the vibrations of the hydrogen bond itself can be measured. Proton nuclear magnetic resonance is a well-developed tool and is, of course, sensitive to the donation of a proton in a hydrogen bond. Ultraviolet spectroscopy measures electronic transitions which are also sensitive to hydrogen bonding. The type of information which can be learned from these methods includes the nature of the bond itself, the nature of the hydrogen-bonded species, and considerable thermodynamic data about bond formation.

The period covered is from 1959 to the present, and while the work is not a comprehensive survey of that period, it is hopefully a representative sampling of the progress which has been made.

II. Infrared Spectroscopy

A. Properties of the Hydrogen Bond

Much of the infrared work on hydrogen-bonded systems involves measurements of the fundamental stretching frequencies of A–H bonds which are proton donors in a hydrogen bond. This fundamental mode is quite sensitive to hydrogen bonding. Work has also been done on the actual hydrogen bond stretching frequency in the far-infrared region. The type of bond involved and how it is reflected in the vibrational spectrum had to be established in order to use these data to determine the species involved and to make thermodynamic calculations.

Theoretical calculations[3,4] of the N–H stretching frequency of amides, alcohols, acids, and secondary amides have been made assuming that hydrogen bonding is essentially an electrostatic interaction. These results were then compared to observed infrared frequencies and the validity of the model was established. The O–H or N–H stretching mode involved in the hydrogen bond was considered independent of the rest of the molecule and was then placed in the electric field of the electron donor. By using the Schrödinger equation with a Morse potential, the following equation was derived:

$$\nu = \frac{a}{\pi c}\left(\frac{D}{2M}\right)^{1/2} - \frac{a^2 h}{4\pi^2 Mc} \tag{1}$$

where D is the dissociation energy, M is the reduced mass, and a is a constant. The dissociation energy had to be corrected because hydrogen bonding changes the unbalanced charges on the atoms in the N–H or O–H bond and

TABLE I

CALCULATED AND OBSERVED O–H AND N–H STRETCHING FREQUENCIES OF HYDROGEN-BONDED SPECIES

Species	$\nu(\mathrm{cm}^{-1})$	
	Observed	Calculated
Methanol	3344	3325[a]
Ethanol	3338	3320[a]
Formic acid	3110	3103[a]
Acetic acid	3035	3025[a]
N-Methylacetamide	3280	3257[a]
Formamide	3210	3228[b]
Acetamide	3205	3234[b]

[a] K. V. Ramiah and P. G. Puranik, *Proc. Indian Acad. Sci., Sect. A* **56**, 155 (1962).

[b] K. V. Ramiah and P. G. Puranik, *Proc. Indian Acad. Sci., Sect. A* **56**, 96 (1962).

affects the electrostatic interaction. A final adjustment was made in the calculations for the acids and amides to take into account resonance structures. New unbalanced charges, new dissociation energies, and, therefore, new frequencies were calculated. The results are listed in Table I. The agreement between observed and calculated frequencies confirmed the validity of the electrostatic model.

Further information about the nature of the hydrogen bond in nonaqueous solvents has been sought by studying the far-infrared region of the spectrum. Where instrumentation will allow, bands below 250 cm^{-1} have been assigned to the hydrogen bond, the most intense band being the stretching mode, ν_σ. Ginn and Wood[5] attempted to answer a question of previous workers[6] concerning the three bands in the O–H stretching region of a gaseous mixture of methanol and trimethylamine. While the central peak at 3350 cm^{-1} was assigned to the O–H stretching mode, the two bands above and below this frequency (3495 and $\sim$3200 cm^{-1}) could be attributed to sum and difference combinations of 3350 cm^{-1} and a hydrogen bond stretching mode at $\sim$140 cm^{-1}. The far-infrared work showed a band at 142 cm^{-1} (ν_σ) which was not present in the unmixed components. Further work by these same authors[7] established hydrogen bond stretching modes for phenol–trimethylamine (143 cm^{-1}), phenol–triethylamine (123 cm^{-1}), and phenol–pyridine (134 cm^{-1} and shoulder at 143 cm^{-1}) complexes. Hall and Wood[8] made further far-infrared studies to investigate the ionic character of the bond in the phenol–

pyridine complex. They found that ν_σ rose in frequency in the more polar solvents. This trend is in the opposite direction from the solvent shift of intramolecular stretching frequencies, but in the same direction as the effect in charge-transfer complexes. Far-infrared spectroscopy has thus been used to emphasize the electrostatic and ionic nature of the hydrogen bond.

The intensity and breadth of a stretching mode involving a proton which is hydrogen-bonded cannot be explained solely on the basis of the combination bands previously mentioned. Hall and Wood[9] made an infrared study of phenol and its derivatives and pyridine and its derivatives to determine the reason for the structure in ν_s, the O–H stretching mode. They made use of Job plots to establish that the concentration regions studied had only 1:1 complexes so that different hydrogen-bonded species could not account for the breadth and structure of the bands. In phenol with six bases the same bands were seen. These were independent of the base, but due only to the complex and could not be due to combination bands with ν_σ. If this were so, they would change in frequency with the base. The best explanation for this phenomenon was the occurrence of Fermi resonance between the fundamental and the overtone or combination bands. Studies with new donors (derivatives of phenol) gave a new pattern of peaks, again independent of the base, but characteristic of the donor, as would be expected if the O–H stretching frequency of the donor was the fundamental involved in Fermi resonance. Further work was done to establish that combinations and overtones of donor fundamentals were, in each case, the other band in resonance.

When hydrogen bonding is present, there is a large increase in intensity and width of ν_s the fundamental stretching mode. In systems where more than one hydrogen-bonded species might be present, the intensity and width of an O–H, N–H, or S–H stretching mode are attributed to contributions from several A–H oscillators with different $R_{\text{A–B}}$ distances. Ažman *et al.*[10] made calculations assuming a linear hydrogen bond in the pyrrole–tetrahydrofuran system and concluded that the increase in intensity of ν_s in weak hydrogen bonding is of the order 1.5 to 2.5.

When strong hydrogen bonding is present in a system, there is an increase in the anharmonicity of the O–H or N–H stretching vibration involved in the hydrogen bond, compared to the free O–H and N–H vibrations. Foldes and Sandorfy[11] found that it was necessary to include off-diagonal terms of the quartic potential constant in order to deal with the anharmonicity of vibrations influenced by hydrogen bonds. Asselin *et al.*[12] studied isopropanol, ethanol, phenol, and 2,6-di-*tert*-butyl-*p*-cresol (DTBC), and measured the frequencies of the O–H and O–D overtones, both free and associated. When the anharmonicity constants and isotopic ratios were computed for the free bands, it was shown

$$X_{12} = \omega_e x_e \simeq X_{23} \tag{2}$$

and the second-order perturbation treatment was shown to be valid. When the same computations were made for the self-association frequencies, the isotopic ratios became higher and the ratios of anharmonic constants were significantly higher. Thus, in the presence of hydrogen bonding anharmonicity constants are not equal for different vibrational levels. In addition, recent work of Asselin and Sandorfy[13] has been completed which indicates that the anharmonicity constant of the O–H stretching mode of a hydrogen-bonded species changes (diminishes) with temperature, whereas the anharmonicity of the free (OH) band is mostly independent of temperature. They concluded that anharmonicity has a significant effect on the shift in the OH stretching frequency on hydrogen bonding. Many workers have attempted to relate this shift in frequency, $\Delta\nu$, to the enthalpy of the hydrogen bond, without considering the effect of anharmonicity.

One question which has arisen about infrared studies of the vibrations of the hydrogen bond is whether the stretching vibration is localized in the O–H···O bond or whether it is a case of the two molecules in the bond vibrating as a whole relative to one another. There are several ways to approach this problem and these methods can be applied to various chemical systems. One can assume that the vibration responsible for the hydrogen bond stretching frequency, ν_σ, is localized and has the frequency of a diatomic oscillator where

$$\nu = \frac{1}{2\pi}\left(\frac{k}{\mu}\right)^{1/2} \tag{3}$$

In a series of acids or alcohols, the different force constants, k, could then be calculated with μ set equal to the reduced mass of O–H···O. If, however, one assumes that the force constant (and anharmonicity) in a series can be considered constant and the vibration is not localized, the frequencies should vary with the reciprocal of the square root of the reduced mass of the two molecules in the dimer. With the assumption that the vibration is highly localized and the force constant unchanging, the product $M\nu_\sigma{}^2$ should vary directly with M, the molecular mass of the compound. Stanevich[14] studied carboxylic acids and found that with the assumption of unchanging force constants and the calculation of all other frequencies from the observed frequency of acetic acid and the reduced masses of the other acids, good agreement was obtained between calculated and observed frequencies. These results are reported in Table II. Lake and Thompson[15] found that in a series of alcohols, if the entire mass is used in calculating frequencies, the relationship $\nu \approx (\mu)^{-1/2}$ does not hold. They also found that $M\nu_\sigma{}^2$ does vary roughly with M, implying localization. However, the trend was not smooth, and they concluded that although the hydrogen bond vibration in alcohols

TABLE II

CALCULATED AND OBSERVED FREQUENCIES OF THE HYDROGEN BOND STRETCHING MODE IN CARBOXYLIC ACIDS AND PHENOLS

Compound	ν_{calc} (cm^{-1})[a]	ν_{obs} (cm^{-1}) at 20°C
CH_3COOH[b]	—	176
CH_3CH_2COOH[b]	158.7	157
$CH_3CH_2CH_2COOH$[b]	144	—
CH_3COOD[b]	176	178
CD_3COOD[b]	171	176
Phenol[c]	—	162
m-Cresol[c]	151	143
p-Cresol[c]	151	178, 124
o-Cresol[c]	151	188, 121
o-Isopropylphenol[c]	135	130

[a] Calculated from observed frequency of acetic acid or phenol using the diatomic harmonic oscillator approximation and the reduced mass of the entire molecule.

[b] A. E. Stanevich, *Opt. Spectrosc.* **16**, 243 (1964).

[c] R. J. Jakobsen and J. W. Brasch, *Spectrochim. Acta* **21**, 1753 (1965).

is localized to some extent, there are other factors, such as the potential function, which are influenced by steric effects and polymer formation.

Jakobsen and Brasch[16] who studied the hydrogen bond in phenols found it useful to calculate the force constant by an independent method and then use it to make further calculations. The force constant can be calculated if the potential function of the hydrogen bonding complex is known. Lippincott and Schroeder[17,18] have suggested a one-dimensional model for the hydrogen bond relating the ratio of the bonded and nonbonded O–H stretching frequency to the force constant, $k_{O-H\cdots O}$; they related this force constant to R, the O $\cdots$ O distance in the complex. Both this distance and the O–H bond length must be known. Unfortunately, R values from the solid or gaseous state must be used in most cases. Stanevich used the force constant computed from his simple diatomic oscillator model to find R, and he found that it was in agreement with the gaseous values for the dimers of formic and acetic acid. This supported his conclusion that the hydrogen bond is delocalized in the carboxylic acids. Lake and Thompson used various values of R and of the O–H bond length from the literature to calculate the force constant. This force constant was then used in the harmonic oscillator approximation with each reduced mass expression to calculate a frequency for phenol. The best agreement was obtained if the bond was considered delocalized and the entire mass of the molecules was used in the calculation. Like Stanevich, these authors used the observed value of phenol and the reduced masses of its derivatives to calculate frequencies for the series. These results are also reported in Table II. Lake and Thompson compared the force constants derived from the Lippincott–Schroeder model with those calculated from the observed values of ν_σ assuming localized hydrogen bonding and found no apparent correlation. They concluded that this model was too simple for the many structures present in alcohols.

It appears that, depending on the system studied, there are differing degrees of localization in hydrogen bonds. The problems inherent in making assumptions about, and calculating, force constants limit the definitive nature of conclusions drawn from these studies. One must try to verify the applicability of potential functions which are generalized, for use in a specific system.

It is often useful in making vibrational studies of hydrogen bonding to substitute deuterated compounds for protonated ones. The question of whether the deuterium bond is very different from the hydrogen bond needed to be answered. Ginn and Wood[19,20] looked at the hydrogen and the deuterium bond directly in the far-infrared region of the spectrum in a series of complexes of phenol and phenol-*d* with bases. They made quantitative intensity studies to show that there was a 1 : 1 complex in each case, and then did a normal coordinate analysis to get the force constants. The frequencies and force constants are shown in Table III. The frequency change could be accounted for by the change in mass, and they used both the localized and delocalized diatomic models of the hydrogen bond to calculate force constants. They concluded that there was no change in force constant or

TABLE III

STRETCHING FREQUENCIES AND FORCE CONSTANTS OF THE HYDROGEN AND DEUTERIUM BONDS

Complex	Frequencies (cm^{-1})	Force constants (mdyne/Å)
phenol–trimethylamine	143[a]	0.271[b]
phenol-*d*-trimethylamine	141[a]	0.274[b]
phenol–triethylamine	123[a]	0.236[b]
phenol-*d*–triethylamine	120[a]	0.220[b]
phenol–pyridine	134[a]	0.229[b]
phenol-*d*–pyridine	130[a]	0.227[b]

[a] S. G. W. Ginn and J. L. Wood, *Chem. Commun.* p. 628 (1965).
[b] S. G. W. Ginn and J. L. Wood, *Spectrochim. Acta, Part A* **23**, 611 (1967).

potential energy on deuteration. Singh and Rao[21] came to a different conclusion using the results of a study of phenol and phenol-*d* with electron donors of differing basicities. Their work differed from that previously discussed in that they studied the O–H and O–D stretching region of the spectrum. They correlated the shift in these stretching frequencies upon hydrogen bonding with the thermodynamic parameters. The enthalpy of formation of a 1:1 complex was always greater with a hydrogen bond than with a deuterium bond. These enthalpy values were calculated from equilibrium constants at different temperatures. The authors did suggest that the enthalpy differences might be due to the longer O–D···Y bond in the deuterated species. It would seem that the direct study in the far-infrared region of the spectrum is perhaps more valid in determining the strength of hydrogen and deuterium bonds.

Infrared spectroscopy is a good tool for studying some of the properties of the hydrogen bond itself. The increased access to the far-infrared region through Fourier transform spectroscopy should increase the amount of information which is necessary to complete the picture of the hydrogen bond.

B. Nature of Self-Association Species

1. Alcohols

The broad, multiband region of the spectrum from 3300 to 3700 cm^{-1} encompasses the various O–H stretching vibrations in alcohols. Most investigators agree that a variety of hydrogen-bonded species contribute to the bands in this region, but there are differences in interpretation. There is the question of whether a particular band can be assigned to monomer, dimer, or higher polymers. An alternative approach would assign a band to an O–H oscillator which is a donor, an acceptor, or both. The assignment of multimers as linear and/or cyclic must also be considered in interpreting spectra. These questions should be answered before the quantitative work using absorption data to calculate equilibrium constants between species can be considered valid.

In most alcohols there is a peak above 3600 cm^{-1} which is concentration dependent and is assigned to the monomer. Another peak somewhere near 3500 cm^{-1} which often increases and then decreases in intensity with increased concentration is assigned to some type of dimer. Bands which are assigned to higher polymers are usually much closer to 3300 cm^{-1}. Some bands of typical alcohols are listed in Table IV. Bellamy and Pace[22] attempted to explain why the dimer band in alcohols is not shifted as far from the monomer band as are higher polymer bands. They suggested that the dimer is linear, and the proton-donating molecule is therefore only a donor,

TABLE IV

FREQUENCIES (CM^{-1}) OF INFRARED BANDS FOUND IN THE O–H STRETCHING REGION OF SOME ALCOHOLS

Species	Monomer	Free OH of dimer	Linear dimer	Cyclic dimer	Polymer
MeOH in CCl_4 [a]	3642	3637	3534	—	3346
Phenol in CCl_4 [a]	3611	3599	3481	—	3352
EtOH in toluene [b]	3640	—	—	3490	3340
		Acceptor	Donor		Donor and acceptor
Phenol in CCl_4 [c]	3611	3599	3481		3393

[a] L. J. Bellamy and R. J. Pace, *Spectrochim. Acta* **22**, 525 (1966).
[b] H. C. Van Ness, J. Van Winkle, H. H. Richtol, and H. B. Hollinger, *J. Phys. Chem.* **71**, 1483 (1967).
[c] A. Hall and J. L. Wood, *Spectrochim. Acta, Part A* **23**, 2657 (1967).

while in higher polymers most species are cyclic, and all the molecules act as both donors and acceptors in hydrogen bonds. This would result in a stronger hydrogen bond, and thus a larger frequency shift because the oxygen atom would be involved in two bonds to protons, weakening the original O–H bond. If dimers are linear, as Bellamy and Pace suggest, there should be an end or terminal proton of the dimer which is more acidic than the monomer proton, and one should be able to distinguish two bands near the monomer frequency. Bellamy and Pace did find two bands in that region in methanol solutions in toluene, carbon tetrachloride, and chloroform. They also found a second band near the monomer band in phenol solutions. In order to further confirm the existence of linear dimers they considered the spectrum of a mixed system of methanol and phenol. Phenol is more acidic than methanol, and in the mixture in carbon tetrachloride a dimer band was observed at 3393 cm^{-1}. The position of this band indicated that this dimer had a stronger hydrogen bond than either of the self-association dimers. By making a concentration study they concluded that this band was most probably due to a dimer, since over a broad concentration range no new bands appeared between 3393 cm^{-1} and the two monomer bands at 3642 and 3611 cm^{-1}. Polymer bands appeared at high concentrations near 3200 cm^{-1}. If, as was assumed, the dimer were linear, phenol would be the proton donor and methanol the base. If the dimer were cyclic, there would be two types of bonds, one for each type of donor–acceptor pair, and two bands would be seen in the dimer region. A further confirmation of the linear-type dimer was the fact that the concentration of the phenol monomer decreased on mixing, whereas that of the methanol monomer did not. The terminal methanol in the dimer would have an O–H stretching vibration included in the monomer band.

Matrix isolation work of Barnes and Hallam[23] on methanol in an argon matrix confirmed the concept of linear dimers in methanol. In the far-infrared region of the spectrum a cyclic dimer would be expected to have an intense hydrogen bond stretching mode, but a weak deformation mode due to the restricted configuration. An open-chain dimer should have stretching and deformation modes of comparable intensity. Two strong bands at 222 and 116 cm^{-1} were reported, supporting the linear dimer model.

Similar work by these authors on ethanol[24] in an argon matrix gave evidence for an open-chain dimer, but extra bands suggested that a second type of dimer, perhaps with a nonlinear hydrogen bond, was present also. The complete infrared spectrum of both these alcohols in matrices was measured and evidence found for a variety of higher polymers both linear and cyclic.

Van Ness *et al.*[25] made an infrared study of ethanol in a variety of solvents and assigned the three bands at 3640, 3490, and 3340 cm^{-1} (in *n*-heptane) to monomer, cyclic dimer, and higher linear polymers, respectively. They found no band near the monomer band to assign to the nondonor end of a linear polymer, and suggested that this O–H oscillator was spectrally similar to the monomer. They cautioned that the intensity of the monomer band should not be used to determine concentrations. Basically, Van Ness *et al.* disagreed with the idea that a hydrogen bond formed by an O–H which is an acceptor as well as a donor is stronger than one formed by an O–H which is a donor only. For this reason they had to postulate a different type of species with different bonding to account for the so-called dimer band at 3490 cm^{-1}. No explanation was given to explain why the cyclic dimer, which they suggested was responsible for this peak, has hydrogen bonds which are weaker than those in polymeric species. The only difference which might account for the differing strength would be the suggestion that the dimer bonds are bent in a cyclic model and are not linear, but it has not been shown that this would have enough effect on the strength of the bond.

Phenol and its derivatives have also been studied by means of infrared spectroscopy. Hall and Wood[26] found three bands in the O–H stretching region of phenol at 3611, 3481, and 3350 cm^{-1}. They saw no second component at 3599 cm^{-1}, but accepted its presence from the evidence of Bellamy and Pace. They attempted to explain this component by the following assignments: monomer, 3611 cm^{-1}; acceptor end group, 3599 cm^{-1}; donor end group, 3481 cm^{-1}; and donor and acceptor, 3393 cm^{-1}. A picture of

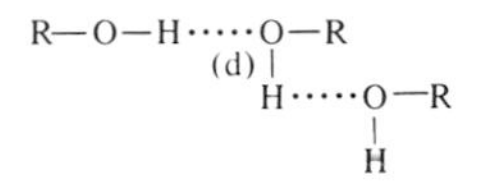

FIG. 1. Alcohol O–H bonds. (a) Monomer, (b) acceptor end group, (c) donor group, and (d) donor and acceptor.

where these bonds would be in a typical alcohol is shown in Fig. 1. These authors are in basic agreement with Bellamy and Pace, and disagreement with Van Ness. The evidence which Bellamy and Pace find of a band which can be assigned to an acceptor but not a donor O–H bond seems to be incontrovertible, and the further interpretation of Hall and Wood is consistent with the positions of the bands in this region of the spectrum. Those bands due to nondonor O–H bonds occur near 3600 cm^{-1}. Those due to donor, but not acceptor O–H bonds, occur near 3500 cm^{-1}, and those due to donor and acceptor O–H bonds occur below 3400 cm^{-1}. In a phenol–pyridine system Hall and Wood found only one band, at 3481 cm^{-1}, and this supports their interpretation that this band is characteristic of a donor end group which is not an acceptor.

Some infrared data has been used to calculate the species in solution without any a priori assumptions about the existence of any species but the monomer. Fletcher and Heller[27] looked at the first overtone region of 1-octanol and 1-butanol in *n*-decane over a broad concentration and temperature range. They did not want to use carbon tetrachloride as the solvent because of its supposed reactivity in hydrogen bonding systems. However, *n*-decane absorbs in the fundamental O–H stretching region, so they used the overtone region. The following equation was fitted both graphically and by computer:

$$A_0 = a_1/\varepsilon_1 + 2K_{1,2}a_1^2/\varepsilon_1^2 + \cdots + nK_{1,n}a_1^n/\varepsilon_1^n \tag{4}$$

A_0 is the total absorbance. They found a predominance of the fourth-order term, and assuming a monomer–tetramer equilibrium calculated enthalpies. Finally, they used all the data to calculate an equilibrium constant, and concluded that, while dimers and trimers may be present in these solutions, they are not present in sufficient amounts to be considered in the material balance equations for the overtone region.

Murty[28] gave reasonable infrared evidence for the existence of dimers in

TABLE V

FREQUENCY SHIFT ON DIMERIZATION OF VARIOUS ALCOHOLS AND PHENOLS IN CARBON TETRACHLORIDE[a]

Alcohol	$\Delta\nu$ (cm^{-1})
Methanol	132
Ethanol	126
Phenol	128
n-Propanol	124
n-Butanol	126
n-Pentanol	127
n-Octanol	121
n-Decanol	128
p-$CH_3C_6H_4OH$	130

[a] T. S. S. R. Murty, *Can. J. Chem.* **48**, 184 (1970).

a variety of alcohols. He found that in 24 alcohols (primary, secondary, and tertiary) and in phenol, where pK_a's ranged from 9.4 to 19.0, there was an O–H band 125 ± 5 cm^{-1} lower than the monomer stretching mode (Table V). This would seem to indicate the presence of dimers. In sterically hindered *tert*-butylcarbinol-*d* no bands were found below 3500 cm^{-1}. This alcohol cannot form cyclic polymers, and its second band must be due to a dimer.

Bufalini and Stern[29] found that the dimer in alcohols was the species most affected by the addition of electrolytes which could form association complexes with the alcohol. They studied the effect of the addition of Bu_4NCl, Bu_4NNO_3, *N,N*-dimethylaniline hydrochloride, $AgClO_4$, and Bu_4NCO_2H on methanol, 1-butanol, and *tert*-butanol in dilute benzene solution. As salt was added, the dimer peak disappeared and a new band at lower frequency appeared due to an association complex, $A^+X^- \cdots HOR$. The anion was important in determining the position of the O–H stretching frequency of the complex. The smaller the anion (greater charge density), the stronger the hydrogen bond and the lower the O–H stretching frequency. Later Raman work of Hester and Plane[30] confirmed the relation between the charge density of anions and the shift in the O–H band. In a study where *tert*-butylammonium salts were added to *tert*-butyl alcohol in carbon tetrachloride, Hyne and Levy[31] found that higher polymers were formed at the expense of the dimer. They postulated that the salt species may serve as a nucleation center for the aggregation of alcohol molecules in such a way that hydrogen bonding is favored.

The assignment of bands in the O–H stretching region to monomer, dimer, or higher polymers, while convenient, must be used with caution. The probability that the monomer band also contains bands due to the end molecules of linear polymers is quite high, and it should not be used quantitatively unless independent studies show that it is related only to the monomer concentration. Although the existence of the dimer in most alcohols (albeit in small amounts) seems to be confirmed by Murty and others, the question of whether it is linear or cyclic seems to depend on the alcohol studied. There is no doubt that in sterically hindered alcohols linear dimers are present, and the matrix studies of methanol seem to indicate that it too has a linear species. However, it does not necessarily hold as a general rule that what is found in one alcohol is found in others. As the length and configuration change, the ability to form cyclic dimers and *n*-mers is greatly enhanced. Distinguishing any further by infrared methods is difficult indeed. The designation of bands in the O–H stretching region according to donor and acceptor roles of the molecules is perhaps a realistic approach, but it does preclude equilibrium studies.

2. Carboxylic Acids

Infrared studies of carboxylic acids have been made in the crystal, in the solution, and in the vapor state. The O–H and C=O regions of the spectrum are both available for investigation as is the far-infrared region where the vibrations of the intermolecular hydrogen bond can be seen. Carboxylic acids are known to form dimers in substantial amounts, and a study of their self-association is somewhat less complex than a study of alcohols. Solvents have some effect on a monomer–dimer equilibrium in solution depending on the polarity and proton-donating or proton-accepting ability of the particular solvent.

A study was made by Lascombe *et al.*[32] of the C=O stretching region of acetic, propionic, butyric, caproic, and benzoic acids in carbon tetrachloride. Two bands were found in each case except in acetic acid. The lower frequency band was assigned to the monomer and the higher frequency band to the antisymmetric stretching mode of the cyclic dimer. Figure 2 shows a typical monomer, a cyclic dimer, and a polymer. The authors found that in strong proton-donating solvents like chloroform the species labeled (d) in Fig. 2 probably was in equilibrium with the dimer, and the lower frequency band did have increased intensity relative to the monomer band in these solvents. In strong proton-accepting solvents like acetonitrile, the species labeled (e) in Fig. 2 probably predominates, since the monomer band increased in intensity relative to the dimer band. When a series of ethers with different proton-acceptor abilities were used as solvents, the concentration of the monomer depended on the basicity of the ethers, decreasing in the order:

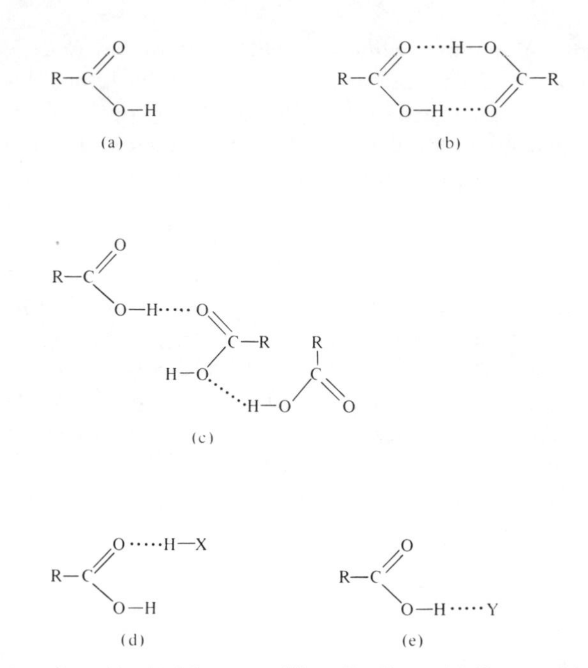

FIG. 2. Carboxylic acids. (a) Monomer, (b) cyclic dimer, (c) linear polymer, (d) complex with proton-donating solvent, and (e) complex with proton-accepting solvent.

dioxane > diethyl ether > isopropyl ether > butyl ether. These are fairly straightforward results, and are best explained in terms of a monomer–dimer equilibrium, with solvent interactions creating new equilibria. In acetic acid three bands were seen in the C=O stretching region in some proton-accepting solvents at some concentrations. This third band appeared in between the two bands assigned to monomer and dimer. If this band were due to a linear dimer, it should disappear with dilution; but it does not, so the authors attributed it to an overtone in Fermi resonance with the C=O stretching fundamental.

Bellamy and Pace[33] looked at the O–H stretching region of anhydrous crystalline oxalic acids. There can be two forms of solid oxalic acid. One is identical to the cyclic dimer (b in Fig. 2), and one is like a linear polymer (c in Fig. 2). The O–H stretching mode for the cyclic molecule is centered at 2890 cm^{-1}, while that for the linear is centered at 3114 cm^{-1}. The hydrogen bond is stronger in the cyclic form. Further studies in the region of the spectrum including ν_{C-O}, $\nu_{C=O}$, and ν_{O-H} confirmed this, but the interpretation was difficult owing to some strong coupling between C–O stretching modes and –OH bending modes.

In a continuing work Bellamy *et al.*[34] studied the spectrum of acetic acid vapor and acetic acid in solution. Because of the complex nature of the ν_{OH} band, they suggested the possibility of an equilibrium between open-chain and cyclic dimers. Each species would give a separate O–H absorption which would be broadened by Fermi resonance effects and by combination with low frequency intermolecular vibrations. The authors examined this band in the vapor and liquid states and in solution, and made a good case for the existence of both types of dimer, but their arguments are not conclusive. The most convincing evidence is that a superposition of the spectra of the linear solid and the cyclic solid oxalic acids gave a pattern similar to that of the vapor and solid states where the existence of both species has been postulated.

Jakobsen *et al.*[35] have attempted to correlate X-ray data of solid carboxylic acids from propanoic to undecanoic with the infrared and Raman spectra in solid, liquid, solution, and vapor. X-ray work shows a cyclic dimer in the solid of propanoic, butanoic, and pentanoic acids. The spectral similarities in the crystal, the liquid, the solutions, and the room temperature vapor indicated that if the crystal is a cyclic dimer, all the other states contain cyclic dimers. For the higher acids no Raman or X-Ray data were available to compare with the infrared data, but they might follow suit. Of course, this work was done in the far-infrared region of the spectrum, whereas the arguments of Bellamy *et al.* are based on the O–H stretching region, and they could both be correct.

Murty and Pitzer[36] studied a system where they tried to shift the equilibrium in favor of linear dimerization. They looked at the ν_{OH} and ν_{CO} region of trifluoroacetic acid. This is a strong acid which promotes solvation of the free O–H group and forces linear dimers. In a carbon tetrachloride solution, they found a strong free ν_{OH} near 3500 cm^{-1} and a free ν_{CO} at 1813 cm^{-1}. In benzene, which is slightly basic, ν_{OH} was split into two bands, 3400 cm^{-1} (terminal O–H of the dimer) and 3455 cm^{-1} (monomer). The band for the cyclic dimer was at 3250 cm^{-1}. In the C=O region of the spectrum two bands were found for the hydrogen-bonded C=O. The band at 1813 cm^{-1} was assigned to the monomer and the free end C=O in the linear dimer. The band at 1792 cm^{-1} was assigned to the terminal acid where C=O is hydrogen-bonded and O–H is free. The band at 1782 cm^{-1} was assigned to the C=O in the cyclic dimer. This type of interpretation might be useful in sorting out the bands encountered by Bellamy in acetic acid, but the spectrum is less complex in these stronger acids.

One good way to determine the structure of these acids in vapor or solution is to do a normal coordinate analysis using the symmetry of the cyclic dimer and comparing the calculated frequencies with the observed frequencies. Nakamoto and Kishida[37] did this for formic and acetic acid, and the agreement was very good, verifying the presence of cyclic dimers in the vapor. Jakobsen *et al.*[38] also did a normal coordinate analysis on these two acids assuming cyclic dimers and found good agreement between calculated frequencies and observed frequencies in both the vapor and solution. It is important to know if both linear and cyclic dimers are present in significant amounts in order to do equilibrium studies. For example, Allen *et al.*[39] made calculations from the spectral data of benzoic acid in solution assuming that only a monomer and a cyclic dimer are in equilibrium. They used the C=O stretching mode for free and bonded acid and found equilibrium constants and values for ΔH and ΔS. Both ΔH and ΔS were negative in the nonpolar solvents, benzene, carbon tetrachloride, and cyclohexane. These thermodynamic parameters were combinations of the ΔH and ΔS of dimerization (both negative) and the ΔH and ΔS of solvation (both positive). The major trend was that the better a solvent was at solvating, the more the positive contribution to ΔH.

Self-association in carboxylic acids is slightly simpler than in alcohols because the cyclic dimer is favored under many conditions without higher polymers interfering. The presence of cyclic dimers in the vapor state, as well as in the liquid and solid states, makes comparisons of spectra of these states more significant and thermodynamic studies more reliable.

3. Others

Infrared studies of self-association of nonaqueous solvents other than alcohols and carboxylic acids are not as numerous, but have been made. Becker[40] found that the integrated absorption coefficient of the C–H stretching band in $CHCl_3$ changed in a linear way with mole fraction and temperature. There was not a large shift in this frequency to indicate strong hydrogen bonding, but the concentration and temperature dependence of the integrated absorption coefficient indicated weak hydrogen bonding. These results cannot be considered conclusive, however, since dipolar interactions might be cited as the reason.

Hydrogen bonding in systems where N–H units are involved has also been studied using infrared spectroscopy. The simplest of these systems is NH_3, and it will be discussed later (*vide infra*), as Raman spectroscopy has become a major tool in its investigation. Amides hydrogen bond through the nitrogen proton to the carbonyl oxygen. Puranik and Ramiah[41] looked at the infrared spectrum of formamide in a variety of solvents and assigned N–H symmetric and antisymmetric stretching modes of a monomer and a trimer species. They attempted to confirm these assignments by attributing the C=O stretching bands to the monomer and trimer also. Kreuger and Smith[42] made an infrared study of fourteen alkyl amides and fifteen benzamides, and in each case found one, two, or three weak bands on the low frequency side of the antisymmetric N–H stretching mode of the free monomer. The positions of these bands were at 3578 ± 3, 3504 ± 4, and $3486 \pm 8\ cm^{-1}$. The intensities depended on concentration and the authors assigned the bands to a cyclic dimer, a cyclic trimer, and a cyclic higher associate, respectively. Although equivalent bands were not found near the symmetric stretching mode, these may have been shifted into the region where they would be in Fermi resonance with the overtone of the NH_2 bending vibration. The amides, like the carboxylic acids, have a good proton donor and good proton acceptor (electron donor) in the carbonyl oxygen which enables them to form association species (especially cyclic ones) easily.

The intensities of infrared bands can be used more quantitatively to determine the fitness of a self-association model. Lady and Whetsel[43] made a study of the overtone of the N–H stretching frequency of aniline in cyclohexane and followed its intensity as it varied with concentration and temperature. They made use of the limiting slope method of Liddel and Becker[44] which assumes dimerization and plots the change in apparent extinction coefficient with concentration. They also tried curve fitting of a monomer–dimer–tetramer model and made use of a stepwise association model. They concluded that the self-association of aniline involves a dimer and a tetramer. This is consistent over a wide range of concentration and temperature. Even though a trimer will fit the model, it is not necessary.

These are only some of the infrared self-association studies which have been made recently. Although none of the models or quantitative methods are exact in fitting the infrared data, the determination of the species present in solution must be made on an overall evaluation of the work done on a particular system. The alcohols remain the most difficult system to assess, especially the smaller, simpler ones because they can form a greater variety of species.

C. Solvent Effects

There has been considerable discussion in the literature in the past fifteen years about the role of the solvent in hydrogen bonding. The stretching frequencies of bonds next to the hydrogen bond often depend on the solvent used in the study. One approach to explaining why this occurs maintains that bulk properties of the solvent such as dielectric constant and refractive index correlate with the frequency shifts. This approach can be generalized to a theory that nonspecific bulk properties of the solvent affect the hydrogen bonding. Considerable work has been done to show that specific interactions between the solvent and hydrogen-bonded species are responsible for the changes in frequency which occur in different solvents.

An early attempt was made by Bellamy and Hallam[45] to rule out bulk dielectric effects on hydrogen-bonded species as being the important factor involved in frequency shifts. They studied the O–H stretching frequency of acetoxime, which forms a cyclic dimer, and diphenylcarbinol, in which the dimer predominates because of steric considerations. They looked at the frequency of the free O–H bond and of the associated O–H bond in a variety of solvents, and found that the band due to the associated O–H bond did not shift, whereas that due to the free OH oscillator did shift by different amounts in different solvents. The band was shifted least by the least hydrogen bonding solvent, *n*-hexane, and the shift increased with the hydrogen bonding ability of the solvent in the following order: CCl_4, $CHCl_3$, C_6H_6, CH_3NO_2, dioxane, and pyridine. In mixed solvents they saw multiple bands at invariant frequencies; a new band appeared with each solvent regardless of the solvent ratios. This seemed to be fairly significant evidence that the dielectric constant which differed in each of the solvents could not alone be responsible for the shifts in the O–H stretching bands. If it were, the bands due to the associated species would shift the same amount as the bands due to the monomer. Only specific interactions between the monomer and the solvent would give these results according to the authors.

A new concept was proposed by Allerhand and Schleyer[46]; it rejected both the dielectric theory nonspecific effects and the specific interaction theory of Bellamy. The rejection of the former was based on evidence similar to that of Bellamy and Hallam, i.e., that changing solvents does not affect the unassociated species and the associated species in the same way. The reasons why Allerhand and Schleyer do not agree with the specific interactions theory are more complex. They found that the stretching frequencies of a hydrogen-bonded complex, RO–H $\cdots$ B, where B is a proton acceptor, are sensitive to the solvent in both self-associated alcohol dimer systems and alcohol–ether systems. In addition, they have found that in mixed complexes the O–H stretching frequency of the RO–H $\cdots$ B complex is sensitive to the concentration of B. If there are specific interactions only between the monomer and the proton-accepting solvent, these changes in frequencies of the associated complex should not be occurring. In a further experiment, the authors looked at solvent shifts of the C=O stretching frequency in benzophenone and compared them to the RO–H $\cdots$ B stretching frequency shifts in the same solvents. They found a linear correlation regardless of the alcohol–ether complex studied and found it difficult to see what similar specific interaction there could be between the solvent and an RO–H $\cdots$ B group and between the solvent and an X=O group.

The somewhat new approach proposed by Allerhand and Schleyer was an empirical linear free energy relationship for the correlation of solvent-sensitive infrared vibrations,

$$(\nu^0 - \nu^s)/\nu^s = aG \tag{5}$$

where ν^0 is the stretching vibration frequency in the vapor phase (either free monomer or hydrogen–bonded complex) and ν^s is the frequency in solution. The symbol a is a function of the particular infrared vibration of a molecule and G is a function of the solvent only. The relationship is similar to the Kirkwood–Bauer–Magat[47] relationship which states

$$\Delta\nu/\nu = C(\varepsilon-1)/(2\varepsilon+1) \tag{6}$$

where C is a constant and ε is the dielectric constant. The Allerhand–Schleyer relationship suggests a new constant G, which cannot be related to a specific property of the solvent, but which must be empirically determined from the best fit of solvent shifts given in the literature for carbonyl and sulfonyl bands. Since a linear relationship has been shown between solvent effects on X=O and RO–H···B stretching vibrations, this was considered to be a valid method of determining the value of G. For the hydrogen-bonded complexes studied, the authors show good linear plots of the hydrogen-bonded O–H stretching frequencies against G. The stretching frequencies of the free O–H bonds do not correlate since they are expected to have specific interactions (weak hydrogen bonds) with the solvents. However, even in the very inert solvents such as hexane and cyclohexane there is no correlation and specific interactions cannot account for this.

The proposed parameter G has no relationship with the basicity, refractive index, or dielectric constant of the solvent. The authors have found it useful in predicting when the O–H stretching frequency of the complex RO–H···B will depend on the concentration of B. When the solvent and the proton acceptor B have similar G values, they have found that this frequency will be independent of solvent concentration; when the G values are very different, it will be concentration dependent.

These authors attempted to refute the specific interactions theory of Bellamy and Hallam[45] by giving a reason for the different bands which appear for each solvent in a mixed solvent study. They suggested that there might be nonequivalent sites in the solvent mixture, i.e., clusters of pure solvents as well as mixed solvents, giving different G values and thence different bands.

Although the above work is perhaps useful in predicting some trends, it does not really explain how the solvent interacts with either the monomer alcohol species (where G values do not correlate with frequencies) or the alcohol dimer or alcohol–ether complexes. Since it was shown that the shift in the carbonyl stretching frequency of benzophenone in different solvents correlated with O–H frequency shifts of hydrogen-bonded species in the same solvents, it is not surprising that G values which were determined using the carbonyl frequencies correlate with the O–H stretching frequencies. This empirical G value is explained somewhat in that it takes into account the total interaction energy between a molecule and its surroundings. This follows the theoretical treatment of Wiederkehr and Drickamer,[48] which resolves this total interaction energy into electrostatic, inductive, dispersive and repulsive contributions. Bellamy *et al.*,[49] while agreeing that these factors are important for systems involving changes from the gas phase to active solvents, tried to show the importance of a single mechanism for the case of transitions from one solvent medium to another. They reputed this mechanism to be dipolar, either hydrogen bonding or similar charge interaction, and attempted to use the results of Allerhand and Schleyer, as well as further results of their own, to substantiate or refute this theory.

While Allerhand attributes the frequency shifts observed on diluting alcohol–ether mixtures with chloroform to a change in base concentration, Bellamy suggests that these changes are due to the formation of a new species, the chloroform–alcohol–ether trimer. The chloroform hydrogen bonds to the oxygen atom of the hydroxyl group and this changes the polarity of the O–H bond and therefore changes the strength of the hydrogen bond to the ether. Previous work of Bellamy[22] already discussed in Section II,B,1 proposed that when this oxygen atom accepts a proton (donates a lone pair of electrons) the proton originally bonded to it becomes more acidic and forms a stronger hydrogen bond. The existence of the trimer or ternary compound would explain why the C=O stretching frequency in benzophenone or acetophenone shifts in solvents in a manner similar to the O–H stretching frequency of an RO–H ··· B complex, since the point of attachment for the solvent is the basic oxygen atom in both cases. This theory implies that the O–H stretching frequency in the alcohol–ether–chloroform system is a summation of two bands, one due to the alcohol–ether complex and one due to the trimer formed with the solvent. Bellamy *et al.* looked at methanol–diethyl ether–chloroform over a wide range of ether/chloroform concentrations and found that the changing band shape of ν_{OH} could be explained by a symmetrical band at 3508 cm^{-1} in a solution of ether in methanol and a symmetrical band at 3444 cm^{-1} in a methanol solution containing chloroform and ether mixed in a 49 to 1 volume ratio. At intermediate ether/chloroform ratios there appeared to be asymmetry. Figure 3 shows that the position of the maximum absorption of the complex does not vary smoothly with the composition of the solvent mixture. This is what would be expected if two species contributed to ν_{OH}.

Although Bellamy's work was not conclusive enough to reject the Allerhand and Schleyer interpretation, it was just as consistent with the experimental data. Additional experiments by Bellamy showed that if the O–H stretching frequency of a phenol–base complex was measured in different solvents and compared with that of the methanol–base in the same solvents, there was no correlation between the two. These results would not be expected according to the Allerhand–Schleyer theory since neither the base nor the solvents were

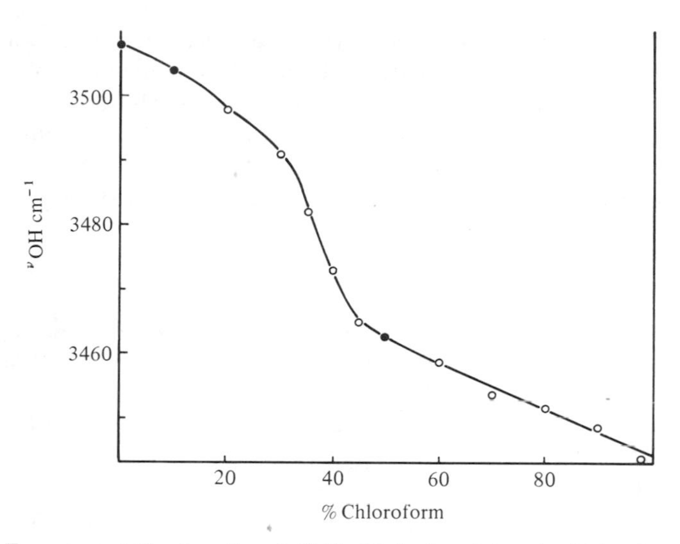

FIG. 3. Frequency shift of methanol (0.05 *M*) hydroxyl in ether/chloroform mixtures. From Bellamy *et al.*[49]

different, only the alcohol. However, this is explained rather well by the Bellamy trimer interpretation since the solvents which are proton donating would complex better with methanol, a stronger base than phenol, and affect its hydrogen bond to the base to a greater extent.

If the Bellamy interpretation is correct, there should be a difference in behavior between the alcohol dimer O–H stretching sensitivity to different solvents and the alcohol–ether O–H stretching sensitivity to the same solvents. The alcohol dimer can form hydrogen bonds with both proton-donating and proton-accepting solvents, whereas the mixed complex can form hydrogen bonds with proton-donating solvents only. Bellamy found that there is a linear relationship between the O–H stretching frequency of RO–H···B in different solvents and the C=O stretching frequency of acetophenone in these solvents much the same as Allerhand and Schleyer did. However, there was no correlation between the O–H stretching frequency of a phenol–methanol complex and the C=O stretching frequency of acetophenone. A great deal of data reported by Bellamy supports this general result. In proton-donating solvents such as chloroform the shifts in ν_{OH} as compared with hexane depend on the basicity of the oxygen atom of the bonded O–H group, which, in turn, depends on the strength of the hydrogen bond, be it with ether ($\Delta\nu_{OH} = 75$ cm^{-1}), dioxane ($\Delta\nu_{OH} = 52$ cm^{-1}), or another alcohol ($\Delta\nu_{OH} = 46$ cm^{-1}). In proton-accepting solvents the complexes and the dimers give very different results as would be expected.

Other workers have also supported the idea of specific interactions between solvent and hydrogen-bonded species as well as between solvents and monomers. Gramstad[50] studied a mixture of phenol or pentachlorophenol with triphenylphosphine oxide in several solvent mixtures. He recorded the $\Delta\nu_{OH}$ (shift between monomer and associated phenol) and calculated the K_{assoc} for different mixtures of CCl_4/CBr_4 and CCl_4/C_6H_{12}. Some of these results are reported in Table VI. He found a linear relation between K_{assoc} and the

TABLE VI

HYDROGEN BONDING OF PHENOL AND PENTACHLOROPHENOL TO TRIPHENYLPHOSPHINE OXIDE: FREQUENCY SHIFT OF PHENOL ON ASSOCIATION, ASSOCIATION CONSTANT, SOLVENT MIXTURE[a]

Phenol		Pentachlorophenol		
$\Delta\nu_{OH}$ (cm^{-1})	$K_{20°C}$ (liter/mole)	$\Delta\nu_{OH}$ (cm^{-1})	$K_{20°C}$ (liter/mole)	Solvent
430	1055.4	578	673.6	100% CCl_4– 0% CBr_4
	836.5		611.3	95% CCl_4– 5% CBr_4
	729.5		514.6	90% CCl_4–10% CBr_4
	643.1		462.8	85% CCl_4–15% CBr_4
	612.6		444.1	80% CCl_4–20% CBr_4
425	500.0	570	390.3	75% CCl_4–25% CBr_4
425	451.6	575	353.0	70% CCl_4–30% CBr_4

[a] T. Gramstad, *Spectrochim. Acta* **19**, 1363 (1963).

CCl_4/CBr_4 composition indicating an interaction between CBr_4 and the base. The P=O stretching mode was also lowered in the presence of CBr_4. This interaction predominated over any solvent effect of CBr_4 on the hydrogen-bonded complex. Rather than forming a trimer, the solvent competes with the phenol for the base.

Cole and Michell[51] performed experiments similar to those done by Bellamy. They used methanol, *tert*-butanol, and phenol in chloroform/diethyl ether, diethyl ether/benzene, and benzene/dichloromethane mixtures, and in all cases but one (*tert*-butanol in benzene/dichloromethane) they could resolve two bands in the O–H stretching region of the associated complex. They attributed these results to specific solvent–solute interactions which appeared to be competitive since the intensities of the bands were not proportional to the relative concentrations of the solvents. While Bellamy *et al.* claimed there were two bands in this system, Cole and Michell showed these bands more clearly as seen in Fig. 4. The Allerhand–Schleyer interpretation does not explain these two distinct bands in the region where the O–H

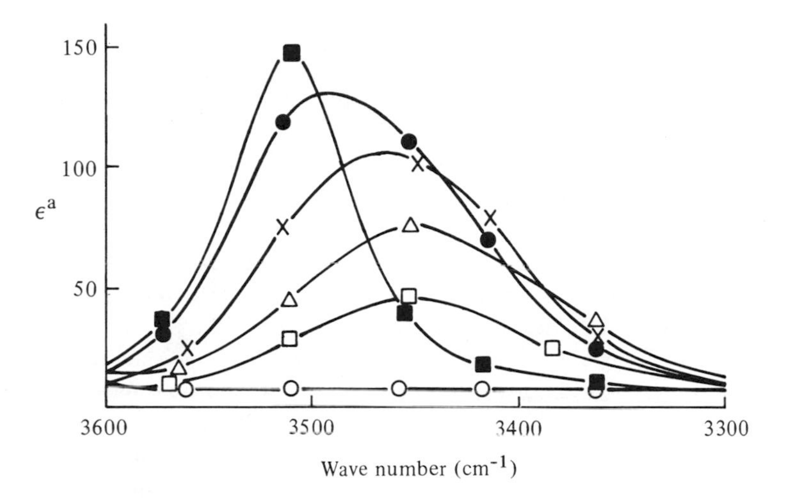

FIG. 4. Spectra of the O–H stretching vibration of methanol in chloroform/ether mixtures. Proportions of chloroform: ○, 1.00; □, 0.89; △, 0.79; X, 0.54; ●, 0.31; ■, 0.00. From Cole and Michell.[51]

stretching mode of an associated complex would be expected. Only a two-species equilibrium which depends on the solvent concentration and thus implies specific interactions can account for it.

Hirano and Kozima[52] in a study which attempted to determine the mechanism of hydrogen bonding in the vapor state and in solution found that there was a correlation between $\Delta\nu_{OH}$, a measure of the strength of the hydrogen bond in a methanol–triethylamine complex, and $1/\varepsilon$, the reciprocal of the dielectric constant of each of the three solvents used. However, this linear relationship could not be extrapolated to the vapor state where $\varepsilon = 1$. These results did not rule out specific interactions, however, since the correlation with dielectric constant involved three fairly similar solvents, all of which are proton donors to some degree.

Further studies have been made more recently by Huong and Lassegues[53] to confirm or further explain the Bellamy ternary complex theory of solvent interaction. They studied the infrared stretching frequency of an A–H oscillator (associated by hydrogen bonding), while varying the nature of the donor (phenol or pyrrole), the base, and the solvent. They found that in a mixture of a polar base and a nonpolar solvent, if the amount of base is increased relative to the amount of donor, the ν_{AH} frequency is decreased. It was suggested that there is a dipolar interaction of the base with the complex, causing a stronger hydrogen bond. In a slightly polar base and a proton-donating solvent, there was a sharp lowering of ν_{AH} of the associated species with increasing concentration of base, and then a frequency increase. This

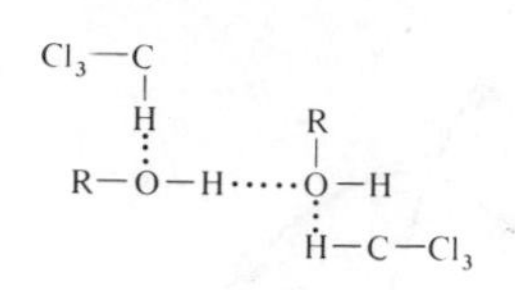

FIG. 5. Quaternary complex.

could be explained first by the formation of the ternary complex suggested by Bellamy, which should cause a lowering of ν_{AH} of the complex. Then, the increase in base could cause quaternary complexes to be formed in which a second solvent molecule donates a proton to the base, as shown in Fig. 5, and the hydrogen bond is weakened. This possibility is a reasonable explanation of the results, and the trends noted do confirm Bellamy's work.

The study of solvent interactions with proton donors, proton acceptors, and hydrogen-bonded complexes led some workers to reevaluate what an "inert solvent" is. Fletcher[54] calculated an association constant, K, for a 1:1 complex between 1-octanol and carbon tetrachloride. He made use of the previously calculated K value for the cyclic tetramer of the alcohol in n-decane to determine the concentration of the free monomer in CCl_4 and of the CCl_4 complex by difference from the total monomer concentration measured directly. He found an enthalpy value ($\Delta H = -0.5$ kcal/mole) for the complex. Christian and Tucker[55] disagreed with this approach and claimed that one cannot use this K value to correct the association constants of alcohols in CCl_4 and that it is hard to say whether specific or nonspecific interactions cause deviation from ideality in carbon tetrachloride. They felt that nonspecific interactions are overwhelming and that equilibrium constants must change with characteristics of the particular solvent. Fletcher[56] answered by arguing that the major characteristic of a solvent which can cause a change in K_{assoc} is the extent to which the solvent has nonbonding electrons causing specific interactions between solute and solvent. In two solvents with the same dielectric constant (n-decane and CCl_4) there is a 15-fold change in equilibrium quotient and only a complex between CCl_4 and the alcohol could account for this.

In another study to examine the interaction of carbon tetrachloride with alcohols, Robinson *et al.*[57] studied the temperature dependence of the molar absorptivity of the fundamental O–H stretching mode of phenols in tetrahydrofuran, benzene, chlorobenzene, carbon tetrachloride, and perfluoromethylcyclohexane. They concluded that the temperature dependence of the molar absorptivity of phenol in carbon tetrachloride results from phenol–CCl_4 hydrogen bonding.

One conclusion which can be drawn concerning solvent effects on hydrogen bonding is that where proton-donating or proton-accepting solvents are concerned, specific interactions are quite likely to occur. These interactions are more important than dielectric properties, in comparing solvents. When comparing vapor phase hydrogen bonding to solution phase hydrogen bonding, both effects must be taken into account. It is also difficult to determine which solvents can be considered inert. Even if solvents have molecules with no overall dipole moment, the bond moments (e.g., in CCl_4) can interact with hydrogen bonding solutes as hydrogen bonding is essentially an ionic interaction. With care, however, solvent systems can be chosen which have minimum interaction with the solutes.

D. Thermodynamic Studies

Various workers studying hydrogen bonding phenomena in nonaqueous systems have attempted to use infrared data to calculate thermodynamic parameters. Relationships between frequency shifts and the enthalpy of the hydrogen bond and between the strength of the hydrogen bond and the acidity of the donor or basicity of the proton acceptor have been proposed. The most significant relationship between ν_s, i.e., the shift in the X–H stretching frequency upon hydrogen bond formation, and the enthalpy of the hydrogen bond was a linear relationship proposed by Badger and Bauer in 1937.[58] Although this relationship will not hold for a wide variety of donors and acceptors, it appears to hold well for one donor and a series of bases which have the same proton-accepting atom, but which change in basicity because of a change in the rest of the molecule.

In most of the work which will be reported in this section, ternary systems (donor, acceptor, and inert solvent) where the concentration of base is much greater than the concentration of donor (alcohol) will be studied. This minimizes the amount of self-association of the proton donor. The association constant, K, for the 1:1 complex is usually calculated from the optical density or absorption of the band due to the monomer. Becker[59] does a typical calculation of the formation of a complex from a monomer and a base:

$$\mathrm{M} + \mathrm{B} = \mathrm{C} \tag{7}$$

If m_0 and b_0 are the initial concentrations of the monomer and the base, m, b, and c are equilibrium concentrations and the equilibrium constant is

$$K = c/mb = \frac{m_0 - m}{m(b_0 - m_0 + m)} \tag{8}$$

It has already been noted that in order to avoid self-association, b_0 must be much greater than m_0 or m. Therefore,

$$K = (m_0 - m)/mb_0 \tag{9}$$

If D, the optical density of the monomer, is equal to αmd (α = molar absorption coefficient and d = path length) and $D_0 = \alpha m_0 d$, then

$$(m_0 - m)/m = (D_0 - D)/D \tag{10}$$

and

$$K = \frac{D_0 - D}{Db_0} \tag{11}$$

Only two measurements are necessary, one of a solution containing B and M, the other containing only M. If performed at two temperatures, $-\Delta H$ and then ΔG and $-\Delta S$ can be calculated. All these parameters increase the information which can be known about these systems.

Becker[59] studied the infrared spectrum of methanol, ethanol, and *tert*-butanol with six bases in carbon tetrachloride. The $\Delta\nu_s$ and thermodynamic parameters are reported in Table VII. He found no trends in $-\Delta H$ for any of the three alcohols with one base. In general, acetone, ethyl acetate, benzophenone, and dioxane have $-\Delta H$ values in the range 2.5 to 3.2 kcal/mole. Dimethylformamide and pyridine have higher $-\Delta H$ values near 3.9 kcal/mole. When a Badger–Bauer type plot of $\Delta\nu_s$ versus $-\Delta H$ was made, all data corresponding to the fifteen O–H···O complexes were fairly close to the line, whereas those representing the three pyridine complexes were way off the line. Later work confirmed this lack of correlation between different types of bases (*vide infra*).

Wimette and Linnell[60] attempted to look at one type of base with pyrrole as the proton donor. Their values for $-\Delta H$ and $-\Delta S$ are listed in Table VII. The addition of a methyl group should increase the basicity of the pyridine and this can be seen when comparing $-\Delta H$ and $-\Delta S$ of pyridine and 2-methylpyridine; stronger hydrogen bonds are formed. However, this trend does not extrapolate to 2,6-dimethylpyridine. In this case, steric factors may influence $-\Delta H$ and $-\Delta S$ and interfere with the trend.

It is important to make these infrared measurements in systems which have a relatively inert solvent. Mitra[61] made a study of alcohols in nitriles both with and without CCl_4 as a solvent. (Results in ternary systems only are reported in Table VII.) He found that the frequency shifts in the binary systems were systematically larger than those in the ternary systems. In the binary systems, where the nitrile was the solvent, he suggested that the bulk dielectric properties of this solvent probably have a significant effect on the hydrogen bond. Acetonitrile has a very high dielectric constant (38.8) compared with CCl_4 (2.24). In the three-component system Mitra did find a linear correspondence between $-\Delta H$ and $\Delta\nu$. It was found that in the series of nitriles the $\Delta\nu$ was smaller for bonding with the unsaturated than the saturated nitrile. This is consistent with the electrostatic nature of the

TABLE VII

INFRARED FREQUENCY SHIFTS AND THERMODYNAMIC PARAMETERS OF ALCOHOL–BASE COMPLEXES

Alcohol	Base	$\Delta\nu$ (cm^{-1})	$-\Delta H$ (kcal/mole)	$-\Delta S$ (e.u.)	Ref.
MeOH	Acetone	112	2.52	7.3	*a*
	Benzophenone	88	2.16	6.5	*a*
	Ethyl acetate	84	2.52	7.8	*a*
	Dioxane	126	2.80	8.6	*a*
	DMF	160	3.72	9.1	*a*
	Pyridine	286	3.88	10.8	*a*
	Acetonitrile	81	2.25	—	*c*
EtOH	Acetone	109	3.46	11.2	*a*
	Benzophenone	84	3.23	10.5	*a*
	Ethyl acetate	80	2.33	7.8	*a*
	Dioxane	123	3.09	10.3	*a*
	DMF	155	3.88	10.5	*a*
	Pyridine	276	3.66	10.5	*a*
tert-Butanol	Acetone	101	2.94	9.8	*a*
	Benzophenone	77	2.67	8.9	*a*
	Ethyl acetate	75	2.92	10.1	*a*
	Dioxane	118	2.94	9.6	*a*
	DMF	143	3.92	11.0	*a*
	Pyridine	252	3.98	12.6	*a*
Phenol	Acetonitrile	160	3.92	—	*c*
	Acrylonitrile	145	3.60	—	*c*
	BrCN	102	2.60	—	*c*
Pyrrole	Acetonitrile	72	1.92	—	*c*
	Pyridine	—	3.2	8.9	*b*
	2-Methylpyridine	—	3.8	10.8	*b*
	2,6-Dimethylpyridine	—	3.4	9.2	*b*
Propan-1-ol	Pyridine	—	4.3	—	*d*
Propan-2-ol	Pyridine	—	6.1	—	*d*
Butan-1-ol	Pyridine	—	5.0	—	*d*
Butan-2-ol	Pyridine	—	4.1	—	*d*

[a] E. D. Becker, *Spectrochim. Acta* **17**, 436 (1961).
[b] H. J. Wimette and R. H. Linnell, *J. Phys. Chem.* **66**, 546 (1961).
[c] S. S. Mitra, *J. Chem. Phys.* **36**, 3286 (1962).
[d] T. J. V. Findlay and A. D. Kidman, *Aust. J. Chem.* **18**, 521 (1965).

hydrogen bond. Phenol–acetonitrile formed the strongest hydrogen bond, while pyrrole–acetonitrile formed the weakest hydrogen bond. Oxygen is more electronegative than nitrogen so the proton of an O–H bond is more acidic than the proton of an N–H bond and the former will form a stronger bond. Zeegers-Huyskens *et al.*[62] found that in a study of hydrogen bonding between alcohols and amines the association constant is higher when the amine is more basic and when the alcohol is more acidic. They carried these results a little further and defined a value D where

$$D = \log K_{\text{assoc}} - (\log K_{\text{i}_{\text{alcohol}}} + \log K_{\text{i}_{\text{base}}}) \qquad (12)$$

K_{i}'s are ionization constants for the alcohol and the amine. They found this value of D to be constant for aliphatic amines and alcohols. These thermodynamic studies helped confirm the theory which treats the hydrogen bond as an electrostatic entity.

Gramstad[63] found similar trends in nitrogen compounds with phenol. There was a correlation between $\log K_{\text{assoc}}$ of the nitrogen base complex with phenol and the heat of mixing of these bases with chloroform. In addition there was a linear relationship between $\log K_{\text{assoc}}$ of pyridines and quinolines with phenol and the pK_a values of the bases. Only tertiary amines did not fit in this picture. Further work by Gramstad[64] studied at 1:1 association of phenol with eight esters, three acid fluorides, sixteen ketones, fourteen aldehydes, and fifteen ethers in carbon tetrachloride. In previous work[65,66] he had studied the ability of organophosphorus compounds and *N,N*-disubstituted amides to hydrogen-bond with phenol, and in the work just mentioned, he looked at this ability in tertiary amines and pyridines. In all these systems there has been a proportionality shown between the log of the association constant of the complex and the frequency shift, $\Delta\nu$, of the O–H band of the proton involved in the hydrogen bond. The ability to form a hydrogen bond with phenol was shown to decrease in the order: organophosphorus compounds > amides > esters > ketones > aldehydes > ethers > pyridines and tertiary amines. In each system studied $\Delta\nu_{\text{OH}}$ varied linearly with $-\Delta H$ and $-\Delta S$. There was a difference between the $\log K_{\text{assoc}}$ vs. $\Delta\nu_{\text{OH}}$ plots and the $-\Delta H$ vs. $\Delta\nu_{\text{OH}}$ plots which has been accounted for by the various negative entropies of association which increased in the order: organophosphorus compounds < amides < esters < ketones < aldehydes < ethers < tertiary amines and pyridines. Gramstad attributed this entropy change to an increase in the polarity of the hydrogen-bonded complex along this series which resulted in a greater orientation of solvent molecules around it, and an increase in the negative entropy of association. Differences in hydrogen bonding within groups as substituents were changed were explained in terms of the influence of mesomeric and inductive effects. The differences in hydrogen bonding ability of compounds containing the carbonyl group were explained in terms of the differences in polarizability of the various proton acceptors and the differences in resonance stabilization or charge delocalization in the complexes.

These studies by Gramstad give a thorough characterization of proton acceptors in hydrogen-bonded complexes. An additional study[67] was added which looked at sulfoxides and nitroso compounds. It was found that the ability of an S=O group to accept a proton lies between amides and esters and that of an N=O group lies between ketones and aldehydes.

Gramstad had determined that if a proton donor is changed, no correlation between $\Delta\nu$ and $-\Delta H$ exists. Findlay and Kidman[68] studied solutions of pyridine with various propanols and butanols (Table VII) and found this to be the case. However, varying the proton donor with the base can give some information about the type of hydrogen bond formed. Zeegers-Huyskens[69] measured the O–H and N–H stretching modes in the infrared spectra of a variety of alcohols with propylamine. She found that the symmetric and antisymmetric stretching modes of the N–H bond of the amine were shifted according to the acidity of the alcohol and that in a solution with *m*-nitrophenol, the O–H stretching frequency disappeared and a large band appeared between 2500 and 2750 cm^{-1}. She attributed this band to an $N–H^{+} \cdots O^{-}$ species, i.e., the proton is actually transferred from the alcohol to the amine, and an ion-pair is formed. Segal[70] had found similar evidence for a strong hydrogen bond between ethylamine and chloroform with the occurrence of a new N–H stretching mode at 4.00 μm (2500 cm^{-1}). Another study of alcohols was made by Motoyama and Jarboe[71] with ethyl and isopropyl ethers as bases. The inductive effect of the extra methyl group on the isopropyl ether was said to be the reason why this ether was a better base in terms of association constants. This is in line with other workers. Naidu[72] made a study of substituted phenols with dioxane in carbon tetrachloride to determine how the $\Delta\nu_s$ was affected by substituents on the phenol. He found that *o*-, *m*-, and *p*-methyl substitution has very little effect and, therefore, the induction of the methyl group on the proton donor is not very important. However, *o*-, *m*-, and *p*-chlorophenols cause large shifts in $\Delta\nu_s$ due to the electron-withdrawing power which causes a net decrease in electron density on the oxygen. This is conducive to the formation of stronger hydrogen bonds in the order: ortho > meta > para.

Gramstad did some more recent infrared work on thioamides, nitriles, and amides with phenol and ethanol, and summed up some of the thermodynamic relationships with spectral data.[73] If a hydrogen-bonded complex is represented by $R'XH \cdots BYR''$, there will be a linear correlation between $-\Delta H$ and $\Delta\nu_{OH}$ with the same R'X, if YR'' changes structurally. However, if B, the proton-accepting atom varies or has different hybridization, this linearity will not hold. Although there are contributions to entropy (and thus to the *K* of association) arising from the polarity of the hydrogen-bonded complex which orders the solvent, the steric effects of the bulky substituents on the species involved in the hydrogen bond also affect the entropy of the system. Bellamy *et al.*[74] tried to differentiate between these steric (entropy) terms and the energy of the hydrogen bond as contributors to the equilibrium quotient, K. They made a study of ortho-substituted phenols with ethers, proceeding from the premise that $\Delta\nu$ is directly related to O···O distances, and these are determined by proton donor and acceptor properties unless they are modified by steric hindrances. In unhindered phenols where steric effects are unimportant, $\Delta\nu/\nu$ of a donor can be plotted against $\Delta\nu/\nu$ of another donor in the same solvents, and a straight line would be obtained depending on the basicity of the solvents. If steric effects interfere, there would be changes in the O···O distances and deviation from the straight line. The $\Delta\nu$ values for unhindered phenols correlated with the basicities of the ethers, even the bulky di-*tert*-butyl ether which is quite basic. When the $\Delta\nu$ values for the ortho-substituted and the 2,6-di-*tert*-butylphenols were measured, di-*tert*-butyl ether moved down the list as an effective acceptor, and K values did not follow the basicities of the ethers. This was attributed to the entropy of the system which is determined by the proportion of collisions which are effective in leading to hydrogen bond formation.

Singh and Rao[75] followed up on the work of Bellamy and found that with hindered phenols and alcohols in a variety of proton-accepting solvents $\Delta\nu$ decreased with an increase in the bulk of the ortho substituent. The $\Delta\nu$ was not a measure of the enthalpy of the hydrogen bond in these cases. Very bulky substituents can thus affect both $\Delta\nu$ which is often used as a measure of the enthalpy and K which is determined by both the enthalpy and the entropy of the system.

The thermodynamic calculations which can be made on hydrogen-bonded species are somewhat limited to the mixed complexes. As has been shown in Section II, B, specific self-association species have not been well characterized in terms of infrared bands, and the absorptions measured cannot be considered proportional to concentrations of particular species. However, the possibilities for information about hydrogen bonding in mixed systems by means of thermodynamic studies are varied and quite good. Such data as acidities, basicities, polarizabilities, and configurations of donors and acceptors can be used effectively in determining more information about hydrogen bonding.

III. Raman Spectroscopy

One of the more interesting solvents which has shown evidence of extensive hydrogen bonding is anhydrous ammonia. It is similar to water in its behavior, but its low freezing and boiling points indicate that it has less hydrogen bonding than water or alcohols. Ammonia has been an interesting subject for study by means of vibrational spectroscopy since it was expected that with less hydrogen bonding, lower polymer species might be identified with spectral bands. While water molecules show evidence for hydrogen bonding in two directions,[76] ammonia was thought to form linear species with only one hydrogen atom per molecule taking part in a hydrogen bond. This theory was put forth by Pimentel *et al.*[77] on evidence of matrix isolation studies of the dimer by infrared spectroscopy. Although there have been infrared studies of liquid ammonia, some of the more recent Raman studies will be discussed in this section as examples of how Raman spectroscopy has been used to investigate hydrogen bonding phenomena. The development of laser Raman technology has made ammonia systems accessible to this type of vibrational study, and Raman spectroscopy can provide additional data in terms of depolarization ratios which infrared spectroscopy cannot.

The Raman spectrum of liquid ammonia exhibits a complex envelope between 3100 and 3500 cm^{-1} composed primarily of N–H stretching bands (Fig. 6). Plint *et al.*[78] recorded the Raman spectrum of ammonia with a mercury arc lamp as the exciting source. They assigned three bands in this region at 3218, 3300, and 3373 cm^{-1} as $2\nu_4$, ν_1, and ν_3, respectively, of an ammonia molecule with C_{3v} symmetry. ν_1 is the symmetric stretching mode,

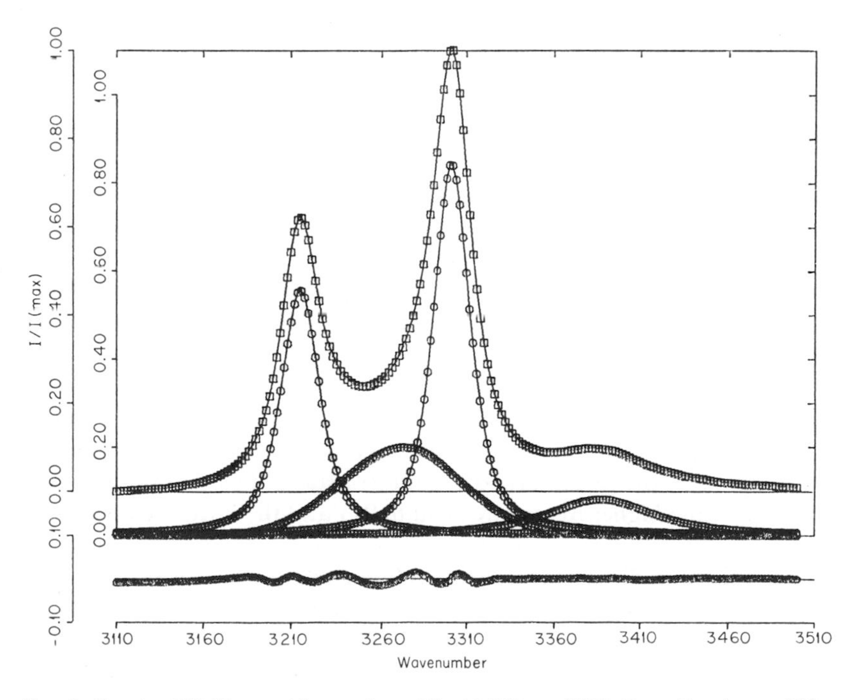

FIG. 6. Resolved N–H stretching region of liquid NH_3 at 25°C. From Lemley *et al.*[83]

spectrum of ammonia at 25°C and −71°C was measured, and the envelope between 3100 and 3500 cm^{-1} was resolved into four bands by computer techniques. Significance tests were made on three-, four-, and five-band fits, and the four-band resolution gave the best fit in the statistical sense. The bands were assigned in a manner somewhat similar to the assignments of Gardiner *et al.*,[82] i.e., the band at 3214 cm^{-1} was assigned to $2\nu_4$ and the other three bands were assigned to fundamental stretching modes of ammonia species. The various assignments are given in Table VIII.

TABLE VIII

RAMAN STRETCHING FREQUENCIES AND ASSIGNMENTS IN NH_3-LIKE SYSTEMS[a]

Symmetry	Mode	ND_3	ND_2H	NH_3 (25°C)	NH_3 (−71°C)
C_{3v}	$2\nu_4$	2347	—	3214	3209
	ν_1	2403	—	3300	3298
	ν_3	2521	—	3385	3380
C_s	ν_2	2373	—	3271	3261
	ν_1	2403	3342, 3367	3300	3298
	ν_5	2521	—	3385	3380

[a] A. T. Lemley, J. H. Roberts, K. R. Plowman, and J. J. Lagowski, *J. Phys. Chem.* **77**, 2185 (1973).

The reported fourth band which occurred near 3270 cm^{-1} in ammonia solutions and near 2373 cm^{-1} in ND_3 solutions indicated that species other than free NH_3 molecules with C_{3v} symmetry were present in solution. It also precluded totally hydrogen-bonded NH_3 molecules which would retain C_{3v} symmetry as seen in the solid state. The characterization of this band as due to another type of NH_3 species was supported by evidence from the spectrum of the species ND_2H. The N–H stretching region of this species would be expected to have only one band if all the molecules were the same. The fact that two bands were clearly present (Fig. 7), both polarized, indicated that two types of ND_2H molecules were present. The two bands were similar in separation and half-widths to the pair of bands at 3271 and 3300 cm^{-1} in the NH_3 spectrum. The only way that ND_2H could have two N–H stretching modes would be for one molecule type to be hydrogen-bonded through the hydrogen atom and the other type to be free. Using this information and the spectrum of NH_3, two models were proposed (Fig. 8). Model I would be expected to have three stretching modes giving rise to three bands. The lowest frequency band (N–H stretching mode of associated hydrogen atoms) should

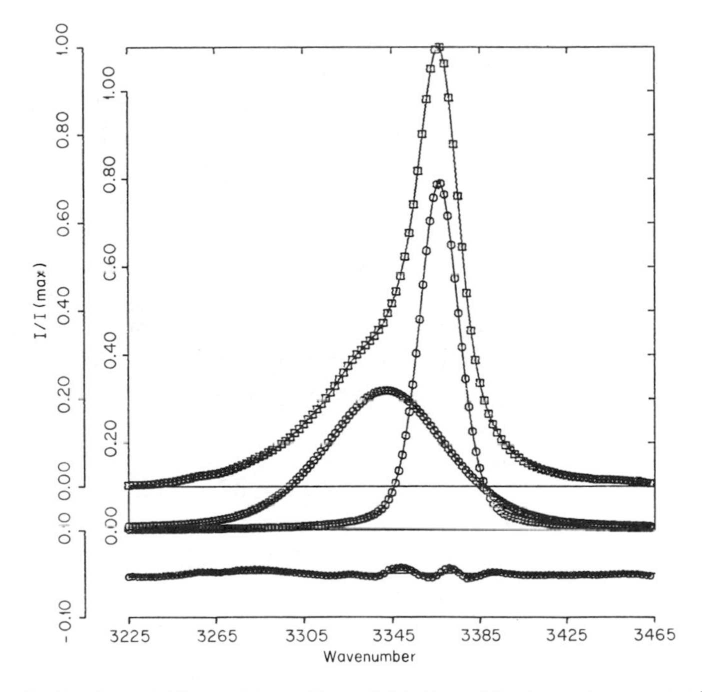

FIG. 7. Resolved N–H stretching of liquid ND_2H at 25°C. From Lemley *et al.*[83]

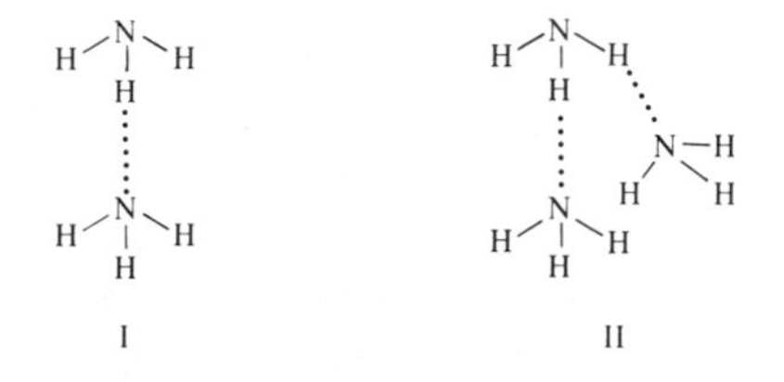

FIG. 8. Models for hydrogen bonding in ammonia.

be polarized. The free N–H bonds of this molecule would give rise to a symmetric and antisymmetric stretching mode and these would probably coincide in frequency with the stretching modes of end molecules of NH_3 with C_{3v} symmetry. These predictions were consistent with experimental data. Three bands in addition to the overtone were observed; the highest frequency one was depolarized, and the two of lower frequency were polarized. Model II would be expected to exhibit two bands (symmetric and antisymmetric) due to the stretching modes of the hydrogen-bonded N–H bonds and one band (symmetric) due to the stretching mode of the free N–H bond. This latter might be expected to coincide in frequency with the symmetric stretching mode of C_{3v} ammonia (3300 cm^{-1}). Thus the band at 3271 cm^{-1} could be assigned to the symmetric mode of the associated N–H bonds, but the 3385 cm^{-1} band would have to be assigned to the antisymmetric stretching mode of the associated N–H bonds *and* to the antisymmetric stretching mode of the free N–H bonds in C_{3v} ammonia. Two depolarized bands would be needed to account for these fundamentals, so Model II was ruled out on this basis and on the work of Pimentel *et al.*[77] with NH_3 dimers.

The authors also pointed out the difficulties inherent in a study similar to that of de Bettignies and Wallart.[80] When more than one type of ammonia species is postulated, the band assigned as $2\nu_4$ must encompass the overtone of the bending modes for these two species. One of these contributions would be in a resonance interaction with the fundamental stretching mode of the same molecule, e.g., the associated species at 3271 cm^{-1}, and the other contribution would be in resonance with the fundamental stretching mode of the NH_3 molecule with C_{3v} symmetry (3300 cm^{-1}). The intensities would not be measures of just one species concentration, so the point at which there would be perfect resonance could not be determined.

In addition, Lemley *et al.* disputed the contention of Gardiner *et al.*[82] that some of the intensity of the band at 3214 cm^{-1} was due to a fundamental stretching mode of an associated species. If this were so, an overtone of this band in the near-infrared region of the spectrum would be expected to be seen because overtones and combinations of the higher frequency fundamentals are of reasonable intensity. The region between 6060 and 6540 cm^{-1} is devoid of bands, however. The most consistent explanation for the 3214 cm^{-1} band appeared to be that it was due only to the overtones of the ν_4 antisymmetric bending modes of all species present. The final interpretation of liquid ammonia suggested that it consists of linear polymeric species incorporating NH_3 molecules associated by hydrogen bonding. The molecule at one end of this entity has C_{3v} symmetry, while the others have C_s symmetry, this perturbation being discernible in the stretching region but not in the bending region of the spectrum.

Following these Raman studies of liquid ammonia, work on salt solutions in ammonia was done in order to investigate the effects of these electrolytes on solvent structure. Gardiner *et al.*[84] looked at the Raman spectra of $LiNO_3$ and NH_4NO_3 solutions over a broad concentration range. They studied both the ammonia fundamental vibrations and the nitrate ion fundamental vibrations. They found that increasing the $LiNO_3$ concentration brought about a splitting of the degenerate antisymmetric stretching mode near 1357 cm^{-1} and of the degenerate symmetric bending mode near 700 cm^{-1}. They also found a broad band centered near 250 cm^{-1}, also present in the spectrum of pure liquid ammonia, and three new bands at 550, 350, and 250

cm^{-1} which became more evident with increasing salt concentration. Some changes were shown to occur in these bands at very high salt concentrations. The spectral changes in the NH_4NO_3 solutions were similar to those found in $LiNO_3$ solutions. The asymmetric stretching region of the nitrate ion was still at least a doublet at very low salt concentrations, however, and the band at 700 cm^{-1} remained a singlet at all concentrations. No bands were observed below 500 cm^{-1}.

These results were interpreted by the authors in terms of the formation first, of solvent separated ion-pairs, and then, of contact ion-pairs in the $LiNO_3$ solutions. The bands at 561 and 361 cm^{-1} were assigned to $[Li(NH_3)_n]^+$ species, while that at 245 cm^{-1} was assigned as a deformation mode of this species. The perturbations of the NO_3^- ion bands would be consistent with this interpretation. It was further postulated that NH_4NO_3 also interacts with ammonia, but not to the same extent owing to the lesser polarizing power of the NH_4^+ cation versus the Li^+ cation. Gardiner *et al.* supported their previous hypothesis that much of the intensity of the 3217 cm^{-1} band is due to NH_3 molecules associated with other NH_3 molecules, arguing that the presence of solutes which can provide alternate bonding possibilities for the NH_3 hydrogen atoms caused a decrease in the intensity of this band. They concluded that the NH_3 molecules which associated with a cation through the one pair of electrons on the nitrogen atom could account for the 3260 cm^{-1} band.

A somewhat different interpretation was made by Lemley and Lagowski[85] of Raman spectra of ammonia solutions of $NaNO_3$, NaSCN, NH_4NO_3, NH_4SCN, $LiNO_3$, and LiSCN. They found that in the sodium salt solutions, the band at 3260 cm^{-1} was shifted to lower frequency with increasing SCN^- ion concentration and remained almost the same or was shifted to slightly higher frequency in the presence of NO_3^- ion. These authors thus continued to attribute this band to NH_3 molecules associated through the hydrogen atom since it was affected by anions. The change in intensity of the 3217 cm^{-1} band was again attributed to changes in the amount of Fermi resonance with differing amounts of hydrogen bonding. It was found that a second band appeared in the NH_3 bending region (1050 and 1120 cm^{-1}) in the SCN^- solutions. (This region was obscured in the NO_3^- solutions.) An interaction through the hydrogen end of the NH_3 molecules with the anion would be expected to hinder this bending mode and cause an increase in frequency. An analysis of the SCN^- ion fundamental modes confirmed that where one would expect only one C–S stretching mode, two were found in all solutions. The positions of these bands indicated some type of bonding through the sulfur end of the ion. The positions were the same (740 and 750 cm^{-1}) in NaSCN and NH_4SCN solutions and by reason of intensity changes with concentration the 740 cm^{-1} band was assigned to an NH_3–SCN^- interaction.

Two bands were also found in the LiSCN solutions at 765 and 740 cm^{-1}. In addition, a second C–N stretching mode at slightly higher frequency was found. This was interpreted as being due to desolvation of the nitrogen end of SCN^- as a result of the highly polarizing Li^+ ion being strongly solvated. In the NaSCN and NH_4SCN solutions this desolvation was not thought to occur because anion interactions with the solvent dominated cation interactions.

The nitrate bands in these solutions were interpreted as follows. The splitting of the antisymmetric stretching mode in NH_4NO_3 solutions, but not the symmetric bending mode, is considered to be due to weak solvent interaction with the nitrate ion. The constant splitting of both these bands in the $NaNO_3$ solutions implies some contact ion-pairing at all concentrations studied. The $LiNO_3$ solutions do not show splitting in the symmetric bending mode region until an 11 to 1 molar ratio is reached. The lithium ion is strongly solvated with ammonia and does not form inner sphere ion-pairs easily. The amount of splitting in the antisymmetric stretching mode changes with concentration, also implying a shift from outer sphere to inner sphere ion-pairing as suggested by Gardiner. However, better spectral resolution showed that inner sphere ion-pairing occurred before the molar ratio of 4 to 1, as Gardiner postulated. The conclusions of this work are that Na^+ and NH_4^+ do not interfere greatly with the structure of bulk ammonia. Li^+ does order NH_3, thereby limiting some NH_3–anion interactions. SCN^- interacts strongly with the solvent (and the cations) and NO_3^- interacts weakly. The model of liquid ammonia postulated in the work of Lemley *et al.*[83] was supported by these results. Later work by Plowman and Lagowski[86] showed that the ~250 cm^{-1} band seen by Gardiner must be assigned to the symmetric stretching mode of $[Li(NH_3)_4]^+$. The other two bands seen by Gardiner were not observed by these authors.

Plowman and Lagowski measured the Raman spectra of ammonia solutions of alkaline earth and some alkali metal salts. They found that the N–H stretching region was not perturbed by these highly polarizing cations. (The anions used were ClO_4^- and NO_3^- which were thought to have weak interactions with the solvent, if any.) However, it was found that the symmetric bending mode of ammonia was shifted to higher frequency and this shift was proportional to the charge density of the cation. This was shown earlier by Corset,[87] and indicated a strong cation–ammonia interaction. Low frequency bands were assigned to the symmetric stretching mode of the solvated cation and these frequencies are reported in Table IX. An electrostatic model taking into account ion-dipole, ion-quadrupole, dipole-dipole, dipole-quadrupole, and quadrupole-quadrupole interactions was used to calculate the interaction energy of the cation in a cage of solvent molecules. A hard sphere model was used to determine R, the distance from the center of the ion to the center of the NH_3 dipole. It was found that an octahedral cage was energetically favored over a tetrahedral cage for both monovalent and divalent cations. By means of normal coordinate analysis calculations and the same hard sphere value for R, a force constant and a frequency were calculated. However, accurate values for R are not available, so it is difficult to compare these values with experimental values. Observed frequencies were used to calculate a value for R which was then used to estimate the enthalpy of solvation. This was compared with the enthalpy determined experimentally by Senozan[88] from heats of solution of salts and metals in liquid ammonia. Enthalpy values were calculated using both tetrahedral and octahedral symmetry, and the one most closely approximating the experimental value was the determining factor in symmetry. It was found that alkali metals have four ammonia molecules in the primary solvation sphere and the alkaline earth metals have six. The observed low frequencies associated with solvated cations in ammonia were consistent with an ionic interaction model if they were assigned as the symmetric stretching mode of the primary solvation sphere of the cation.

TABLE IX

RAMAN FREQUENCIES OF SYMMETRIC STRETCHING MODES OF SOLVATED CATIONS[a]

Cation	ν (cm^{-1})
Li^+	241
Na^+	194
Mg^{2+}	328
Ca^{2+}	266
Sr^{2+}	243
Ba^{2+}	215

[a] K. R. Plowman and J. J. Lagowski, *J. Phys. Chem.* **78**, 143 (1974).

While Raman spectroscopy has long been used as a tool to determine information about hydrogen bonding in water and other solvents, these recent examples of its use in studying the structure of liquid ammonia are typical of the way it can be used. The ammonia system seems to be somewhat simpler than water or alcohols assuming that linear hydrogen bonding is a correct model, and knowledge about its structure and spectral characteristics should aid in the study of these other systems.

IV. Nuclear Magnetic Resonance Spectroscopy

A. Application of NMR Techniques

Resonance shifts reflect the extent of hydrogen bonding if they are measured in hydrogen bonding solvents diluted by inert solvents over an extensive concentration range. The association shift, which is always presented as a negative shift, is the proton resonance of the pure liquid just above the melting point less the proton resonance of the gas. The fact that these shifts are negative indicates a decrease in diamagnetic shielding around the proton upon hydrogen bonding. In general, the greater the shift, the stronger the hydrogen bond. Huggins *et al.*[89] showed that in phenols, the limiting behavior at low concentration is dominated by a monomer–dimer equilibrium. The observed shift δ is equal to a weighted mean

$$\delta \propto \delta_{\mathrm{M}} + (1-\alpha)\delta_{\mathrm{D}} \tag{13}$$

The limiting value of δ and $d\delta/dx$ at zero concentration are

$$\begin{aligned} \delta_0 &= \delta_{\mathrm{M}} \\ (d\delta/dx)_0 &= 2K(\delta_{\mathrm{D}}-\delta_{\mathrm{M}}) \end{aligned} \tag{14}$$

where x is the apparent mole fraction of the phenol and K is the association equilibrium constant in mole fraction units. Many studies have been made which attempt to relate chemical shifts to association constants, generally for the monomer–dimer equilibrium. Muller and Reiter[90] suggested that when studies are made correlating resonance shifts caused by changing temperature with the dissociation of hydrogen bonds, that results might be altered owing to the fact that the shift of a particular hydrogen bond can change with a change in temperature. Equation 14 would not hold then, because all the changes in the observed shift would not be due to shifting equilibria. They proposed that a large part of the shift is due to changes in the effective length of the hydrogen bond, which are, in turn, due to the anharmonicity and low frequency of the hydrogen bond stretch. This problem can be overcome, however, by using dilution techniques instead of temperature changes to alter the amount of hydrogen bonding. Creswell and Allred[91] pointed out another pitfall when an aromatic is used as the inert solvent. There is a high field shift of a donating proton with aromatics due to the magnetic anisotropy arising from the induced circulation of π electrons. This gives a secondary field which opposes the applied one in the area of the symmetry axis of the ring and augments the field near the edge of the ring.

Other types of NMR measurements have been used to study hydrogen bonding. Sato and Nishioka[92] made proton spin-lattice relaxation studies of chloroform and acceptor solvents. By determining T_1 for each of these pairs of donor and acceptor, the authors were able to say something about the order of association of the solvents. In one group, T_1 was not affected by the solvent, in another group it was slightly affected, and in a third group (dimethyl sulfoxide, DMSO), the spin-lattice relaxation rate increased by a factor of four with a small amount of solvent. The order of the bases as proton acceptors as determined by this work corresponded with association constants determined by other methods. Other data to be derived from relaxation times are the lifetimes of complexes.

The ^{13}C chemical shift of C=O in acetone was studied by Maciel and Ruben,[93] and it was found to be sensitive to the environment of proton donors. ^{15}N Chemical shifts have also been used and will be discussed later (*vide infra*).

B. Self-Association Studies

Unlike infrared data, NMR data do not show a different peak for each species present. One resonance shift is observed and the behavior of this shift with changes in concentration determines what species are in equilibrium. Saunders and Hyne[94] measured the NMR spectra of methanol, *tert*-butanol, and phenol and attributed the data to an equilibrium between the monomer and one other species (either a trimer or a tetramer). They plotted log C vs. δ and found a zero slope at low concentrations, ruling out the dimer. However, Becker[95] replotted this data as concentration vs. chemical shift and found a nonzero slope at zero concentration for phenol and *tert*-butanol. This indicated that dimers are present because of the relationship in Eq. 14. Becker concluded that trimers and tetramers might predominate, but that the dimer is present in amounts up to 10 or 15% at the most favorable concentrations. Davis *et al.*[96] used the same limiting slope method in alcohols and found K at different temperatures, and calculated apparent enthalpies for dimerization. These are reported in Table X where they are compared with some infrared values. The NMR values are quite a bit higher for methanol and ethanol. The authors point out that they were calculated assuming that the shift was independent of temperature, and as Muller and Reiter[90] showed, this assumption is not always valid. Dixon[97] measured the proton resonance shift in methanol in a variety of solvents and found that a monomer–tetramer equilibrium gave the best fit to his data. He did not rule out the presence of dimer or trimer in small amounts.

Rao *et al.*[98] looked at the chemical shift of OH, NH_2, and SH protons as a function of concentration, and found large shifts in phenol which have been interpreted as monomer–dimer–polymer equilibria. Aniline showed less of a shift and thiophenol had a nearly linear variation in shift, perhaps due to a

TABLE X

APPARENT ENTHALPIES FROM NMR AND IR DATA

	$-\Delta H$	
Alcohol	NMR	IR
In CCl_4		
Methanol	9.4 ± 2[a]	4.59[b]
Ethanol	7.6 ± 2[a]	3.6 ± 0.8[c]
Isopropanol	7.3 ± 3[a]	—
tert-Butanol	4.4 ± 2[a]	5.3 ± 0.5[d]
In C_6H_2		
Ethanol	5.1 ± 1[a]	—

[a] J. C. Davis, Jr., K. S. Pitzer, and C. N. R. Rao, *J. Phys. Chem.* **64**, 1744 (1960).
[b] R. Mecke and H. Nückel, *Naturwissenschaften* **31**, 248 (1943).
[c] U. Liddel and E. D. Becker, *Spectrochim. Acta* **10**, 70 (1957).
[d] E. G. Hoffman, *Z. Phys. Chem., Abt. B* **53**, 179 (1943).

monomer–dimer equilibrium over the entire concentration range. These results are consistent with hydrogen bonding phenomena as interpreted by infrared spectroscopy. Yamaguchi[99] studied ring proton, methyl proton, and OH proton shifts of *p*-cresol in carbon tetrachloride, benzene, dioxane, acetone, and pyridine. Although the OH proton shifts showed a different concentration dependence in the various solvents, the methyl and ring proton did not show much dependence on the solvent, and were probably not involved in hydrogen bonding. These shifts did have a linear decrease with dilution which was probably due to the diamagnetic anisotropy effect of the aromatic rings. The plots of the chemical shifts of the OH protons vs. mole fraction of solute (Fig. 9) indicated that in benzene, chloroform, and carbon tetrachloride the self-association of *p*-cresol was decreased upon dilution. The curves for dioxane, acetone, and pyridine suggested that new hydrogen bonds were formed between the solute and the solvents, and the pyridine probably formed the strongest bonds. This type of plot is a good qualitative map of some of the hydrogen bonding interactions taking place.

A similar observation was made by Somers and Gutowsky[100] in a system containing hindered phenols which were diluted by carbon tetrachloride, ethanol, and dioxane. Steric factors limited species larger than dimers. It was found that on dilution with carbon tetrachloride there was either an upfield shift or a negligible shift indicating a decrease in hydrogen bonding. On dilution with ethanol or dioxane, there was a downfield shift greater in

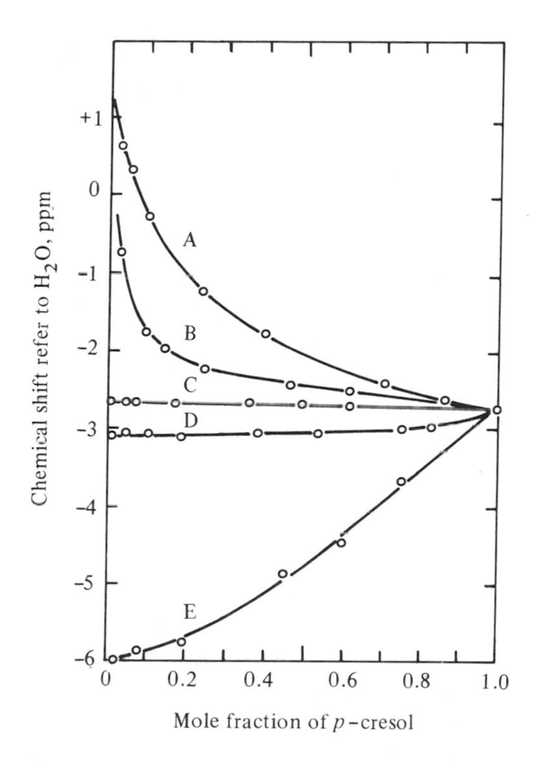

FIG. 9. Concentration dependence of OH proton chemical shift for *p*-cresol in various solutions. A, Benzene; B, CCl_4; C, dioxane; D, acetone; and E, pyridine. From Yamaguchi.[99]

ethanol than in dioxane. It was concluded that these two solvents formed better hydrogen bonds with these phenols than the bond involved in self-association. However, the limiting slope method relates this shift to an equilibrium constant and this must reflect the entropy of the system. The authors pointed out that the phenols are bulky in self-association, but that a small molecule like ethanol will hydrogen-bond easily. Again, it is difficult to make any statements about the strengths of these bonds from this data.

Feeney and Sutcliffe[101,102] made two studies of the NMR spectra of monoethylamine, diethylamine, and isobutylamine in carbon tetrachloride. They first used the limiting slope method and found a zero slope at concentrations less than 0.1 *M*, and postulated a monomer–tetramer equilibrium in order to interpret the data from progressive dilution. They then found that the association shift was the same for all three amines. This would necessitate the same ratio of hydrogen-bonded to non-hydrogen-bonded amine protons for each compound. This would be more likely if cyclic *n*-mers were present, and they made some calculations of theoretical curves assuming this, finding a good fit with experimental results for a monomer–tetramer model. However, their curve fitting techniques were not sensitive to n-mers where $n > 4$.

In general, NMR self-association studies appear to have been quite helpful in determining whether a dimer is present in solution. The limiting slope method if used properly is a good test for dimerization and for determining equilibrium constants. However, the extension of this to other thermodynamic parameters, particularly the enthalpy of the bond, is questionable. Studies must be made at different temperatures in order to relate K to $-\Delta H$, and there is no assurance that the resonance shift of a hydrogen-bonded proton is constant over a wide temperature range.

C. Mixed Complexes

The same limiting slope method applied to dimers can be applied to 1:1 complexes of proton donor and acceptor. A large number of the studies made have used chloroform as the proton donor and they will be discussed at some length. Other types of donors have also been investigated and a few examples will be mentioned. Kanekar *et al.*[103] compared the proton-donating ability of CH_3OH, $CHCl_3$, and C_6H_5SH in dimethylformamide (DMF) and dimethylacetamide (DMA). They found that all proton resonances shifted downfield on dilution with the solvent, indicating the formation of hydrogen bonds. When the same proton donor was used, DMA formed stronger bonds than DMF. This is in line with the greater basicity of DMA. For the same acceptor, the order of proton-donating ability was OH > SH > CH. Again, this would be the expected order of proton acidity. Johnston *et al.*[104] made an NMR study of CH_2Cl_2, $CHCl_3$, $CHCl_2CN$, and C_6H_5OH in methylbenzenes using heptane as an inert solvent. The donors were thought to bond to the π electrons of the ring and the order of proton-donor ability was $C_6H_5OH > CHCl_2CN > CHCl_3 > CH_2Cl_2$. The acceptor strength was mesitylene > toluene > benzene. None of these were as good acceptors as dioxane, however, implying that the lone pair of electrons on oxygen are better acceptors than the π electrons of an aromatic ring. Lin and Tsay[105] made a PMR study of chloroform with some of these same solvents. They attempted to test if 1:1 complexes were formed, and to determine the association constants. They found that 1:1 complexes were formed and the K's were of the same order as found by Johnston *et al.*[104] They concluded too that the bond between $CHCl_3$ and the benzene ring was electrostatic in character, but was less stable than the more common hydrogen bond.

Chloroform has been widely used as a proton donor in NMR studies of hydrogen bonding. For example, McClellan *et al.*[106] considered a mixture of chloroform and dimethyl sulfoxide and plotted the chemical shift vs. the mole fraction of chloroform. This data did not fit the calculated curve for a 1:1 complex, even allowing for some self-association of chloroform. When the complex 1:2 DMSO–$CHCl_3$ was postulated and some dimer formation was allowed for, the calculated curve gave a good fit with the experimental data, and a $-\Delta H$ value for this bond of 3.3 kcal/mole was determined.

Berkeley and Hanna[107–109] measured the chemical shift of chloroform with nitrogen bases. They suggested a model to account for correspondence between the shift, Δ, which is the difference between the proton shift of the complex and the monomeric chloroform in an inert medium, and the basicity of the nitrogen bases. They also calculated the association constants for the bases pyridine, *N*-methylpyridine, $CH_3CH{=}NCH(CH_3)_2$, and acetonitrile. These values are listed in Table XI along with the values of Δ. One can relate

TABLE XI

ASSOCIATION CONSTANTS FOR CHLOROFORM COMPLEXES FROM NMR DATA[a]

Base	*K* (kcal/mole)	Δ (ppm)
Pyridine	0.69	−3.90
N-Methylpyridine	2.2	−2.05
$CH_3CH{=}NCH(CH_3)_2$	5.2	−1.98
Acetonitrile	3.2	−0.63

[a] P. J. Berkeley, Jr. and M. W. Hanna, *J. Phys. Chem.* **67**, 846 (1963).

Δ to the basicity of the nitrogen bases if one takes into account the aromaticity of pyridine and the paramagnetic contribution of acetonitrile. A model was proposed that as the chloroform molecule approaches the base, the H–C bond dipole interacts with the electric field produced by the electric asymmetry of the base, and there is a lowering of energy in the system. This only continues until the van der Waals radius of H plus N is reached. The electric field of the base also causes polarization of the H–C bonding electrons toward the carbon atom resulting in further stabilization and reduction in the screening constant of the proton. This causes a shift in the magnetic resonance downfield. Using both the Buckingham electric field effect and the neighbor anisotropy effects, the authors[109] obtained the magnitudes of these two effects as functions of various parameters. These two effects, which are the principal contributions to the shift on hydrogen bonding,[110] were found insensitive to the magnetic anisotropy of the bond, the electric field chemical shift proportionality constant, and the type of orbitals used to represent the lone-pair electron distribution. They were found to be sensitive to the hydrogen bond length. The authors used the chemical shift information to determine bond lengths, and found them to be in agreement with X-ray crystallographic data. The lengths increased as hybridization of the lone pair went from sp^3 to sp^2 to sp. The conclusion of the work was that the shift on hydrogen bond formation is a good criterion for the basicity of proton acceptors for weak hydrogen bonds.

The NMR technique has been used quite extensively then to determine thermodynamic data in weak hydrogen-bonded systems. However, much of this data is not precise for a variety of reasons, most of which have been dealt with in a recent work by Wiley and Miller.[111] They have obtained precise thermodynamic data for the hydrogen bonding of chloroform with twelve proton acceptors in cyclohexane, and have compared their results with those obtained by others. The discrepancies in these data are numerous and raise a question about their reproducibility which the authors attempt to answer.

The first point made by Wiley and Miller is with regard to plotting data. The limiting slope method of plotting is open-ended, and values for K are determined from only a small range of association. They have derived the following relationship:

$$(\delta - \delta_A)/B_0 = K(\Delta - \Delta_{obsd}) \tag{15}$$

where it is assumed that B_0, the concentration of the base, is much greater than A_0, the concentration of chloroform, $\Delta \equiv \delta_c - \delta_A$, and $\Delta_{obsd} \equiv \delta - \delta_A$. δ is the experimental chemical shift; δ_c is the shift due to the complex; and δ_A is the shift in monomeric chloroform. A plot of this equation is linear and can cover a wide range of concentrations. This analysis would break down if there were dimerization of chloroform or if 1:2 or 2:1 complexes were formed. Another concept whose importance was emphasized by these authors was the fraction of chloroform associated known as the saturation factor, $s = C/A_0$; in terms of chemical shifts:

$$s = \Delta_{obsd}/\Delta \tag{16}$$

This factor has a limiting value of 0.0 when there is no base present and 1.0 at complete complexation. Deranleau[112] has pointed out that the range where equilibrium constants are most reliable is $s = 0.2$–0.8. These authors have stated, however, that broad, middle s ranges are often impossible to achieve experimentally, especially if K is small or large, or if concentrations are limited by solubility, etc. In Table XII thermodynamic parameters for chloroform complexes and the s ranges for these studies for a variety of workers are reported. The results of Wiley and Miller have greater precision and in most cases a better s range. The association constants were determined from Eq. 15 and then by an iterative method they were fitted to an equation which does not assume B_0 greater than A_0. They are reasonably confident that

TABLE XII

THERMODYNAMIC PARAMETERS FROM PMR FOR ASSOCIATION OF CHLOROFORM WITH PROTON ACCEPTORS

Base	Δ(Hz)	K (liter/mole)	−ΔH (kcal/mole)	−ΔS⁰ (e.u.)	S
$(CH_3)_2CO$[a]	58.31	0.751	2.34	8.37	0.15–0.80
$(CH_3)_2CO$[b]	57.8	0.21	3.6	—	0.1–0.36
$CH_3COOC_2H_5$[a]	50.51	0.674	2.51	9.03	0.05–0.78
$CH_3COOC_2H_5$[b]	40.7	0.4	3.8	—	0.12–0.5
1,4-Dioxane[a]	40.53	0.582	2.56	9.64	0.17–0.79
1,4-Dioxane[c]	72	1.4[e]	2.0	—	—
$(C_2H_5)_2S$[a]	55.95	0.219	1.7	8.65	0.01–0.65
$(C_2H_5)_2S$[b]	46.2	0.11	2.3	—	0.06–0.23
$(n\text{-}C_4H_9)_2O$[a]	53.29	0.243	2.35	5.32	0.02–0.40
$(n\text{-}C_4H_9)_2O$[d]	—	1.33[e]	1.9	—	—

[a] G. R. Wiley and S. I. Miller, *J. Amer. Chem. Soc.* **94**, 3287 (1972).
[b] F. L. Slejko, R. S. Drago, and D. G. Brown, private communication, in Wiley and Miller[a].
[c] R. Kaiser, *Can. J. Chem.* **41**, 430 (1963).
[d] J. R. Baker, I. D. Watson, and A. G. Williamson, *Aust. J. Chem.* **24**, 2047 (1971).
[e] The units are mole fraction $^{-1}$.

dimerization of chloroform or formation of 2:1 complexes does not occur in significant amounts. However, 1:2 complexes or base dimers might be present and cannot be accounted for. The authors also were aware of the temperature dependencies of the chemical shift which were given by Muller and Reiter.[90] They found that they could combine a linear temperature dependence of the chemical shift with the concentration dependence and obtain a chemical shift relation which by computer fit would yield ΔH^0 and ΔS^0. Although this was not the method used to obtain the values in Table XII, they have pointed out the future usefulness of this approach. This type of method, as well as the way K values were determined and the s range covered, combines reliable data with a statistical approach to using the NMR technique.

Litchman *et al.*[113,114] made ^{15}N NMR studies on NH_3 to determine information about hydrogen bonding in that solvent. In the first work[113] they looked at the resonance signal in NH_3 and ND_3 as the temperature was changed, and found that both species have downfield shifts with decreasing temperature, which could be due to increased hydrogen bonding. They urge caution however as there might be some temperature dependence of the shift due to changes in the N–H bond length.

The other study made by these workers[114] considered the ^{15}N resonance signal for NH_3 in thirteen solvents over a considerable concentration range. Since the plots of shift vs. concentration were smooth, they were extrapolated to zero NH_3 concentration to give a set of infinite dilution shifts which did not coincide with the gaseous value. A model was proposed to interpret these shifts which considered that two types of interactions contributed to them. The first interaction was that of the lone pair of electrons on nitrogen with the solvent molecules. This might be hydrogen bond formation with the solvent protons or an interaction between the lone pair and the solvent R groups. The second type of interaction was hydrogen bond formation between solvent electron donor sites and the NH_3 protons. The infinite dilution shifts were given as a linear combination of six interaction parameters which took into account all the possible interaction sites on the solvents. The coefficients for each of the thirteen equations depended on the number of such sites on the solvent molecule. The calculated values fitted quite well with the observed values (see Table XIII), but did not necessarily prove the model. It is nevertheless a worthwhile approach to studying hydrogen bonding in that it gives

TABLE XIII

Calculated and Observed Dilution Shifts of $^{15}NH_3$ in Various Solvents[a]

Solvent	δ_{calc} (ppm)	δ_{obsd} (ppm)
Me_2O	−7.8	−7.7
Me_3N	−9.0	−10.0
Me_4C	−11.1	−11.1
$MeNH_2$	−12.9	−12.3
MeOH	−13.8	−13.4
EtOH	−15.4	−16.6
Me_2NH	−10.9	−10.1
$EtNH_2$	−14.0	−13.7
Et_2O	−11.1	−10.8
Et_2NH	−13.1	−12.7
NH_3	−14.9	−15.9

[a] W. M. Litchman, M. Alei, Jr., and A. E. Florin, *J. Amer. Chem. Soc.* **91**, 6574 (1969).

some insight into the way hydrogen bonding involving the lone pair affects the paramagnetic term in the chemical shift.

The increased use of statistical fits of data by means of computer techniques, and the availability of nuclear magnetic resonance using a variety of nuclei have added to the value of this technique in studying hydrogen bonding. Used in conjunction with vibrational spectroscopy, good thermodynamic data can be calculated and structural information can be obtained.

V. Ultraviolet Spectroscopy

Electronic absorptions are affected by hydrogen bonding, especially if that portion of the molecule most concerned in the given transition is involved in the hydrogen bond. Frequency shifts of $n \rightarrow \pi^*$ and $\pi \rightarrow \pi^*$ transitions of electrons on bases are both affected. In the first transition electron density is shifted away from the basic group making it a poorer hydrogen bonder in the excited state and, thus, the transition undergoes a shift to higher energy, a blue shift. A $\pi \rightarrow \pi^*$ transition enhances electron density in the outer parts of the molecule, making electrons more available for hydrogen bonding and stabilizing the excited state. This causes a shift to lower energy (red shift) in the transition. As with other techniques, electronic absorption spectroscopy can be used to study self-association in hydrogen-bonded solvents by varying concentration with an inert solvent or to study complex formation in a series of proton donors and/or bases.

Dearden and Forbes[115] showed the concentration dependence of the UV band in phenol and alkyl-substituted phenols, confirming hydrogen bonding. They found less dependence in aniline and *N,N*-dimethylaniline. The bond in phenol was weaker than that in benzoic acids which they had also studied.[116] In that work they measured the concentration dependencies of substituted benzoic acids and from noting the maximum concentration at which this dependence began, they were able to discuss relative strengths of the dimeric bond. They found that *m*-methyl, *p*-methyl, and *p*-hydroxy substituents strengthened the hydrogen bond and *o*-methyl, *o*-hydroxy, and *o*-methoxy substituents weakened the hydrogen bond.

Ito[117] also studied hydrogen bonding in phenol. He observed the band at 270 mμ which shifted to higher energy with increased concentration in *n*-hexane. He made a control study with anisole in *n*-hexane and did not get a similar shift, concluding that hydrogen bonding caused the shift in phenol. He also postulated a cyclic dimer since only one band was observed, meaning that all oxygen atoms were both hydrogen bond donors and acceptors. He argued that linear dimers should produce two bands. These arguments are similar to those made about infrared data and help to confirm or deny some models. Kaye and Poulson[118] found UV evidence for association in methanol. By comparing the absorptivity of the 195 mμ band with that of the first overtone of the O–H stretching mode at 1.4 μm they were able to confirm that the O–H bond was the source of the ultraviolet band.

A considerable amount of work has been done measuring the ultraviolet spectra of proton donors and acceptors which form hydrogen-bonded complexes. From optical density measurements equilibrium constants can be calculated. Table XIV shows a variety of shift data and calculated equilibrium constants. In most cases studies were conducted on a series of complexes

TABLE XIV

ULTRAVIOLET FREQUENCY SHIFTS AND EQUILIBRIUM CONSTANTS FOR HYDROGEN-BONDED COMPLEXES

Donor	Acceptor	λ (mμ)	K (liter/mole)
n-Propanol	Pyridazine	340 → 315	3.05[a]
	Diethylnitrosoamine	366 → 350	2.16[a]
	Benzophenone	348 → 335	0.62[a]
n-Butanol	Pyridazine	340 → 315	3.38[a]
	Diethylnitrosoamine	366 → 350	2.25[a]
	Benzophenone	348 → 335	0.64[a]
Isopropanol	Pyridazine	340 → 335	1.72[a]
	Diethylnitrosoamine	366 → 350	1.44[a]
	Benzophenone	348 → 335	0.39[a]
Diphenylamine	Diethylnitrosoamine	366 → 350	28.2[b]
	Quinoxaline	360 → 350	17.4[b]
	Benzophenone	348 → 335	14.4[b]
p-Chloroaniline	Diethylnitrosoamine	366 → 350	4.4[b]
	Quinoxaline	360 → 350	2.9[b]
	Benzophenone	348 → 335	1.8[b]
Methylaniline	Diethylnitrosoamine	366 → 350	4.2[b]
	Quinoxaline	360 → 350	2.7[b]
	Benzophenone	348 → 335	1.4[b]
p-Chlorophenol	Tetrahydrofuran	290 → 296	26.01[c]
	1,4-Dioxane	290 → 296	15.59[c]
	Diethyl ether	290 → 296	13.88[c]
p-Hydroxydiphenyl	Tetrahydrofuran	256 → 264	24.00[c]
	1,4-Dioxane	256 → 264	14.58[c]
	Diethyl ether	256 → 264	12.90[c]
α-Naphthol	Tetrahydrofuran	320 → 322	21.50[c]
	1,4-Dioxane	320 → 322	13.39[c]
	Diethyl ether	320 → 322	11.11[c]
β-Naphthol	Tetrahydrofuran	326 → 333	20.20[c]
	1,4-Dioxane	326 → 333	12.87[c]
	Diethyl ether	326 → 333	10.58[c]
Phenol	Tetrahydrofuran	276 → 281	19.43[c]
	1,4-Dioxane	276 → 281	12.34[c]
	Diethyl ether	276 → 281	10.52[c]
Diphenylamine	Triethylamine	—	0.85[d]
Pyrrole	Triethylamine	—	2.0[d]
2-Methylindole	Triethylamine	—	2.2[d]
Carbazole	Triethylamine	—	2.75[d]

[a] A. K. Chandra and S. Basu, *Trans. Faraday Soc.* **56**, 632 (1960).
[b] B. B. Bhowmik and S. Basu, *Trans. Faraday Soc.* **58**, 48 (1962).
[c] B. B. Bhowmik and S. Basu, *Trans. Faraday Soc.* **59**, 813 (1963).
[d] A. B. Sannigrahi, *Indian J. Chem.* **4**, 532 (1966).

with varying basicities or in different solvents. The hydrogen bonding ability of a donor or acceptor is reflected in the equilibrium constant, but both the strength of the bond and the entropy factor contribute to it. Rao *et al.*[119] studied a series of aliphatic C=O derivatives in several solvents. The $n \rightarrow \pi^*$ transition showed a blue shift with increasing polarity of the proton acceptor or with increasing hydrogen bonding ability of the solvent.

Ito[120] measured the ultraviolet absorption spectra of a variety of complexes. The $\pi \rightarrow \pi^*$ transition of phenol showed a red shift as ethanol was added to a solution of phenol in *n*-hexane. The proposed linear dimer with phenol as the donor compound is in accord with infrared work of Bellamy and Pace[22] discussed in Section II,B,1. The $\pi \rightarrow \pi^*$ transition of aniline showed a large red shift in ethyl ether, and from the number of bands seen, both a 1:1 and a 1:2 complex were proposed. There was a red shift in the diphenylamine $\pi \rightarrow \pi^*$ transition in the presence of ethanol. In this case Ito explained why diphenylamine must be considered the proton donor. *N*-methyldiphenylamine showed no shift in any bands in an ethanol mixture. It would be expected to be the same type of acceptor as diphenylamine, and the shift in the band of diphenylamine has to be due to proton donation. Finally, the $n \rightarrow \pi^*$ transition of acetone showed an expected blue shift as ethanol was added, confirming hydrogen bonding.

Balasubramanian and Rao[121] were able to correlate the shift in the $n \rightarrow \pi^*$ transition band of C=O, C=S, NO_2, and N=N groups to the proton-donating ability of the solvent. The blue shifts in the alcohols decreased in the order 2,2,2-trifluoroethanol > methanol > ethanol > isopropanol > *tert*-butanol. This order also approximates the decreasing electron-withdrawing power of the alkyl groups and hence the decreasing acidity. However, the ability of these alcohols to self-associate is exactly the opposite of this order, implying that their abilities as acceptors are equally or more important than as donors for dimerization.

Basu *et al.*[122–124] made a series of studies of hydrogen bonding by means of ultraviolet spectroscopy. In the first study[122] they calculated equilibrium constants for pyridazine, diethyl nitrosoamine, and benzophenone with five alcohols from the $n \rightarrow \pi^*$ blue shift in the spectrum of the base. They found that the equilibrium constant decreased in the order primary > secondary > tertiary alcohol. The proton donor character of alcohols also follows this order so the change in K in the alcohol series is largely due to an enthalpy change in the bond. The K values also appeared to indicate that alcoholic groups formed better hydrogen bonds with nitrogen than with oxygen. However, higher alcohol concentrations had to be used to get good spectral measurements in the benzophenone system, and there was most probably a competing self-association reaction of the alcohol which would decrease the amount of complex formed. The second study in this series[123] was concerned with the proton donor ability of the N–H proton in amines. The $n \rightarrow \pi^*$ blue shifts of diethylnitrosoamine, benzophenone, and quinoxaline were used, and linear Beer's law type plots indicated the existence of 1 : 1 complexes in all systems studied. The authors found that if they plotted $\log K$ of the calculated constant against pK_b of the amine, a linear relationship existed. This pK_b value refers to the proton donor character of the conjugate acids of the amines, but even as an approximate measure of proton donor ability it shows that the formation of these hydrogen bonds depneds on the acidity of the proton.

The third study in the series[124] examined phenols and their proton donor abilities. The red shift of the $\pi \rightarrow \pi^*$ transition of the phenols was measured in mixtures with tetrahydrofuran, dioxane, and diethyl ether. The fact that the UV band measured was associated with the proton donor made the comparison of the K values with the pK_a's of the phenols more valid than the correlations in the previous systems. It was found that $\log K$ for each phenol with a given base had a linear relationship with the pK_a's of the phenols, confirming that the hydrogen bond, which is electrostatic in character depends on the acidity of the proton donor if steric hindrance is not important.

Sannigrahi[125] looked at various proton donors with triethylamine, and calculated equilibrium constants for hydrogen bonded complexes. These are listed in Table XIII along with other results from ultraviolet work. These are just a few representative studies of the many systems in which ultraviolet absorption spectroscopy has been used to obtain information on hydrogen bonding trends and thermodynamic parameters. The usefulness of this method is inherent in the great variety of transitions which are sensitive to hydrogen bonding and can be monitored as concentration, temperature, or solvent are changed. New bands for each type of association are often seen, and results can be correlated with both infrared and NMR data for a more thorough understanding of hydrogen bonding phenomena.

References

1. G. C. Pimentel and A. L. McClellan, "The Hydrogen Bond." Freeman, San Francisco, California, 1960.
2. G. C. Pimentel and A. L. McClellan, *Annu. Rev. Phys. Chem.* **22**, 347 (1971).
3. K. V. Ramiah and P. G. Puranik, *Proc. Indian Acad. Sci., Sect. A* **56**, 96 (1962).
4. K. V. Ramiah and P. G. Puranik, *Proc. Indian Acad. Sci., Sect. A* **56**, 155 (1962).
5. S. G. W. Ginn and J. L. Wood, *Nature (London)* **200**, 467 (1963).
6. D. J. Millen and J. Zabicky, *Nature (London)* **196**, 889 (1962).
7. S. G. W. Ginn and J. L. Wood, *Proc. Chem. Soc., London*, p. 370 (1964).
8. A. Hall and J. L. Wood, *Spectrochim. Acta, Part A* **24**, 1109 (1968).
9. A. Hall and J. L. Wood, *Spectrochim. Acta, Part A* **23**, 1257 (1967).
10. A. Ažman, B. Borštnik, and O. Hadži, *J. Mol. Struct.* **8**, 315 (1971).

11. A. Foldes and C. Sandorfy, *J. Mol. Spectrosc.* **20**, 262 (1966).
12. M. Asselin, G. Bélanger, and C. Sandorfy, *J. Mol. Spectrosc.* **30**, 96 (1969).
13. M. Asselin and C. Sandorfy, *J. Mol. Struct.* **8**, 145 (1971).
14. A. E. Stanevich, *Opt. Spectrosc.* **16**, 243 (1964).
15. R. F. Lake and H. W. Thompson, *Proc. Roy. Soc., Ser. A* **291**, 469 (1966).
16. R. J. Jakobsen and J. W. Brasch, *Spectrochim. Acta* **21**, 1753 (1965).
17. E. R. Lippincott and R. Schroeder, *J. Chem. Phys.* **23**, 1099 (1955).
18. R. Schroeder and E. R. Lippincott, *J. Phys. Chem.* **61**, 921 (1957).
19. S. G. W. Ginn and J. L. Wood, *Chem. Commun.* p. 628 (1965).
20. S. G. W. Ginn and J. L. Wood, *Spectrochim. Acta, Part A* **23**, 611 (1967).
21. S. Singh and C. N. R. Rao, *Can. J. Chem.* **44**, 2611 (1966).
22. L. J. Bellamy and R. J. Pace, *Spectrochim. Acta* **22**, 525 (1966).
23. A. J. Barnes and H. E. Hallam, *Trans. Faraday Soc.* **66**, 1920 (1970).
24. A. J. Barnes and H. E. Hallam, *Trans. Faraday Soc.* **66**, 1932 (1970).
25. H. C. Van Ness, J. Van Winkle, H. H. Richtol, and H. B. Hollinger, *J. Phys. Chem.* **71** 1483 (1967).
26. A. Hall and J. L. Wood, *Spectrochim. Acta, Part A* **23**, 2657 (1967).
27. A. N. Fletcher and C. A. Heller, *J. Phys. Chem.* **71**, 3742 (1967).
28. T. S. S. R. Murty, *Can. J. Chem.* **48**, 184 (1970).
29. J. Bufalini and K. H. Stern, *J. Amer. Chem. Soc.* **83**, 4362 (1961).
30. R. E. Hester and R. A. Plane, *Spectrochim. Acta, Part A* 23, 2289 (1967).
31. J. B. Hyne and R. M. Levy, *Can. J. Chem.* **40**, 692 (1962).
32. J. Lascombe, M. Haurie, and M.-L. Josien, *J. Chim. Phys.* **59**, 1233 (1962).
33. L. J. Bellamy and R. J. Pace, *Spectrochim. Acta* **19**, 435 (1963).
34. L. J. Bellamy, R. F. Lake, and R. J. Pace, *Spectrochim. Acta* **19**, 443 (1963).
35. R. J. Jakobsen, Y. Mikawa, and J. W. Brasch, *Spectrochim. Acta, Part A* **25**, 839 (1969).
36. T. S. S. R. Murty and K. S. Pitzer, *J. Phys. Chem.* **73**, 1426 (1969).
37. K. Nakamoto and S. Kishida, *J. Chem. Phys.* **41**, 1558 (1964).
38. R. J. Jakobsen, Y. Mikawa, and J. W. Brasch, *Spectrochim. Acta, Part A* **23**, 2199 (1967).
39. G. Allen, J. G. Watkinson, and K. H. Webb, *Spectrochim. Acta* **22**, 807 (1966).
40. E. D. Becker, *Spectrochim. Acta* 743 (1959).
41. P. G. Puranik and K. V. Ramiah, *J. Mol. Spectrosc.* **3**, 486 (1959).
42. P. J. Krueger and D. W. Smith, *Can. J. Chem.* **45**, 1611 (1967).
43. J. H. Lady and K. B. Whetsel, *J. Phys. Chem.* **68**, 1001 (1964).
44. U. Liddel and E. D. Becker, *Spectrochim. Acta* p. 1070 (1957).
45. L. J. Bellamy and H. E. Hallam, *Trans. Faraday Soc.* **55**, 220 (1959).
46. A. Allerhand and P. von R. Schleyer, *J. Amer. Chem. Soc.* **85**, 371 (1963).
47. E. Bauer and M. Magat, *J. Phys. Radium* **9**, 319 (1938).
48. R. R. Wiederkehr and H. G. Drickamer, *J. Chem. Phys.* **28**, 311 (1958).
49. L. J. Bellamy, K. J. Morgan, and R. J. Pace, *Spectrochim. Acta* **22**, 535 (1966).
50. T. Gramstad, *Spectrochim. Acta* **19**, 1363 (1963).
51. A. R. H. Cole and A. J. Michell, *Aust. J. Chem.* **18**, 102 (1965).
52. E. Hirano and K. Kozima, *Bull. Chem. Soc. Jap.* **39**, 1216 (1966).
53. P. V. Huong and J. C. Lassegues, *Spectrochim. Acta, Part A* **26**, 269 (1970).
54. A. N. Fletcher, *J. Phys. Chem.* **73**, 2217 (1969).
55. S. D. Christian and E. E. Tucker, *J. Phys. Chem.* **74**, 214 (1970).
56. A. N. Fletcher, *J. Phys. Chem.* **74**, 216 (1970).
57. E. A. Robinson, H. D. Schreiber, and J. N. Spencer, *J. Phys. Chem.* **75**, 2219 (1971).
58. R. M. Badger and S. H. Bauer, *J. Chem. Phys.* **5**, 839 (1937).
59. E. D. Becker, *Spectrochim. Acta* **17**, 436 (1961).

60. H. J. Wimette and R. H. Linnell, *J. Phys. Chem.* **66**, 546 (1961).
61. S. S. Mitra, *J. Chem. Phys.* **36**, 3286 (1962).
62. T. Zeegers-Huyskens, L. Lamberts, and P. Huyskens, *J. Chim. Phys.* **59**, 521 (1962).
63. T. Gramstad, *Acta Chem. Scand.* **16**, 807 (1962).
64. T. Gramstad, *Spectrochim. Acta* **19**, 497 (1963).
65. T. Gramstad, *Acta Chem. Scand.* **15**, 1337 (1961).
66. T. Gramstad and W. J. Fugelvik, *Acta Chem. Scand.* **16**, 1369 (1962).
67. T. Gramstad, *Spectrochim. Acta* **19**, 829 (1963).
68. T. J. V. Findlay and A. D. Kidman, *Aust. J. Chem.* **18**, 521 (1965).
69. T. Zeegers-Huyskens, *Spectrochim. Acta* **21**, 221 (1965).
70. L. Segal, *J. Phys. Chem.* **65**, 697 (1961).
71. I. Motoyama and C. H. Jarboe, *J. Phys. Chem.* **71**, 2723 (1967).
72. P. R. Naidu, *Aust. J. Chem.* **19**, 2393 (1966).
73. T. Gramstad and J. Sandström, *Spectrochim. Acta, Part A* **25**, 31 (1969).
74. L. J. Bellamy, G. Eglinton, and J. F. Morman, *J. Chem. Soc., London* p. 4762 (1961).
75. S. Singh and C. N. R. Rao, *J. Amer. Chem. Soc.* **88**, 2142 (1966).
76. W. A. P. Luck and W. Ditter, *J. Phys. Chem.* **74**, 3687 (1970).
77. G. C. Pimentel, M. O. Bulanin, and M. Van Thiel, *J. Chem. Phys.* **36**, 500 (1962).
78. C. A. Plint, R. M. B. Small, and H. L. Welsh, *Can. J. Phys.* **32**, 653 (1954).
79. T. Birchall and I. Drummond, *J. Chem. Soc., A* p. 1859 (1970).
80. B. de Bettignies and F. Wallart, *C. R. Acad. Sci., Ser. B* **271**, 640 (1970).
81. J. H. Roberts, A. T. Lemley, and J. J. Lagowski, *Spectrosc. Lett.* **5**, 271 (1972).
82. D. J. Gardiner, R. E. Hester, and W. E. L. Grossman, *J. Raman Spectrosc.* **1**, 87 (1973).
83. A. T. Lemley, J. H. Roberts, K. R. Plowman, and J. J. Lagowski, *J. Phys. Chem.* **77**, 2185 (1973).
84. D. J. Gardiner, R. E. Hester, and W. E. L. Grossman, *J. Chem. Phys.* **59**, 175 (1973).
85. A. T. Lemley and J. L. Lagowski, *J. Phys. Chem.* **78**, 708 (1974).
86. K. R. Plowman and J. J. Lagowski, *J. Phys. Chem.* **78**, 143 (1974).
87. J. Corset, Ph.D. Thesis, University of Bordeaux, 1967.
88. N. M. Senozan, *J. Inorg. Nucl. Chem.* **35**, 727 (1973).
89. C. M. Huggins, G. C. Pimentel, and J. N. Shoolery, *J. Phys. Chem.* **60**, 1311 (1956).
90. N. Muller and R. C. Reiter, *J. Chem. Phys.* **42**, 3265 (1965).
91. C. J. Creswell and A. L. Allred, *J. Phys. Chem.* **66**, 1469 (1962).
92. K. Sato and Z. Nishioka, *Bull. Chem. Soc. Jap.* **44**, 1506 (1971).
93. G. E. Maciel and G. C. Ruben, *J. Amer. Chem. Soc.* **85**, 3903 (1963).
94. M. Saunders and J. B. Hyne, *J. Chem. Phys.* **29**, 1319 (1958).
95. E. D. Becker, *J. Chem. Phys.* **31**, 269 (1959).
96. J. C. Davis, Jr., K. S. Pitzer, and C. N. R. Rao, *J. Phys. Chem.* **64**, 1744 (1960).
97. W. B. Dixon, *J. Phys. Chem.* **74**, 1396 (1970).
98. B. D. N. Rao, P. Venkateswarlu, A. S. N. Murthy, and C. N. R. Rao, *Can. J. Chem.* **40**, 963 (1962).
99. I. Yamaguchi, *Bull. Chem. Soc. Jap.* **34**, 1602 (1961).
100. B. G. Somers and H. S. Gutowsky, *J. Amer. Chem. Soc.* **85**, 3065 (1963).
101. J. Feeney and L. H. Sutcliffe, *Proc. Chem. Soc., London* p. 118 (1961).
102. J. Feeney and L. H. Sutcliffe, *J. Chem. Soc., London* p. 1123 (1962).
103. C. R. Kanekar, C. L. Khetrapal, K. V. Ramiah, and C. A. Indirachory, *Proc. Indian Acad. Sci., Sect. A* **66**, 189 (1967).
104. M. D. Johnston, Jr., F. P. Gasparro, and I. D. Kuntz, Jr., *J. Amer. Chem. Soc.* **91**, 5715 (1969).
105. W. Lin and S. Tsay, *J. Phys. Chem.* **74**, 1037 (1970).

106. A. L. McClellan, S. W. Nicksic, and J. C. Guffy, *J. Mol. Spectrosc.* **11**, 340 (1963).
107. P. J. Berkeley, Jr. and M. W. Hanna, *J. Phys. Chem.* **67**, 846 (1963).
108. P. J. Berkeley, Jr. and M. W. Hanna, *J. Chem. Phys.* **4**, 2530 (1964).
109. P. J. Berkeley and M. W. Hanna, *J. Amer. Chem. Soc.* **86**, 2990 (1964).
110. J. A. Pople, J. J. Bernstein, and W. G. Schneider, "High Resolution Nuclear Magnetic Resonance." McGraw-Hill, New York, 1959.
111. G. R. Wiley and S. I. Miller, *J. Amer. Chem. Soc.* **94**, 3287 (1972).
112. D. A. Deranleau, *J. Amer. Chem. Soc.* **91**, 4044 and 4050 (1969).
113. W. M. Litchman, M. Alei, Jr., and A. E. Florin, *J. Chem. Phys.* **50**, 1031 (1969).
114. W. M. Litchman, M. Alei, Jr., and A. E. Florin, *J. Amer. Chem. Soc.* **91**, 6574 (1969).
115. J. C. Dearden and W. F. Forbes, *Can. J. Chem.* **38**, 896 (1960).
116. W. F. Forbes, A. R. Knight, and D. L. Coffen, *Can. J. Chem.* **38**, 728 (1960).
117. M. Ito, *J. Mol. Spectrosc.* **4**, 125 (1960).
118. W. Kaye and R. Poulson, *Nature (London)* **193**, 675 (1962).
119. C. N. R. Rao, G. K. Goldman, and A. Balasubramanian, *Can. J. Chem.* **38**, 2508 (1960).
120. M. Ito, *J. Mol. Spectrosc.* **4**, 106 (1960).
121. A. Balasubramanian and C. N. R. Rao, *Spectrochim. Acta* **18**, 1337 (1962).
122. A. K. Chandra and S. Basu, *Trans. Faraday Soc.* **56**, 632 (1060).
123. B. B. Bhowmik and S. Basu, *Trans. Faraday Soc.* **58**, 48 (1962).
124. B. B. Bhowmik and S. Basu, *Trans. Faraday Soc.* **59**, 813 (1963).
125. A. B. Sannigrahi, *Indian J. Chem.* **4**, 532 (1966).

3

Redox Systems in Nonaqueous Solvents

MICHEL RUMEAU

Faculté des Sciences et des Techniques
Centre Universitaire de Savoie, Chambery, France

I. Generalities about Redox Systems in Nonaqueous Solvents

A. Introduction

In nonaqueous solvents redox reactions have been studied far less than acid–base reactions. Although numerous results have been obtained from acid–base reactions in protic solvents analogous to water (alcohol, acid, etc.) and in dipolar aprotic solvents (acetonitrile, propylene carbonate, nitromethane), very little work has been carried out on solvents having a low dielectric constant, such as hydrocarbons or their derivatives. The main reason for this is that it is extremely difficult to make electrochemical measurements, which is by far the best method of studying these reactions.

For acid–base reactions spectrophotometric methods are generally employed to study these solvents; however these methods have remained practically unused in redox systems. Nevertheless, both spectrophotometric and electrochemical methods can be used for most of these solvents without too much trouble.

When using spectrophotometric methods to study redox reactions, there are far fewer redox indicators than acid–base indicators. For electrochemical methods there is the difficulty of finding an electrolyte which provides sufficient conduction in these solvents.

The different methods used in studying redox systems in nonaqueous solvents will be discussed in Section I, E.

In the special case of hydrocarbons and their slowly dissociating aprotic solvent derivatives, the acid–base reactions are well known, and their mechanisms, constants, and representations were perfected a long time ago.

On the other hand, there has been very little study of redox reactions in these solvents owing to the absence of appropriate measuring and representation methods.

At present a start is being made to study and to use just a few of these solvents for redox reactions employing electrochemical methods.

B. Exchange of Electrons

An acid–base reaction corresponds to a proton transfer; in the same way a redox reaction corresponds to an electron transfer. Nevertheless, the behavior of the different solvents is not the same for these two particles. In the most usual solvents (water, alcohols, ethers, ketone, etc.) the proton may be found in the solvated state and one can then have a measurable concentration of "free" protons that are symbolized by HS^+, where S represents the solvent.

So it is possible to determine a scale of acidity directly linked to this concentration, itself related to the acid–base properties of the solvent. Conversely, in these same solvents the electron is not generally found in the "free state". The electron transfer cannot therefore be made except directly from a reductant to an oxidant. No redox scale can then be directly associated with the concentration of the solvated electron or with the redox properties of the solvent.

The electron exchange in the usual solvents is therefore comparable with the proton exchange in slowly basic aprotic solvents.

The classification of the different solvents for redox reactions will have to be different from the classification decided on for acid–base reactions, as we shall see in Section I, C.

The exchange reactions of protons in these solvents have been fully studied, particularly by Marion MacLean Davis.[1,*] We know that in these solvents when the constants of acid–base reactions do not allow the transfer of protons there exists a hydrogen bond between these two species capable of reacting (acid and base). That is to say, although there is no transfer of protons, there exists an attraction, however weak, for the latter on the part of the base.

In particular, the species produced by an acid–base reaction are linked between themselves by a hydrogen bond, and one can write the acid–base reactions in aprotic solvents according to the scheme:

$$B + HA \rightleftharpoons BH^{+} \cdots A^{-}$$

where B and HA represent, respectively, the base and the reaction acid, and $BH^{+} \cdots A^{-}$ the ion-pair associated by a hydrogen bond resulting from the acid–base reaction.

In much the same way, even if the electron transfer is impossible, there is usually an association between donors and acceptors in most solvents, especially if the latter have a weak dielectric constant and are not themselves electron acceptors. This kind of association allows the formation of numerous complexes linked to the donor or acceptor part of the reacting species. Of course, these associations are not just as simple as this and are more than mere hydrogen bonds; but this concept is nevertheless necessary if the redox reactions are to be clearly understood.

The electron-donor property of these species is characterized by their "donicity."[2] See Table I for some solvent "donicity" values.

It must be noted that the associations which arise under these conditions are not only of an electrostatic nature. They can also be compared to the associations by hydrogen bonds,† and there are associations between electron

* See Vol. III.

† See Chapter 2 of this volume.

TABLE I

DONICITY NUMBER OF VARIOUS POLAR SOLVENTS[2]

Solvents	Donicity number	Dielectric constant
Nitromethane	2.7	35.9
Nitrobenzene	4.4	34.8
Benzonitrile	11.9	25.2
Sulfolane	14.8	38.0
Acetone	17.0	20.7
Diethyl ether	19.2	4.3
Tetrahydrofuran	20.0	7.6
Pyridine	33.1	12.3
Hexamethylphosphotriamide (HMPT)	38.8	30.0

donors and acceptors just as there are between proton acceptors and donors. The bridge is not purely of an electrostatic nature and there is no particle or charge transfer.

Usually, donor–acceptor interactions are stronger in solvents having a low dielectric constant because hydrogen bonds are stronger in these solvents.

1. Oxidants and Reductors

Every substance capable of receiving one or more electrons is called an oxidant. Every substance capable of giving one or more electrons is called a reductant. A redox reaction will correspond to an exchange of electrons between two oxidant and reductant couples. This exchange may be represented by the most general equilibrium:

$$n_1\mathrm{Ox}_2 + n_2\mathrm{Red}_1 \rightleftharpoons n_1\mathrm{Red}_2 + n_2\mathrm{Ox}_1 \tag{1}$$

By applying the law of action of mass to this equilibrium, we obtain

$$K = \left(\frac{[\mathrm{Ox}_1]}{[\mathrm{Red}_1]}\right)^{n_2} \cdot \left(\frac{[\mathrm{Red}_2]}{[\mathrm{Ox}_2]}\right)^{n_1} \tag{2}$$

2. Electrochemical Reactions

The donation or acceptance of electrons can be made at an electrode surface. Then there is an electrochemical reaction for reduction or oxidation as shown in Eqs. 3 and 4, respectively.

$$\mathrm{Ox}_1 + n_1 e\,(\mathrm{electrode}_1) \rightleftharpoons \mathrm{Red}_1 \tag{3}$$

$$\mathrm{Red}_2 - n_2 e\,(\mathrm{electrode}_2) \rightleftharpoons \mathrm{Ox}_2 \tag{4}$$

By applying Nernst's law to the first electrode a potential may be defined:

$$E_1 = E_1{}^0 + \frac{2.3RT}{n_1 \mathscr{F}} \log \frac{(\mathrm{Ox}_1)}{(\mathrm{Red}_1)} \tag{5}$$

and at the second electrode:

$$E_2 = E_2{}^0 + \frac{2.3RT}{n_2 \mathscr{F}} \log \frac{(\mathrm{Ox}_2)}{(\mathrm{Red}_2)} \tag{6}$$

where $E_1{}^0$ and $E_2{}^0$ are the normal potentials of the redox couples 1 and 2 determined in relation to the same reference, R is the ideal gas constant, $\mathscr{F}$ the Faraday $= Ne$ (Avogadro number multiplied by the charge of the electron), T the absolute temperature, 2.3 the conversion factor of Naperian logarithms into decimal logarithms, and n_1 and n_2 are the number of electrons exchanged.

At equilibrium $E_1 = E_2$ and under these conditions we have

$$E_2{}^0 - E_1{}^0 = \frac{2.3RT}{\mathscr{F}} \log K \tag{7}$$

Thus the difference in the normal potentials between two redox couples is linked to the redox equilibrium constant between these two systems. This observation will be of use to us later on in establishing redox molecular scales. It will be possible to set up scales of redox reactivity just as well from the potentials as from the logarithms of the equilibrium constants. The potentials or the logarithms of the constants will then be determined with the same system chosen as reference.

C. Influence of the Solvent Characteristics

We shall not distinguish here between aprotic solvents and others; this distinction has, in fact, only a secondary influence on the redox reactions. However, all solvents do not have the same behavior in redox reactions. The essential characteristic, which will influence the method of interpretation of these reactions, is the presence or absence of free ions in solution. It is known that the concentration of free ions is essentially due to the dielectric constant of the solvent affecting the value of the dissociation constants of ion-pairs.* A solvent with a small dielectric constant will lead to relatively little dissociation of ion-pairs and our interpretation of redox phenomena will be more difficult, as we shall see in Section II,A. In addition, the electrochemical measurements will be more difficult in view of the slight conductivity of the

* See Chapter 1 of this volume.

solutions. The solvating power of the solvent likewise has an influence on the dissociation of the ions, but in this case there cannot be residual hydrogen bonds, so this property of the solvent is, in general, less important than for the acid–base reactions.

According to their behavior in redox reactions, we shall distinguish three main types of solvents.

(1) Very poorly oxidizing solvents capable of solvating the electron. In a small number of solvents, ammonia or the amines, for example, the electron may be found free in a solvated form.[3–5]* If an alkaline metal is dissolved in an amine, the latter is dissociated in a reaction such as the following.

$$\mathrm{Na} + \mathrm{S} \rightleftharpoons \mathrm{Na}^+ + \mathrm{S}e^- \tag{8}$$

where S stands for the solvent and $\mathrm{S}e^-$ the solvated electron; here we have taken no account of the eventual solvation of the cation. This dissociation is comparable to the dissociation of acids in water or other solvents of the same type. In these instances it is possible to define pe as $-\log(e)$ in much the same way as one does for pH in solvents such as water and alcohols, where one may be faced with solvated protons.

The dielectric constant of the solvent also has considerable influence on the solvation of electrons. The latter is stabilized by a lowering in the dielectric constant.[6,7]

(2) Dissociating solvents where the electron is not found in a free state. This category includes the largest number of solvents; water, in particular, is found in this group. The exchange of electrons is made directly between ions or between a molecule and an ion. These are the most typical redox reactions of the type:

$$\mathrm{Ox} + ne \rightleftharpoons \mathrm{Red} \tag{9}$$

At equilibrium, the potential is given by Nernst's law

$$E = E_0 + \frac{RT}{n\mathcal{F}} \log \frac{(\mathrm{Ox})}{(\mathrm{Red})} \tag{10}$$

E is the redox potential measured by the electrode and E_0 is the standard potential of the system.

We shall not dwell on this type of redox reaction which is by far the most familiar.

(3) Poorly dissociating solvents where the reactions occur between molecules or ion-pairs. Solvents with a small dielectric constant are generally poorly solvating. This is the category which is our main concern here because the hydrocarbons and their derivatives are classified in this group of solvents.

* See also Vol. II.

Redox reactions of this type have been much less studied. Electrical measurements are difficult to make because of the low conductivity of the solutions and spectrophotometric measurements are not generally very useful for studying redox reactions.

The behavior of these solvents in redox reactions is comparable to the behavior of slightly dissociating aprotic solvents in acid–base reactions. Poorly solvating aprotic solvents such as the hydrocarbons and their derivatives generally behave in the same way for redox reactions and acid–base reactions. In fact, in these practically inert solvents all the reactions take place directly between molecules or ion-pairs with little or no influence by the solvent.

Redox reactions, like acid–base reactions, cannot be compared with each other except relative to an arbitrary reference which may be quite unrelated to the solvent.

D. Diagrammatic Representations

The limit of the acid–base scales for most solvents is known, but there also exists a limit for the redox scales. Redox and acidity limits, moreover, may be interrelated; in water and numerous other solvents the potential limits depend on pH. It is possible to establish such relationships in diagrammatic form; generally the limits of the redox reactions are fixed on the ordinate and the limits of the acidity scale on the abscissa.

1. Acidity Scales

Two methods of representation for acidity scales are possible depending on the solvent type.

a. *pH Scale.* This scale is used for dissociating protic solvents where the proton can be found in the solvated state. In these solvents the acidic and basic limits are imposed by the basic and acid properties of the solvent. This method of representation is the commonest and most practical one; it is particularly applicable to water and alcohols.

b. *Molecular Acidity Scale.* In all the aprotic solvents the pH scale cannot be utilized; a scale of comparison in relation to the properties of the solvent is not available.[8] Under such conditions it is necessary to choose an arbitrary reference and to determine all the constants of acid–base reactions in relation to this reference.

On the other hand, these solvents generally have a very low dielectric constant and the reactions occur between molecules or ion-pairs. The classification of the bases is then made using the same acid as a reference. It is accepted that all the ion-pairs are in the same size range.

2. Redox Scales

In the preceding section we have defined three types of solvents for redox reactions. For each type of solvent there is a corresponding method for representing redox scales.

a. *pe Scale.* In the case of solvents of the first type, where solvated electrons exist, one can make use of a scale defined as $pe = -\log e$, similar to the definition of pH scales. In addition, this method of representation has been used by Buvet and his collaborators[9] for the study of the properties of organic semiconductors which were considered as solid solvents.[10]

b. *Potential Scale.* Potential scales are well known in water and many other solvents; they permit the prediction of the course of redox reactions much as pH scales permit predictions concerning acid–base reactions. The potential chosen as a reference may be absolutely arbitrary; nevertheless, it must be carefully chosen if we wish to make a comparison between different solvents. Indeed, the very first consideration must be to establish a relationship between the redox scales in different solvents and an absolute scale; it is therefore necessary to choose a reference which is, as far as possible, independent of the solvent. Different redox couples have been proposed, in particular the system Rb/Rb^+, which has the disadvantage of being impossible to use in solvents that are too oxidizing. The ferrocene–ferricinium system also has been found useful. These systems appear to fulfill the conditions cited previously; indeed, the large size and the low charge of ions present limit the effects of solvation which appears to be the essential factor when one considers a variation of the solvent.* Other redox systems have been proposed as references, and all are based on the same principle of weakly solvated reacting species.

Such scales are used for all the aprotic solvents whether they are polar or not, and also for poorly dissociating protic solvents such as acetic acid. In this type of solvent the two methods of representation may be used. We shall not refer again to these methods of representation which are well known and are not the concern in this chapter.[1,8]

* See Vol. I.

c. *Redox Molecular Scales.* In the poorly dissociating solvents where reactions take place between molecules or ion-pairs it is impossible to bring into play the simple ratio of (oxidant):(reductant) because at least one of the two species has a charge. It is no longer possible to give a simple representation of the redox reactions from a potential scale.[11] The redox reactions have to be compared to each other by determining their reaction constants using the same redox couple as a reference. We shall discuss the establishment of these scales in Section II, A.

3. Different Types of Diagrams

There are, theoretically, six different types of diagrams possible for relating the acid–base properties of the solvent to its redox properties. In fact, however, only three types can be used, the most common of which is the potential–pH diagram used in water[12] or similar solvents, i.e., protic, solvating, and dissociating solvents (Fig. 1). The second type of diagram which has been employed uses pe on the ordinate axis and pH on the abscissa; this has been used in the study of organic semiconductors.[13] Finally, the third type of display, which has been employed for aprotic, poorly dissociating solvents such as hydrocarbons and their derivatives, involves the molecular redox scale on the ordinate axis and the molecular acidity scale on the abscissa. If

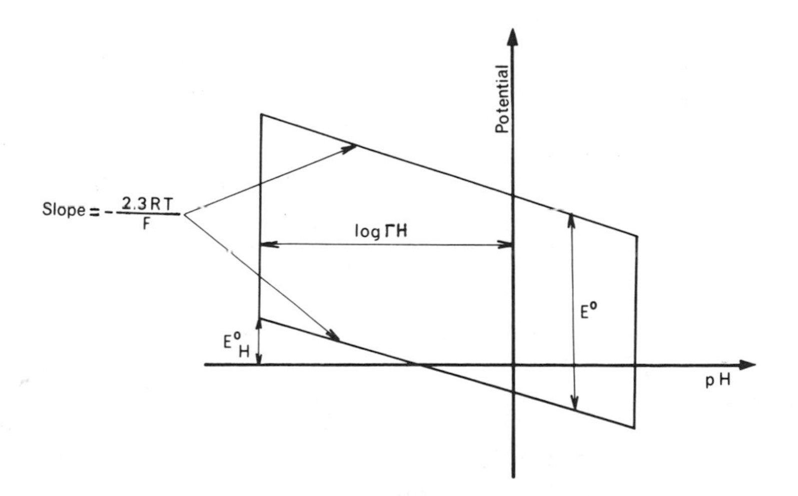

FIG. 1. Determination of potential–pH diagram for a protic solvent. $E_H{}^0$ is the potential of a normal hydrogen electrode vs. an internal reference electrode; ΓH is the difference in solvation coefficient between the solvent and water; and E^0 the potential range for the solvent. The reference for pH is $[H_3O^+]=1$ in water and the reference for the potential is an arbitrary internal reference electrode.

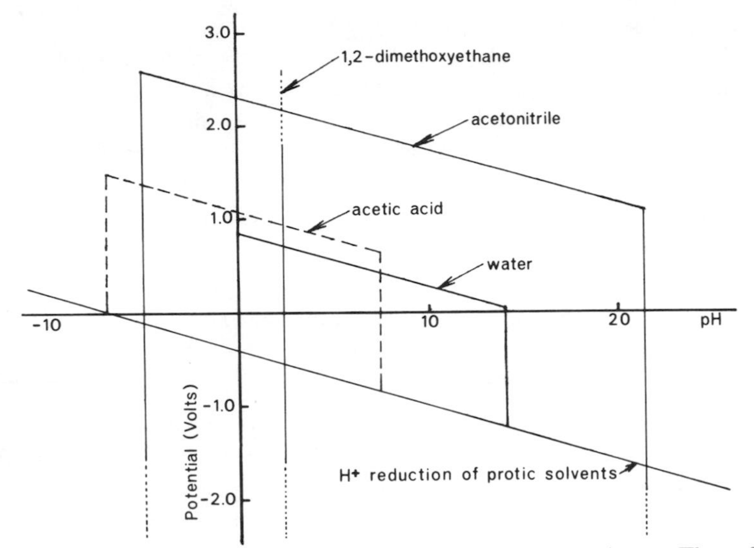

FIG. 2. Potential–pH diagrams of various protic and aprotic solvents. The reference potential is the ferrocene–ferricinium system.

we take, for example, the diagram shown in Fig. 2 which is the classic potential–pH diagram, we see that the potential of the ferrocene–ferricinium system is the reference and the pH reference is that established in water.

In the case of protic solvents the potential of the normal hydrogen electrode related to the reference couple makes it possible to determine the reduction limit of redox scale protons at unit molar concentration. Furthermore, this potential makes it possible to determine the solvation activity coefficient of the proton, and to use the scales $R_0(\mathrm{H})$ defined by Strehlow[14] (Eq. 11).

$$\log \Gamma\mathrm{H} \simeq R_0(\mathrm{H}) = \frac{\mathscr{F}}{2.3RT}(E_{\mathrm{S}}{}^0 - E_{\mathrm{W}}{}^0) \tag{11}$$

where $\Gamma\mathrm{H}$ is the solvation activity coefficient for the proton, $R_0(\mathrm{H})$ is Strehlow's redox function, $\mathscr{F}$ is the Faraday, R is the gas constant, T is the absolute temperature, $E_{\mathrm{S}}{}^0$ is the potential of the hydrogen electrode in the solvent, and $E_{\mathrm{W}}{}^0$ is the potential of this same electrode in water; the index $_0$ indicates that the reductant of the reference couple (ferrocene) has no charge, and H refers to the fact that the system under consideration is the couple H_2/H^+. The definition of pH in a solvent will be dependent on the potential of the normal hydrogen electrode relative to the reference chosen.

Thus, for protic solvents we can use one point to establish the diagram relating redox phenomena to acid–base phenomena. In effect, all we need is to plot the experimental length of the redox scale on the ordinate axis in order to obtain the limit of potential at pH = 0. The limit of the pH scale can be obtained by plotting, e.g., 14 for water; 14.5 for acetic acid; and 27 for acetonitrile.[8] Furthermore, we know that in protic solvents such as water or acetic acid, the reduction limit of the potential varies as $-(2.3RT/\mathscr{F})$pH, and, similarly the oxidation limit varies as $E-(2.3RT/\mathscr{F})$pH, where E is the potential of oxidation at pH = 0 of the solvent when compared to the normal hydrogen electrode. Diagrams of the types shown in Fig. 2 are easy to establish.

The reduction potential limit is not known for acetonitrile, an aprotic solvent; on the other hand, the oxidation limit is established by the reaction shown in Eq. 12. This solvent's oxidation potential varies as

$$2CH_3CN \rightarrow CN\text{-}CH_2\text{-}CH_2\text{-}CN + 2H^+ + 2e^- \qquad (12)$$

$E-(2.3RT/\mathscr{F})$pH. Neither the redox limit nor the upper pH limit for 1,2-dimethoxyethane are known.[15]

Such methods of representation make it possible not only to determine the redox and pH limits of a solvent, but also to show the properties of the different dissolved species. Other types of similar diagrams have also been used, particularly for fused salts, where the theory of solvoacidity[16,17] is often used.

E. Methods Used for the Study of Redox Reactions in Nonaqueous Solvents

1. Electrochemical Methods

Electrochemical methods are generally the best for the study of redox reactions.[18,19] We shall discuss potentiometric methods, nonstationary diffusion methods, and electrolyses.

It should be pointed out that these methods are readily applicable in dissociating solvents (solvents of the first two categories seen in Section I, C, but applications to the poorly dissociating solvents (the third category) are generally more difficult. In such solvents spectrophotometric methods have been preferred.

a. *Potentiometry.* In theory this is the simplest method and, in any case, the most direct one for the study of redox reactions. Potentiometric methods allow thermodynamic data to be obtained, only, however, if the electrochemical systems are reversible. Many electrochemical systems which are nonreversible at a rotating electrode (voltametry) may be reversible at the stationary electrodes used in potentiometry. Under voltametric conditions electrochemical reactions may not be limited by the rate of electron exchange between the electrode and the solution, but by the slower diffusion rate of the electroactive ion toward the electrode. In certain cases, potentiometry may be more advantageously used than voltametry.

The potentiometric methods are most often used in titration schemes[20]; they can provide much more information and are often easier to use than other electrochemical methods.[21,22] However, potentiometric methods are not generally applicable to the study of complex reactions which occur at electrodes.

The verification of the reversibility of an electrochemical system potentiometrically can be done by establishing the validity of the Nernst equation (Eq. 10) corresponding to the equilibrium shown in Eq. 9. The redox properties of cyclopentadienyl–metal complexes are of particular interest in establishing potential scales; we shall use as an example the oxidation of bicyclopentadienyliron (ferrocene) by iodine or bromine.[11,23] Figures 3 and 4 show the potentiometric titrations of ferrocene (Fc) with iodine and bromine. In Fig. 3 the titration reaction corresponds to the equilibrium shown in Eq. 13.

$$2Fc + I_2 \rightleftharpoons 2Fc^+I^- \tag{13}$$

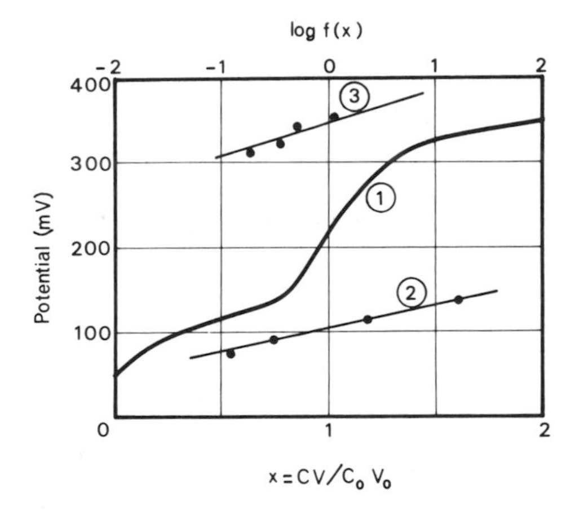

Fig. 3. Titration of ferrocene with iodine.

(1) Titration curve $E = f(x)$

(2) For $x < 1$

$$E = f\left(\log \frac{x}{(1-x)^2}\right)$$

(3) For $x > 1$

$$E = f\left(\log \frac{(x-1)}{(3-x)^3}\right)$$

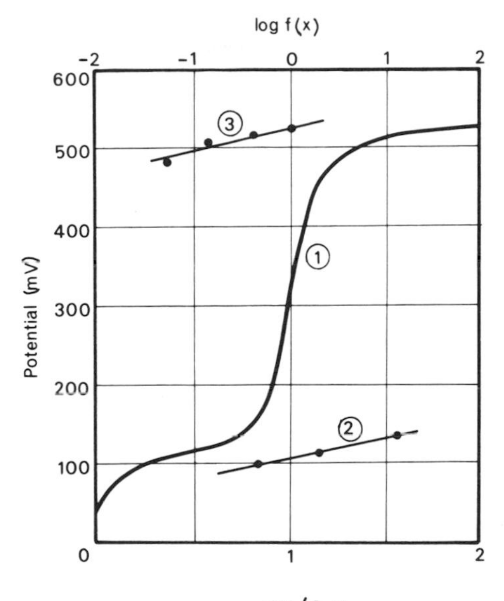

FIG. 4. Titration of ferrocene with bromine.

(1) Titration curve $E = f(x)$

(2) For $x < 1$

$$E = f\left(\log \frac{x}{(1-x)^2}\right)$$

(3) For $x > 1$

$$E = f\left(\log \frac{(x-1)}{(3-x)^3}\right)$$

With an excess of iodine, the ferricinium iodide formed produces the triiodide complex according to reaction 14.

$$Fc^+I^- + I_2 \rightleftharpoons Fc^+I_3^- \tag{14}$$

The potential before ($x < 1$) and after ($x > 1$) the end point ($x = 1$) of the titration is, therefore, associated with the equilibria shown in Eqs. 15 and 16.

$$Fc \rightleftharpoons Fc^+ + e^- \tag{15}$$

$$3I^- \rightleftharpoons I_3^- + 2e^- \tag{16}$$

Before the end point ($x < 1$), the expression of potential which corresponds to Eq. 15 is given by Eq. 17.

$$E_{eq} = E^0 + \frac{RT}{\mathscr{F}} \log \frac{[Fc^+]}{[Fc]} \tag{17}$$

The Fc^+ ion is associated with the I^- anion and the following equations are obtained (ignoring the concentration of free ions).

$$C_0 = [Fc] + [Fc^+I^-] \tag{18}$$

$$[Fc^+I^-] = xC_0 \tag{19}$$

The electroneutrality condition in this system is given by Eq. 20.

$$[Fc^+] = [I^-] \tag{20}$$

It follows that

$$[Fc] = C_0[1-x] \tag{21}$$

$$[Fc^+] = [k_d \times C_0]^{1/2} \tag{22}$$

where k_d is the dissociation constant of the ion-pair Fc^+I^-. The equation for the equilibrium potential is given by Eq. 23.

$$E_{eq} = E^0 - \frac{RT}{2\mathscr{F}} \log \frac{C_0}{k_d} + 0.029 \log \frac{x}{(1-x)^2} \tag{23}$$

After the endpoint ($x > 1$), the equilibrium potential is given by Eq. 24.

$$E_{eq} = E_1{}^0 + \frac{RT}{2\mathscr{F}} \log \frac{[I_3{}^-]}{[I^-]^3} \tag{24}$$

The anions $I_3{}^-$ and I^- are associated with the ferricinium ion Fc^+. If we designate $k_d{}'$ and k_d, respectively, as the constants of dissociation of corresponding ion-pairs, we may write Eqs. 25 and 26.

$$[I^-] = k_d \frac{[Fc^+I^-]}{[Fc^+]} \tag{25}$$

$$[I_3{}^-] = k_d{}' \frac{[Fc^+I_3{}^-]}{[Fc^+]} \tag{26}$$

A consideration of the mass balance gives Eqs. 27 and 28 for $(I_2) \ll C_0$.

$$C_0 = [Fc^+I^-] + [Fc^+I_3{}^-] \tag{27}$$

$$[Fc^+I_3{}^-] = (x-1)\frac{C_0}{2} \tag{28}$$

The electroneutrality condition dictates Eq. 29

$$[Fc^+] = [I_3{}^-] + [I^-] \tag{29}$$

and we obtain

$$[I_3{}^-] = \frac{k_d{}'(C_0/2)(x-1)}{[I^-] + [I_3{}^-]} \tag{30}$$

Considering that the dissociation constants k_d and k_d' are of the same magnitude ($k_d' \simeq k_d$), we can easily deduce

$$E_{eq} = E_1{}^0 - \frac{RT}{\mathscr{F}} \log \frac{k_d C_0}{2} + \frac{RT}{2\mathscr{F}} \log \frac{(x-1)}{(3-x)^3} \tag{31}$$

Such an analysis of potentiometric titration curves enables us to determine the constants for redox reactions, even in poorly dissociating solvents, but the values found depend on dissociation constants of ion-pairs. In the case of molecular acidity scales, it is necessary to assume that all dissociation constants of ion-pairs are of the same magnitude to obtain useful results.

b. *Steady State Diffusion Methods: Conventional Voltametry and Polarography.* For our purposes we shall consider voltametry as a general term to describe the process which measures a current flow at a microelectrode to which a potential has been imposed. Polarography is a particular method of voltametry in which a dropping mercury microelectrode is used. In nonaqueous solvents, the classic arrangement with three electrodes is indispensable to compensate for the ohmic drop between the working electrode and the auxiliary electrode. Indeed, in many of these solvents the dielectric constant is low, the salts weakly dissociated, and, thus, the solutions are only mildly conductive. Furthermore, to obtain state diffusion, stirring at the electrode surface is necessary (solid revolving electrodes or dropping mercury electrodes). As we have already seen, such agitation increases the rate of transfer, but has the disadvantage of rendering certain systems less reversible. Indeed, if the transfer rate is too rapid, the electron exchange rate limits the rate of the overall reaction. These methods also require a fairly large concentration of supporting electrolyte to assure good conductivity, because the properties of the solvent are usually affected.

In spite of their disadvantages, steady state diffusion methods have been used for the study of redox reactions in halogen derivatives of the hydrocarbons, and in particular, chloroform and methyl chloride. The limits of electroactivity in chloroform at a platinum electrode and a dropping mercury electrode, in the presence of various electrolyte supports, are shown in Fig. 5.

To make the interpretation of voltametric curves possible, it is indispensable that the electrochemical kinetics of the systems in question be rapid. The nature and properties of solvents have an important influence on such kinetics, which is specifically related to the molecular refraction of the solvent.[26,27]

Different types of electrodes have been used. Usually platinum or mercury are preferred, but graphite[28] and vitreous carbon[29] have been employed to study organic,[30–33] organometallic,[34] and inorganic[35–37] species.

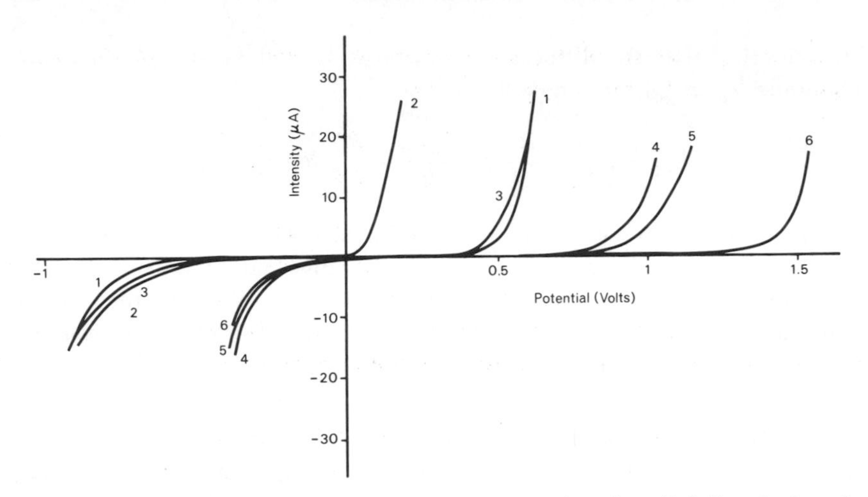

Fig. 5. Electroactivity range in chloroform with various electrolytes (1, 2, 3) at the dropping mercury electrode and (4, 5, 6) at the bright platinum electrode. Electrolytes: 1, 4; 1 *M* triethylammonium sulfate–0.5 *M* sulfuric acid; 2, 5, tetraphenylarsonium chloride; 3, 6, 0.25 *M* piperidine–0.75 *M* piperidinium perchlorate.

c. *Non-steady-State Diffusion Methods.* The most widely used of such methods are chronoamperometry, cyclic voltametry,[38,39] oscillographic polarography,[40] and chronopotentiometry.[41] These methods which have been widely used in polar solvents[42] (e.g., acetonitrile) for the study of organic reactions can also be applied to the study of redox reactions in hydrocarbons and their derivatives.[41,43]

d. *Electrolysis.* Apart from their role in the quantitative determination of certain substances, electrolysis experiments permit the study of certain redox reactions by determining the stoichiometry or the number of electrons exchanged.[44–47] Such experiments enable us to determine the nature of the products of a redox reaction. For example, it has been demonstrated that in chloroform certain quinones are reduced to semiquinones with the exchange of an electron; if the reduction is continued, an irreversible destruction of the molecule occurs.[36] Coulometry using electrogenerated species has been successfully performed in this solvent.[48]

2. Spectrophotometric Methods

a. *Direct Method.* Under certain conditions in a redox reaction, either the oxidized or reduced species is capable of absorbing light; such reactions are easily followed in a redox titration by spectrophotometric methods. At the equivalence point, the difference between the observed stoichiometry and

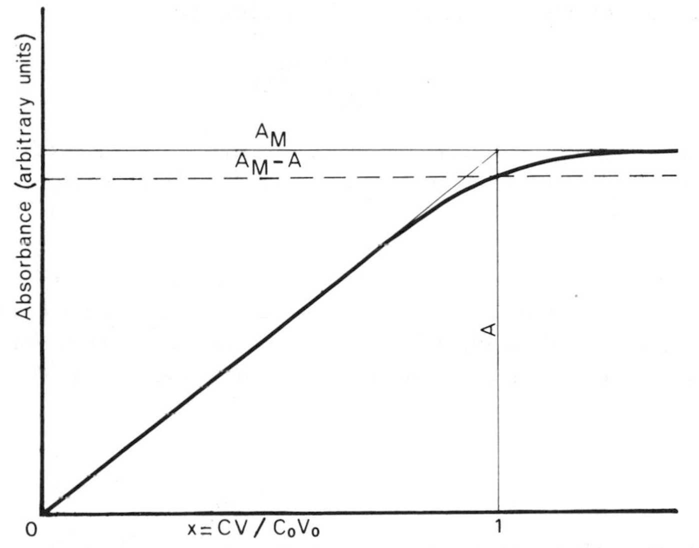

FIG. 6. Spectrophotometric titration of tetraethylammonium iodide with iodine at $\lambda = 295$ mm; variation of absorbance with the addition of iodine.

that expected is related to the redox equilibrium constant. Figure 6 represents the results of the titration of iodine (4.2×10^{-3} M) by tetraethylammonium iodide in chloroform.[11] The stoichiometric discrepancy at the equivalence point indicated by the quantity $(A_M - A)$ makes it possible to determine the equilibrium constant.

$$K = \frac{[Et_4NI_3]}{[Et_4NI][I_2]} \tag{32}$$

For $x = 1$ we have, in effect,

$$[I_2] = [Et_4NI] = C_0 - [Et_4NI_3] \tag{33}$$

where C_0 represents the concentration of iodine at the beginning of the titration.

If we call the maximal absorbance due to Et_4NI_3 when the latter reaches the concentration C_0, A_M, and A the absorbance at the equivalent point, it follows that

$$[Et_4NI_3] = \frac{A\ C_0}{A_M} \tag{34}$$

$$[I_2] = [Et_4NI] = \frac{A_M - A}{A_M} C_0 \tag{35}$$

The equilibrium constant expressed in Eq. 32 becomes

$$K = \frac{A \times A_{\mathrm{M}}}{C_0 (A_{\mathrm{M}} - A)^2} \tag{36}$$

This method is simple and rapid, but requires that at least one of the reacting substances or products of the reaction be absorbent in an easily measurable wavelength range.

b. *Indirect Method Using Spectrophotometric Redox Indicators.* As in the case for the pH determinations, the potential of a solution may be determined by means of colored indicators. If, for example, we call the oxidized form of this indicator I, and its reduced form after electron exchange has occurred $\mathrm{I}e_{\mathrm{n}}$, Eq. 37 is obtained (where the electron charges are not shown).

$$E = E^0 + \frac{RT}{n\mathscr{F}} \log \frac{[\mathrm{I}]}{[\mathrm{I}e_{\mathrm{n}}]} \tag{37}$$

Equation 37 allows us to determine redox potentials in relation to the normal potential of an indicator by processes similar to those used in determining acid–base constants. For the redox reaction given in Eq. 38,

$$\mathrm{Ox} + me \rightleftharpoons \mathrm{Red} \tag{38}$$

the Nernst expression is given by Eq. 39

$$E = E_{\mathrm{i}}{}^0 + \frac{RT}{n\mathscr{F}} \log \frac{(\mathrm{Ox})}{(\mathrm{Red})} = E^0 + \frac{RT}{n\mathscr{F}} \log \frac{(\mathrm{I})}{(\mathrm{I}e_{\mathrm{n}})} \tag{39}$$

and we deduce

$$E^0 - E_{\mathrm{i}}{}^0 = \frac{RT}{nm\mathscr{F}} \log \frac{(\mathrm{Red})^n}{(\mathrm{Ox})^n} \times \frac{(\mathrm{I})^m}{(\mathrm{I}e_{\mathrm{n}})^m} \tag{40}$$

The [Red]/[Ox] ratio is fixed experimentally and spectrophotometric measurements enables us to determine the relationship $[\mathrm{I}]/[\mathrm{I}e_{\mathrm{n}}]$. The potential difference can be determined provided that the relations [Red]/[Ox] and $[\mathrm{I}]/[\mathrm{I}e_{\mathrm{n}}]$ are neither too large nor too small to obtain acceptable precision. Furthermore, it is important that the total concentration of the indicator be negligible compared with that of the redox couple being studied.

Using such techniques, it is also possible to determine the coefficients of the redox reaction, and we shall see in Section II,A that such information is very important in the study of redox reactions which occur in solutions of hydrocarbons and their derivatives. In certain cases, the electrochemical and spectrophotometric measurements can be performed simultaneously when it is possible to determine the redox potential of an indicator.[49] Some values for redox potentials relative to the normal hydrogen electrode in water as well as the redox potentials of some organometallic compounds which can be used as indicators, are given in Table II.

TABLE II

POTENTIAL OF VARIOUS ORGANOMETALLIC COMPOUNDS IN WATER

Compound	$E_H{}^0$	Ref.
Dicyclopentadienyliron (ferrocene)	0.40	(8)
	0.81	(18)
Dicyclopentadienylcobalt (cobaltocene)	−0.92	(8)
Tris(*o*-phenanthroline sulfonate-6)Fe^{II}	1.22	(8)
Tris(*o*-phenanthroline)Fe^{II} (ferroin)	1.13	(8)
Dixylenechromium(I)	−1.31	(18)
Dihexamethylbenzenechromium(I)	−1.77	(18)

II. Redox Systems in Hydrocarbons and Their Derivatives

A. Molecular Scales Redox

It is impossible to establish the strength of acids and bases in poorly solvating solvents with a low dielectric constant using the same ideas associated with the pH scale as developed for water or other solvents with a high dielectric constant. For similar reasons, it also appears impossible to establish a classification using a redox potential in solvents of low dielectric constant and the arguments developed for aqueous systems. Nonetheless, redox molecular scales similar to modes of representation for the molecular scales of acidity or basicity can be established using similar approximations; that is, we shall consider that all dissociation constants for ion-pairs are of the same order of magnitude in a given solvent. Two modes of classification are then possible: (1) Classification relative to an oxidant or a reductant at a given concentration and (2) classification relative to a redox couple.

One can classify the strength of several reductants relative to an oxidant. For example, if we choose bromine as an oxidant reference, the classification of redox systems can be established using the following argument. On a scale of $pBr_2 = \log Br_2$ (the reference is $[Br_2] = 1.0\ M$), we can place values of pBr_2 which correspond to redox couples such as [oxidant] = [reductant]. The values of pBr_2 are obtained by using redox reaction constants between bromine and the different species studied.

A better method of classification can be obtained by using a redox couple as a point of reference. The choice of the system of the reference is very important; one of the best systems currently available, which may also be used with other solvents as a reference, is the ferrocene–ferricinium couple introduced by Strehlow.[50] Taking this redox couple as an example, we can establish a scale based on Eq. 41

$$pX = \log \frac{[Fc^+][A^-]}{[Fc]} \tag{41}$$

where Fc^+ represents the ferricinium ion, Fc ferrocene, and A^- any anion, assuming that the nature of this anion has only a negligible influence on the dissociation constants of the ion-pair. This scale is similar to the redox potential scale. We can employ either redox couples or a given oxidant (or reductant) at a given concentration. Such a classification enables us to predict easily the possible reactions between oxidants and reductants in solvents of low dielectric constant, as well as the elaboration of redox buffer solutions using the systems discussed. Several results obtained using potentiometry in chloroform[11] are given in Table III. The molecular scales of acidity shown in Figs. 7 and 8 have been established from the data given in Table III.

TABLE III

REDOX REACTION AND COMPLEX FORMATION CONSTANTS IN CHLOROFORM[11]

Redox reaction[a]	$\log K$
$2Cc + Cc^+I_3^- \rightleftharpoons 3Cc^+I^-$	20.5
$2Fc + Fc^+I_3^- \rightleftharpoons 3Fc^+I^-$	8.3
$2Fc + Fc^+Br_3^- \rightleftharpoons 3Fc^+Br^-$	14.5
Complex formation	**$\log K$**
$R_4N^+I^- + I_2 \rightleftharpoons R_4N^+I_3^-$	6.7–7.1
$R_4N^-I^- + Br_2 \rightleftharpoons R_4N^+Br_3^-$	8.3

[a] Cobaltocene (Cc); cobalticinium (Cc^+); ferrocene (Fc); ferricinium (Fc^+).

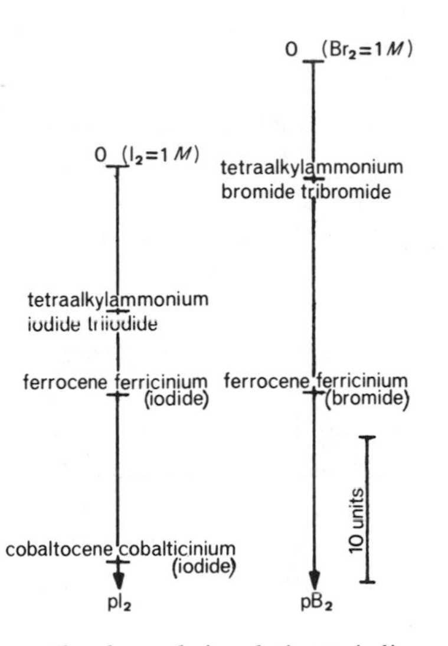

FIG. 7. Redox molecular scale in relation to iodine or bromine.

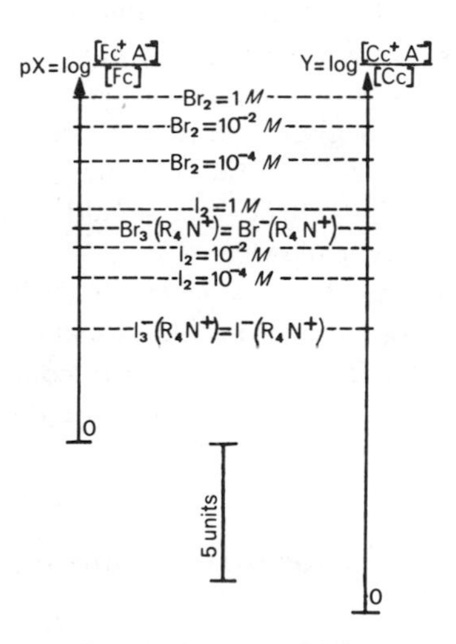

FIG. 8. Redox molecular scale in relation to ferrocene–ferricinium system or cobaltocene–cobalticinium system.

B. Comparison of Scales between Different Solvents

Although the problems associated with comparisons of phenomena between solvents have already been dealt with in Vol. II of this treatise by Meek[51] and Strehlow,[14] we wish to point out that a comparison using potential scales is possible only with sufficiently polar and dissociating solvents. We have seen in Section II,A that only redox molecular scales can be used in the less ionizing solvents. It is, of course, possible to choose the same or similar reference systems for all solvents, but we shall not elaborate on that problem again here. We have seen that it is possible to establish three types of scales of comparison corresponding to three types of solvents discussed in Section I,C: (1) $pe = -\log e$ scale, which is similar to pH scales and is used in solvents supporting a solvated electron; (2) potential scales which are the most widely used for sufficiently dissociating solvents[8]; and (3) redox molecular scales for poorly dissociating solvents where redox reactions occur between molecules or ion-pairs. This type of scale is comparable to the molecular scales of acidity used in the same type of solvents.[11]

In this section we shall discuss the molecular scale, which is the only one that can be used in poorly dissociating solvents, where the direct influence of the solvent (by solvation) will be negligible. Yet there may exist a variation in scale between various solvents arising essentially from differences between the dissociation constants of ion-pairs from one solvent to another. Indeed, interactions between solutes are very strong in such solvents, as the solutes are not strongly associated with the solvent molecules. In "inert" solvents, the effects of solvation are negligible and the essential factor which varies from one solvent to another is the difference between dissociation constants of ion-pairs. To suppress that effect, one should use redox couples of the type $(Ox^+)/(Red^-)$ in this type of solvent.

At this time, comparisons of redox scales in inert solvents do not seem valid, considering the large variations in dissociation constants observed from one solvent to another. Indeed, in order to compare the scales in each of the inert solvents we must take into account the divergences between dissociation constants.

C. Influence of Other Chemical Reactions on Redox Reactions

1. Influence of Acidity

Each time a redox reaction involves an exchange of protons, the acidity (or the basicity) of the solvent has an influence on the values of the redox reaction potentials or constants. This influence can be clearly illustrated by using the

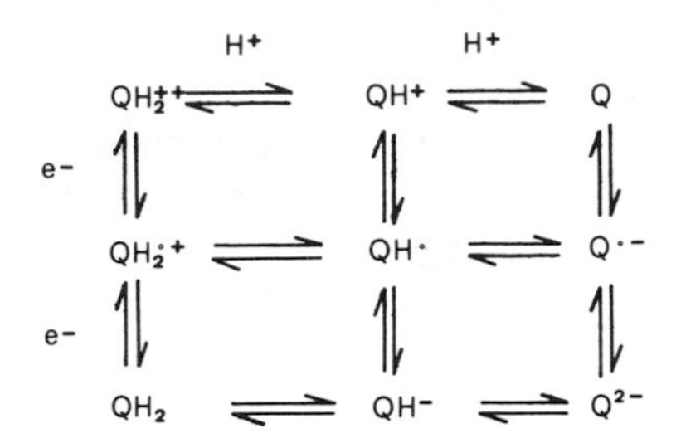

FIG. 9. Scheme of various possible reduction and protonation processes of quinones.

diagrams described in Section I, D. The exchange of protons and electrons can be simultaneous or sequential. Schemes in which all reaction possibilities are revealed can be established after Manning[52] in the form of a diagram; the exchange of protons is shown on the lines and the exchange of electrons in the columns. See Figs. 9 and 10 for a representation of this scheme and the corresponding diagram for quinone. It is obvious that not all the species shown in these diagrams are stable in all the solvents, but a scheme such as this does make it possible to foresee the complete range of reaction possibilities, which facilitates the experimental study of the system. Thus, as the acidity of the system influences redox reactions, it is usually necessary to employ buffers to study such redox reactions.[53] Such influences have been studied in aprotic polar solvents by Farsang,[49,54] Cauquis,[55] and their collaborators.

The influence of acidity on redox systems is well known for quinones, aromatics,[56] cyclopentadienyl–metal complexes, and all species which act as bases.[57,58] Such effects are observed when electron and proton exchanges occur simultaneously.

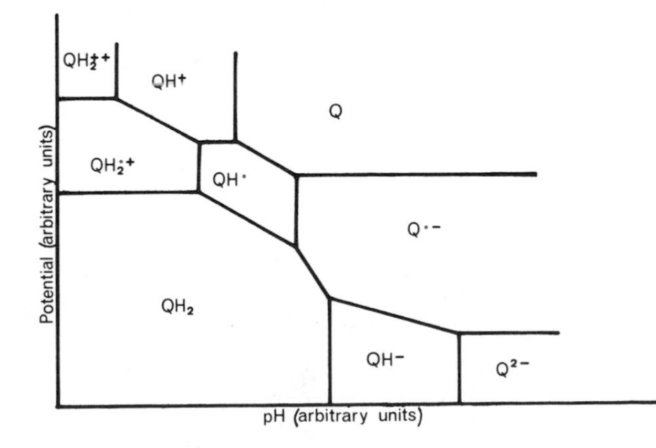

FIG. 10. Complete theoretical potential–acidity diagram of quinones.

2. Influence by the Formation of Complexes

As a general rule solutes in the poorly solvating solvents tend to form stable complexes, given that the different species are not associated with the solvent. Under these conditions the redox reactions are easily influenced by modifications in the solution. In electrochemical studies, particularly, the choice of electrolyte is very important; no electrolytes are really inert in the presence of the solutes generally studied.

Given the relatively great stability of the complexes in this kind of solvent, the redox reactions between two given solutes can be very different from those observed in more highly solvating solvents. For example, the oxidation of tetraethylammonium iodide by the bromine in chloroform takes place in two steps according to the equilibria shown in Eqs. 42 and 43.

$$6Et_4N^+I^- + 2Br_2 \rightleftharpoons 2Et_4N^+I^- + 4Et_4N^+Br^- \quad (42)$$

$$2Et_4N^+I^- + 4Et_4N^+Br^- + 4Br_2 \rightleftharpoons 3Et_4N^+(I_2Br)^- + 3Et_4N^+Br_3{}^- \quad (43)$$

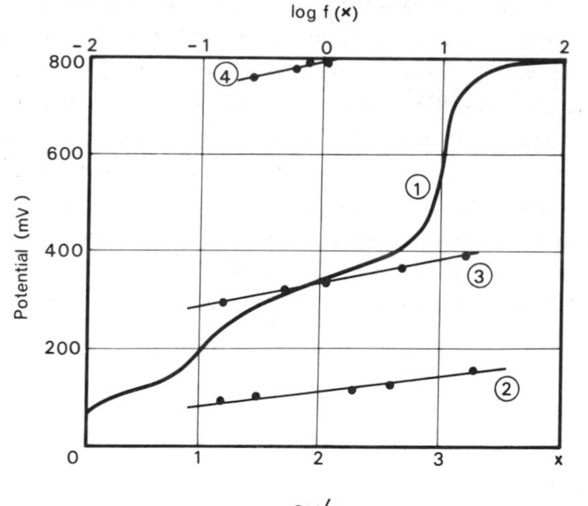

Fig. 11. Titration of tetraethylammonium iodide with bromine.

(1) Titration curve $E = f(x)$

(2) For $x < 1$

$$E = f\left(\log \frac{x}{(1-x)^2}\right)$$

(3) For $1 < x < 3$

$$E = f\left(\log \frac{(x-1)}{(3-x)^3}\right)$$

(4) For $x > 3$

$$E = f(\log(x-3))$$

Figure 11 shows that the reaction for the formation of the complex gives a greater variation in potential than the redox reaction proper. The iodine–iodide complexes have also been studied in other solvents.[58]

Numerous complexes can form during redox reactions in inert solvents[59,60] between the products of the reaction and different cations[61] and, in particular, those of electrolytes used in electrochemical studies.[62–64] Tetraalkylammonium ions also give complexes with the halogen derivatives of hydrocarbons.[65]

Among the solvents which are practically inert, chloroform plays a special role because of its acid behavior[66] (its pK_{H_2O} is about 24). As a result it easily forms hydrogen-bonded complexes,[67] particularly with amines.[68,69]

D. Redox Reactions in Various Solvents

1. Chloroform

Among the halogen derivatives of hydrocarbons, chloroform has been studied most. Because of its relatively highly solvating power, electrolytes are more dissociated than in other similar solvents, and the solutions are better conductors. These characteristics have made it possible to perform electrochemical studies in this solvent, especially potentiometry,[11,59] voltametry on a platinum electrode,[37,70] and polarography.[36,59,71] Redox studies have also been carried out using spectrophotometry,[71] in particular, the photochemical oxidation of aromatic amines.[72]

a. *Potentiometric Titrations.* Potentiometric titrations using chloroform as a solvent (Figs. 3 and 4) have led to the determination of redox equilibrium constants and formation constants for complexes (Fig. 11). Potentiometric titrations in chloroform solutions of dicyclopentadienyl–metal complexes by halogens (I_2, Br_2) have been reported.

Results obtained by potentiometry appear in Table III. These data permit us to establish redox molecular scales (Figs. 7 and 8). The reference electrode used for these measurements consists of a platinum wire dipped in a mixture of 0.1 *M* tetrabutylammonium iodide and 0.1 *M* tetrabutylammonium triiodide separated from the working solution by a glass frit.

b. *Voltametry and Polarography.* These methods usually employ chloroform[36,59,68,69,70] as solvent. The following reference electrode systems have been used: silver/silver chloride in tetrabutylammonium chloride,[71] silver/10^{-3} *M* silver nitrate in 0.25 *M* tetrabutylammonium perchlorate,[70] silver/silver iodide in 0.05 *M* tetrabutylammonium perchlorate,[59] and tetrabutylammonium iodide/triiodide.[11,36,37] The potential of the

TABLE IV

POTENTIAL OF FERROCENE–FERRICINIUM SYSTEM VS. VARIOUS REFERENCE ELECTRODES IN CHLOROFORM

Reference electrodes	Potential of Fc/Fc^+ system (V)	Ref.
$Ag/AgCl–But_4NCl$	~ +0.1	71
$Ag/10^{-3}$ M Ag^+	−0.35	70
$Ag/AgI–0.05$ M But_4NI	+0.42	59
But_4NI_3/But_4NI	−0.58	36

ferrocene–ferricium couple determined with reference to these electrodes is given in Table IV.

Numerous salts and mixtures have been used as electrolytes, namely, tetrabutylammonium perchlorate,[59,70,71] tetraphenylarsonium chloride,[37] piperidine–piperidinium perchlorate,[36] and triethylamine–triethylammonium sulfate.[37]

Numerous organic and inorganic substances have been studied by voltammetry and/or polarography; the half-wave potentials of various oxidants or the equilibrium potentials of various redox couples are summarized in Table V. Figure 12 gives an example of the quantitative determination of vitamin K in chloroform using these techniques.

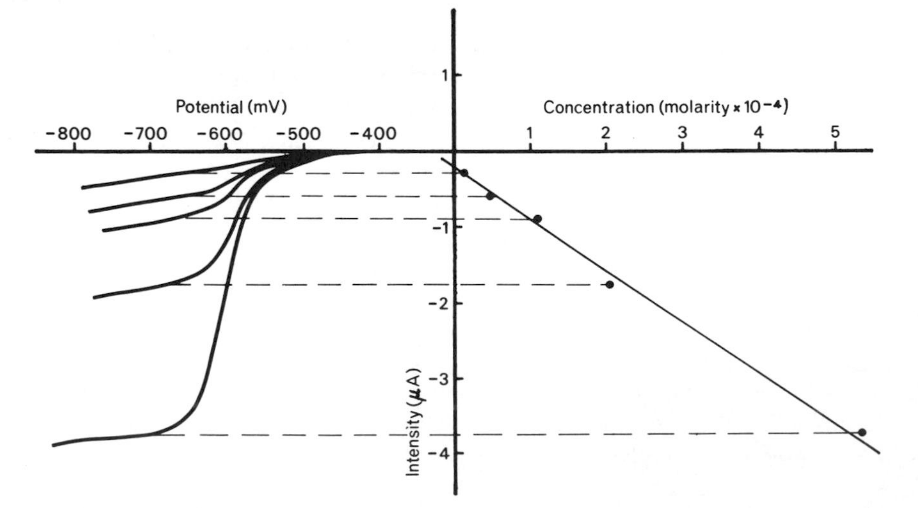

FIG. 12. Polarography of vitamin K in chloroform at various concentrations with piperidine–piperidinium perchlorate electrolyte.

TABLE V

EQUILIBRIUM OR HALF-WAVE POTENTIALS OF VARIOUS REDOX COUPLES OR OXIDANTS[a]

Redox couple	E_{eq} (V)	Reference electrode[d]	Working electrode[e]	Electrolyte[f]	Ref.
Quinhydrone	−0.32	I	DME	Pip/PPC (0.25/0.75 M)	36
Chloranil/tetrachlorohydroquinone	−0.25	I	DME	Pip/PPC (0.25/0.75 M)	36
Naphthoquinone/naphthalenediol	−0.50	I	DME	Pip/PPC (0.25/0.75 M)	36
Anthraquinone/semianthraquinone[b]	−0.65	I	DME	Pip/PPC (0.25/0.75 M)	36

Oxidant	$E_{1/2}$ (V)	Reference electrode[d]	Working electrode	Electrolyte[f]	Ref.
1,4-Naphthoquinone 2-methyl vitamin K	−0.60	I	DME	Pip/PPC (0.25/0.75 M)	36
Cu^{I}[c]	−0.47	I	DME	ϕ_4AsCl	36
V^{V}	−0.35	I	DME	ϕ_4AsCl	37
Hg^{II}	−0.20	I	DME	ϕ_4AsCl	36
Tetrahydroxy-1,2,5,8-anthraquinone	−0.91	I	DME	ϕ_4AsXl (0.5 M)	37
Tetrahydroxy-1,2,5,8-anthraquinone	−0.88	—	Pt	H_2SO_4/Et_3N (1.5/1 M)	37
Anthraquinone	−0.72	—	DME	ϕ_4AsCl (0.5 M)	37
Anthraquinone	−0.78	—	Pt	H_2SO_4/Et_3N (1.5/1 M)	37
Phenazothionium	−0.18	—	—	H_2SO_4/Et_3N (1.5/1 M)	37

(Continued)

TABLE V—*continued*

Oxidant	$E_{1/2}$ (V)	Reference electrode[d]	Working electrode	Electrolyte[f]	Ref.
TMPD	−0.64	II	BPt	n-Bu_4NClO_4 (0.25 *M*)	70
DMDPPD	−0.43	II	BPt	n-Bu_4NClO_4 (0.25 *M*)	70
DPPD	−0.43	II	BPt	n-Bu_4NClO_4 (0.25 *M*)	70
TPPD	−0.25	II	BPt	n-Bu_4NClO_4 (0.25 *M*)	70
HDPA	−0.26	II	BPt	n-Bu_4NClO_4 (0.25 *M*)	70
MPPD	−0.51	II	BPt	n-Bu_4NClO_4 (0.25 *M*)	70
2,3-Dichloro-5,6-dicyanoquinone	0.81	III	DME	n-Bu_4NClO_4 (0.5 *M*)	59
Tetracyanoethylene	0.53	III	DME	n-Bu_4NClO_4 (0.5 *M*)	59
Chloranil	0.53	III	DME	n-Bu_4NClO_4 (0.5 *M*)	59

[a] *N,N,N',N'*-tetramethyl-*p*-phenylenediamine (TMPD), *N,N'*-dimethyl-*N,N'*-diphenyl-*p*-phenylenediamine (DMDPPD), *N,N'*-diphenyl-*p*-phenylenediamine (DPPD), *N,N,N',N'*-tetraphenyl-*p*-phenylenediamine (TPPD), 4-hydroxydiphenylamine (HDPA), *N*-(*p*-methoxyphenyl)-*p*-phenylenediamine (MPPD).

[b] The coulometric reduction of anthraquinone gives a semianthraquinone with only one electron transferred.

[c] In the presence of piperidine, Cu^{II} is very easily oxidized. Only Cu^{I} is stable in this condition.

[d] Electrode systems: I, 0.1 *M* Bu_4NI–0.1 *M* Bu_4NI_3; II, Ag/10^3 *M* Ag; III, Ag/AgI in 0.05 *M* Bu_4NI.

[e] DME, Dropping mercury electrode; BPt, bright platinum.

[f] Data in parentheses are concentration values. Pip, Piperidine; PPC, piperidium perchlorate.

2. Methylene Chloride

Methylene chloride is becoming more recognized as a useful solvent for the electrochemical study of aromatic hydrocarbons. In certain cases[73] these studies are carried out at low temperatures, making it possible to obtain more reproducible measurements. Potential measurements in this solvent have been compared to those obtained in other solvents[42,74] but, because of the great influence of the dissociation constants of the ion-pairs on such measurements, the comparisons are not too useful.

TABLE VI

Oxidation Potentials of Various Aromatic Hydrocarbons in Methylene Chloride

Aromatic hydrocarbon	Oxidation potential (V)	Reference electrode[a]	Ref.
Rubrene	0.17	Ag/Ag^+	73
Perylene	0.35	Ag/Ag^+	74
Tetracene	0.35	Ag/Ag^+	74
Tetracene	1.0	SCE_{aq}	74
Ferrocene	0.37	SCE_{aq}	74
Thianthrene	0.64	Ag/Ag^+	73
1,4-Dimethoxybenzene	0.70	Ag/Ag^+	73
Anthracene	0.73	Ag/Ag^+	73, 74
Hexamethylbenzene	1.05	Ag/Ag^+	73
Phenanthrene	1.13	Ag/Ag^+	73, 74
9,10-Diphenylanthracene	1.22	SCE_{aq}	74

[a] Silver/saturated silver nitrate reference electrode in methylene chloride (Ag/Ag^+); saturated calomel electrode in aqueous solution (SCE_{aq}).

The methods used to study redox phenomena in this solvent have been polarography,[41,59,74] chronopotentiometry,[42] and cyclic voltametry.[73] The most frequently used reference electrode systems are silver/saturated silver nitrate,[73] and silver/silver iodide in 0.05 *m* tetrabutylammonium iodide.[59] Tetrabutylammonium perchlorate[59] and tetrabutylammonium fluoroborate[73] have been used as electrolytes.

The oxidation potentials of various aromatic systems have been determined in these solvents (Table VI).

3. Benzene

Few studies have been carried out in this solvent, but some measurements of half-wave potentials for organomercurials have been reported in 1:1 benzene–ethanol mixtures.[75] Table VII contains the results of the polarographic reduction of organomercury halides shown in Eq. 44.

TABLE VII

Half-Wave Potential for the Polarographic Reduction of Organomercury Compounds[75]

Organomercuric compounds	Potential (V)
C_3H_5HgCl	−0.16
C_3H_5HgBr	−0.26
C_3H_5HgI	−0.39

[a] In 1 : 1 ethanol–benzene mixtures.

$$RHgX + e^- \rightarrow RHg + X^- \tag{44}$$

A second reduction wave sometimes occurs in these systems which corresponds to the cleavage of a mercury–carbon bond.

$$H^+ + RHg + e^- \rightleftharpoons RH + Hg \tag{45}$$

4. Other Essentially Inert Solvents

Polarographic studies have been conducted on solutions of benzoyl fluoride.[76] Various metallic cations (Tl, Zn, Cd, Mn, Co, Ni), as well as ferrocene, have been investigated. A few measurements attempting to establish the polarity of solvents have also been made by spectrophotometry, with "phenol blue" as a suitable indicator.[77] The main problem with most inert solvents is finding an electrolyte which is sufficiently soluble and dissociated to assure sufficient conductivity to make electrochemical measurements possible. Whenever this is not possible it is necessary to employ spectrophotometric methods.

III. Conclusion

It is impossible to use either the redox potential scales or the pH scales in solutions of hydrocarbons and their derivatives. However, for such solvents it is possible to establish redox molecular scales which are similar to the molecular scales of acidity used with this kind of solvent. Two types of scale are useful: one consists of an oxidant (or a reductant) at a given concentration as the reference and the other (preferable) uses a redox couple as a reference. It is necessary to use a reference couple which is independent of the solvent if a comparison of redox molecular scales is to be made among poorly ionizing solvents. Systems of the type $(Ox^+)/(Red^-)$ would seem the best suited for solutions of hydrocarbons and their derivatives. In effect, the most important factor involved in changing from one solvent of this type to another is not solvation phenomena (because none solvate well), but the variation in the dissociation constant of ion-paired species. The use of an appropriate reference couple of the type $(Ox^+)/(Red^-)$ suggests that the dissociation constants are the same for all ion-pairs for a given solvent. Under such conditions the reference couple can be assumed to be independent of the solvent, as long as the latter solvates poorly. Preferential solvation of the anion or cation may take place in solvents that are more polar, in which case the suggested reference system is no longer valid.

The experimental study of redox reactions is without a doubt more difficult in poorly ionizing solvents than in polar solvents. Electrochemical measurements are not always possible because electrolytes are not sufficiently soluble and dissociated. Spectrophotometric measurements are a less direct method of studying redox reactions. Recently, advances in electrochemical methodology have permitted a detailed study of redox reactions in chloroform and methylene chloride. One advantage of these solvents is their slight interaction with the solutes; under these conditions the study of redox reactions between two solutes is not affected by the solvent, which makes it much easier to study the reactions. Furthermore, the influence of solutes or even of the presence of other solvents can be studied more systematically by eliminating competing interactions. It is felt that the study of redox reactions in such solvents will facilitate an understanding of the mechanisms of these reactions, particularly in organic chemistry, as well as of the influence of the different substances capable of catalyzing reactions.

References

1. M. M. Davis, *in* "The Chemistry of Non-aqueous Solvents" (J. J. Lagowski, ed.), Vol. 3 p. 2. Academic Press, New York, 1970.
2. M. Szwarc, "Ions and Ions Pairs in Organic Reactions." Wiley (Interscience), New York, 1972.
3. G. A. Kenney and D. C. Walker, *in* "Electroanalytical Chemistry" (A. J. Bard, ed.), p. 2. Dekker, New York, 1971.
4. G. Howat and B. C. Webster, *Phys. Chem.* **75**, 626 (1971).

5. R. F. Gould, "Solvated Electron." Amer. Chem. Soc., Washington, D.C., 1965.
6. A. S. Davydov, *Zh. Eksp. Teor. Fiz.* **18**, 913 (1948).
7. B. J. Brown, N. T. Barker, and D. P. Sangster, *J. Phys. Chem.* **75**, 3639 (1971).
8. G. Charlot and B. Trémillon, Les réactions chimiques dans les solvants et les sels fondus." Gauthier-Villars, Paris, 1963.
9. R. Buvet, S. Desacher, M. Jozefowicz, J. Perichon, and L. T. Yu, *Electrochim. Acta* **13**, 1441 (1968).
10. L. T. Yu, *Inform. Chim. Anal.* **3**, 23 (1970).
11. M. Rumeau, *Analusis* **2**, 501 (1973).
12. M. Pourbaix, "Atlas d'équilibres électrochimiques," Gauthier-Villars, Paris, 1963.
13. R. Buvet, S. Desacher, M. Jozefowicz, J. Perichon, and L. T. Yu, *Rev. Gen. Elec.* **75**, 1023 (1966).
14. H. Strehlow, *in* "The Chemistry of Non-aqueous Solvents" (J. J. Lagowski, ed.), Vol. 1, p. 29. Academic Press, New York, 1966.
15. A. Caillet, Thesis, Paris, 1974.
16. J. Goret and B. Trémillon, *Electrochim. Acta* **12**, 1065 (1967).
17. R. G. Doisneau, Thesis, Paris, 1973.
18. C. K. Mann and K. K. Barnes, "Electrochemical Reactions in Non-aqueous Systems." Dekker, New York, 1970.
19. A. J. Bard, "Electroanalytical Chemistry." Dekker, New York, 1971.
20. P. K. Agasyan and M. A. Sirakanyan, *Zh. Anal. Khim.* **26**, 1599 (1971).
21. G. A. Harlow and D. H. Morman, *Anal. Chem.* **38**, 485R (1966).
22. L. M. Pozin, I. E. Flis, and V. L. Khejfec, *Zh. Prikl. Khim.* (*Leningrad*) **44**, 1784 (1971).
23. S. P. Gubin, S. A. Smirnova, L. I. Denisovitch, and A. A. Lubovitch, *J. Organometal. Chem.* **30** (2), 243 (1971); **57** (1), 87 (1973).
24. A. S. Gorokhovskaya, *Elektrokhimiya* **8**, 644 (1972).
25. O. Duscheck and V. Gutmann, *Z. Anorg. Allg. Chem.* **394**, 243 (1972).
26. S. V. Gorbachev, N. D. Kalugina, and M. I. Anisimova, *Mosk. Khim. Teknol. Inst.* **67**, 242 (1970).
27. S. V. Gorbachev and M. I. Anisimova, *Zh. Fiz. Khim.* **46**, 1039 (1972).
28. J. Besenhard, H. Jwergen, H. P. Fritz, and P. Heinz, *Z. Naturforsch. B* **27**, 1294 (1972).
29. G. Cauquis and D. Serve, *J. Electroanal. Chem.* **34**, **1** (1972).
30. G. De La Fuente and P. Federlin, *Tetrahedron Lett.* **15**, 1497 (1972).
31. L. Ya. Kheifets, V. D. Bezuglyi, and L. I. Dmitrievskaya, *Zh. Obshch. Khim.* **41**, 68 (1971).
32. A. L. Woodson and D. E. Smith, *Anal. Chem.* **42**, 242 (1970).
33. A. J. Bard, *Pure Appl. Chem.* **25**, 379 (1971).
34. L. I. Denisovitch and S. P. Gubin, *J. Organometal. Chem.* **57**, 109 (1973).
35. J. A. Friend and N. K. Roberts, *Aust. J. Chem.* **11**, 104 (1958).
36. S. M. Golabi, M. Rumeau, and B. Trémillon, in preparation.
37. S. Lagache and M. Rumeau, unpublished results (1973).
38. K. S. V. Santhanam, *Z. Phys. Chem.* (*Leipzig*) **250**, 145 (1972).
39. F. Magno and G. Bontempelli, *J. Electroanal. Chem.* **36**, 389 (1972).
40. T. A. Kowalski and P. J. Lingane, *J. Electroanal. Chem.* **71**, 1 (1971).
41. T. Matsumoto, M. Sato, and A. Ichimura, *Bull. Chem. Soc. Jap.* **44**, 1720 (1971).
42. R. Andruzzi, M. E. Cardinali, and A. Trazza, *Electroanal. Chem.* **41**, 67 (1973).
43. P. H. Plesch and A. Stasko, *J. Chem. Soc., B* **10**, 2052 (1971).
44. J. K. Chambers, P. R. Moses, R. N. Shelton, and D. L. Coffen, *J. Electroanal. Chem.* **38**, 245 (1972).
45. I. N. Rozhkov, A. V. Bukhtiarov, E. G. Gal'pern, and I. L. Knunjanc, *Dokl. Akad. Nauk SSSR* **199**, 369 (1971).

46. E. J. Rudd, M. Finkelstein, and S. D. Ross, *J. Org. Chem.* **37**, 1763 (1972).
47. M. Sainsbury and R. F. Schinazi, *J. Chem. Soc., Chem. Commun.* **12**, 718 (1972).
48. P. Wuelfing, E. A. Fitzgerald, and H. H. Richtol, *Anal. Chem.* **42**, 299 (1970).
49. G. Farsang, M. Kovacs, T. M. H. Saber, and L. Ladanyi, *J. Electroanal. Chem.* **38**, 127 (1972).
50. H. M. Koepp, H. Wendt, and H. Strehlow, *Z. Elektrochem.* **64**, 483 (1960).
51. D. W. Meek, *in* "The Chemistry of Non-Aqueous Solvents" (J. J. Lagowski, ed.), Vol. 1, p. 1. Academic Press, New York, 1966.
52. L. Jeftic and G. Manning, *J. Electroanal. Chem.* **26**, 195 (1970).
53. G. Cauquis and D. Serve, *C. R. Acad. Sci.* **273**, 1715 (1971).
54. T. M. Saber, G. Farsang, and L. Ladangi, *Microchem. J.* **17**, (2), 220 (1972); **18** (1), 66 (1973).
55. J. Bressard, G. Cauquis, and D. Serve, *Tetrahedron Lett.* **35**, 3103 (1970).
56. M. Fujimira, H. Suzuki, and S. Hayano, *J. Electroanal. Chem.* **33**, 393 (1971).
57. J. Cognard, Thesis, Grenoble, France, 1971.
58. C. Barraqué, J. Vedel, and B. Trémillon, *Anal. Chim. Acta* **46**, 263 (1969).
59. E. Peover, *Trans. Faraday Soc.* **60**, 417 (1964).
60. R. Scott, D. De Palma, and S. Vinogradov, *J. Phys. Chem.* **72**, 2192 (1968).
61. K. Sasaki, T. Matsumoto, and A. Kitani, *Nippon Kagaku Kaishi* **6**, 1039 (1972).
62. A. Lasia and M. K. Kalinowski, *J. Electroanal. Chem.* **36**, 511 (1972).
63. T. Fusinaga, K. Izutsu, and T. Nomura, *J. Electroanal. Chem.* **29**, 203 (1972).
64. U. Takaki, T. E. Hogen Esch, and J. Smid, *J. Phys. Chem.* **76**, 2152 (1972).
65. S. G. Majranovskij and T. Ya. Rubinskaya, *Elektrokhimiya SSSR* **8**, 424 (1972).
66. Z. Margolin and F. A. Long, *J. Amer. Chem. Soc.* **95**, 2757 (1973).
67. M. Rumeau, *Analusis* **2**, 420 (1973).
68. J. A. Friends and N. K. Roberts, *Aust. J. Chem.* **11**, 104 (1958).
69. L. Michaelis and E. S. Hill, *J. Amer. Chem. Soc.* **55**, 1481 (1933).
70. G. Cauquis and D. Serve, *Anal. Chem.* **44**, 2222 (1972).
71. R. D. Holm, W. R. Carper, and J. A. Blancher, *J. Phys. Chem.* **71**, 3960 (1967).
72. E. A. Fitzgerald, P. Wuelfing, and H. H. H. Richtol, *J. Phys. Chem.* **75**, 2737 (1971).
73. L. Byrd, L. L. Miller, and D. Pletcher, *Tetrahedron Lett.* **24**, 2419 (1972).
74. D. Bauer and J.-P. Beck, *Bull. Soc. Chim. Fr.* p. 1252 (1973).
75. L. I. Denisovich and S. P. Gubin, *J. Organometal. Chem.* **57**, 99 (1973).
76. V. Gutmann and V. Cechak, *Monatsh. Chem.* **103**, 1447 (1972).
77. O. W. Kolling and J. L. Goodnight. *Anal. Chem.* **45**, 160 (1973).

4

Tetramethylurea

Barbara J. Barker

Department of Chemistry
Hope College, Holland, Michigan

AND

Joseph A. Caruso

Department of Chemistry
University of Cincinnati, Cincinnati, Ohio

I. Introductory Comments

Tetramethylurea (TMU), a urea derivative which is a liquid at room temperature, has been of considerable interest and importance as a non-aqueous medium for many chemical reactions and for a number of recent fundamental and applied chemical investigations. A variety of acids and bases have been titrated successfully in TMU. The conductance behavior of an extensive series of alkali metal, ammonium, and quaternary ammonium salts in this solvent has been determined. Also investigated have been conductance–time effects of iodine–TMU solutions. This chapter is a review of tetramethylurea and of its use as a solvent for these investigations.

II. Solvent Preparation and Purification

Although tetramethylurea has been available commercially from various chemical suppliers for the past several years, Lüttringhaus and Dirksen in their authoritative article have reported[1] several methods of preparing the solvent. They indicate that an especially useful method is the aminolysis of aromatic carbonates such as diphenylcarbonate.

$$(C_6H_5O)_2CO + HN(CH_3)_2 \rightarrow C_6H_5O\text{—}CO\text{—}N(CH_3)_2 + C_6H_5OH$$

$$C_6H_5O\text{—}CO\text{—}N(CH_3)_2 + HN(CH_3)_2 \rightarrow (CH_3)_2N\text{—}CO\text{—}N(CH_3)_2 + C_6H_5OH$$

The reaction apparently proceeds well at 200°C and the phenol is removed as a phenoxide.[1]

Pure TMU is obtained best by vacuum distillation through a packed column from some suitable drying agent such as granular barium oxide.[2] Such solvent typically contains $<0.01\%$ water.

III. Physical Properties

Some of the most important physical properties of tetramethylurea as a solvent are listed in Table I. TMU had a wide liquid range which is unusual for ureas and is a liquid at room temperature. The usually strong association of amides due to hydrogen bonding through the amide nitrogen is absent in TMU. The high boiling point does indicate, however, a strong dipole association. This is consistent with the relatively high dipole moment observed. The combination of high dipole moment, moderate dielectric constant, low

TABLE I

STRUCTURE AND PHYSICAL PROPERTIES OF TETRAMETHYLUREA

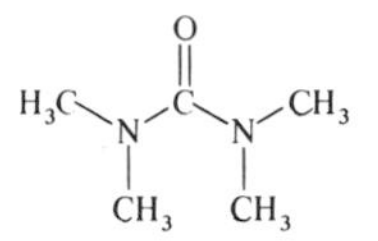

1,1,3,3-Tetramethylurea, tetramethylurea, temur, TMU

Property and unit	Data
Molecular weight	116.16
Molecular formula	$C_5H_{12}N_2O$
Boiling point[a] (°C) [760 mm Hg]	176.5
Melting point[a] (°C)	−1
Density[b] (g/ml) [25°C]	0.9619 ± 0.0001
Viscosity[b] (poise) [25°C]	0.01401 ± 0.00003
Dielectric constant[b] (1 MHz) [25°C]	23.45 ± 0.06
Specific conductance[b] ($ohm^{-1}cm^{-1}$) [25°C]	$<2 \times 10^{-8}$
Basicity[c] (aqueous pK_B)	12
Dipole moment[c] (Debye units)	3.37
Infrared absorption[c] (cm^{-1}) [ν(C=O)]	1640
Ultraviolet absorption[c] (nm) [$a_{max} = 1940$]	217.5

[a] "Technical Bulletin on Tetramethylurea." Ott Chem. Co., Muskegon, Michigan, 1968.
[b] B. J. Barker and J. A. Caruso, *J. Amer. Chem. Soc.* **93**, 1341 (1971).
[c] A. Lüttringhaus and H. W. Dirksen, *Angew. Chem., Int. Ed. Engl.* **3**, 260 (1964).

viscosity and low specific conductance, makes TMU a useful solvent for studying the behavior of electrolytes in solution. Additional discussion of the physical properties of TMU is given by Lüttringhaus.[1]

IV. SOLUBILITY CHARACTERISTICS

Table II lists the solubilities of inorganic compounds in TMU.[1] For strongly ionic salts TMU has only a modest dissolving power of about the same strength as acetone.[1] Alkali and alkaline earth chlorides are sparingly soluble, the bromides are moderately soluble, while the iodides are quite soluble. (Some chemical reactions occur between iodides and TMU which cause the solutions to turn deep yellow.) Not shown in Table II are alkali metal perchlorates and tetraphenylborates which also are quite soluble.[2]

TABLE II

SOLUBILITIES OF INORGANIC COMPOUNDS IN TETRAMETHYLUREA[a]

Compound	Solubility (g/100 g)	Compound	Solubility (g/100 g)
NaCl	0.14 (75°C)	NH_4Cl	0.72
NaBr	5.8	NH_4NO_3	32
NaI	92	$CuSO_4$	0.22
NaCN	0.27	$AgNO_3$	Good
$NaNO_3$	3.8	$CaCl_2$	2.0
$Na_2S_2O_3 \cdot 5H_2O$	0.47	$ZnSO_4 \cdot 7H_2O$	0.39
Na acetate·$3H_2O$	0.26	$Cd(NO_3)_2 \cdot 4H_2O$	>60
KCl	0.15 (75°C)	H_3BO_3	12
KBr	0.11	$MnCl_2 \cdot 4H_2O$	85
KI	14.2	$FeCl_3$	6.9
KCN	0.25	$Co(NO_3)_2 \cdot 6H_2O$	>60
KOCN	0.16	$Ni(NO_3)_2 \cdot 6H_2O$	74
KSCN	29.4	$NbCl_5 \cdot 2DMF$	Good
$K_2S_2O_5$	0.45	TMU·HCl[b]	Good
K acetate	0.21	TMU·HBr[b]	Moderate

[a] Data given in g/100 gm of tetramethylurea at 22°C. Reproduced from A. Lüttringhaus and H. W. Dirksen, *Angew. Chem., Int. Ed. Engl.* **3**, 263 (1964) with permission of the copyright holder, Verlag Chemie.

[b] Formation of a compound.

TABLE III

SOLUBILITIES OF ORGANIC COMPOUNDS IN TETRAMETHYLUREA[a]

Compound	Solubility[b] (g/100 g)	Compound	Solubility[b] (g/100 g)
	Aliphatic and cycloaliphatic derivatives		
Acetylene (18 atm)	50	Sebacic acid	36
Petroleum ether	c.m.	Citric acid	12.3
n-Octadecane	c.m.	Acetamide	44.7
Cyclohexane	c.m.	Dimethylformamide	c.m.
Methanol	c.m.	*N*-Methylpyrrolidone	c.m.
Ethanol	c.m.	Caprolactam	81
1,10-Decanediol	26.4	Urea	2.5
Ethyl acetate	c.m.	Glucose	1.1
Ethyl acetoacetate	c.m.	Carbon tetrachloride	c.m.
Diethyl sebacate	c.m.	Chloroform	c.m.
Acetone	c.m.	Methylene chloride	c.m.
Paraformaldehyde	0.05	Carbon disulfide	c.m.
Urotropin	0.32	Thiourea	7.2
Acetic acid	c.m.		

TABLE III—*continued*

Compound	Solubility[b] (g/100 g)	Compound	Solubility[b] (g/100 g)
	Aromatic derivatives		
Benzene, toluene, xylenes	c.m.	2,4-Dinitrotoluene	s.
Naphthalene	s.	1,3,5-Trinitrobenzene	s.
Phenanthrene	50	2,4,6-Trinitrotoluene	s.
Anthracene	3.8	Benzoic acid	67
Pyrene	73	2-Aminobenzoic acid	85
Biphenyl	77	4-Aminobenzoic acid	80
Acenophthene	26	1,2-Diaminobenzene	46
Phenol	c.m.	2-Aminophenol	40
o-Cresol	c.m.	1,3-Diaminobenzene-4-sulfonic acid	0.3
Hydroquinone	40	Diphenyl carbonate	79
4-Iodotoluene	v.s.	Benzoquinone	60
1-Iodonaphthalene	v.s.	Anthraquinone	0.25
Nitrobenzene	c.m.	Diphenylene oxide	72
4-Nitrotoluene	115	Azobenzene	65
1,3-Dinitrobenzene	s.		
	Heterocyclic derivatives		
Tetrahydrofuran	c.m.	Caffeine	1.7
Dioxane	c.m.	Benzimidazole	50
Pyridine	c.m.	Antipyrine	28
Quinine	39	Benzotrithione	20
	Amino acids		
Glycine	0.02	Phenylglycine	0.25
DL-Valine	0.03	Phenylalanine	0.66
DL-Leucine	0.13	DL-Tyrosine	0.08
Cysteine·HCl	3.5	DL-Tryptophan	0.19
Cystine	0.07	L(−)-Histidine	s.s.
DL-Methionine	0.13	*N*-Dinitrophenylglycine	s.
Lysine (crude)	1.1	*N*-Dinitrophenylalanine	s.
Glutamic acid	0.38		

[a] Data given in g/100 g of tetramethylurea at 22°C. Reproduced from A. Lüttringhaus and H. W. Dirksen, *Angew. Chem., Int. Ed. Engl.* **3**, 264 (1964) with permission of the copyright holder, Verlag Chemie.

[b] c.m., Completely miscible; v.s., very soluble; s., soluble; s.s., sparingly soluble.

TMU is miscible with most types of organic liquids as shown by Table III; it also is miscible with water. These solubility characteristics are expected when the tetramethylurea model discussed by Lüttringhaus and Dirksen is considered.[1] The model shows that the carbonyl oxygen is free to hydrogen bond, while at the same time the model's surface is formed by methyl groups which accounts for the high dissolving power for hydrocarbons. It also may be noted from Table III that generally TMU has excellent dissolving power for aromatic hydrocarbons and many of their derivatives. Lüttringhaus and Dirksen[1] postulate that this excellent dissolving power is due to: "... the van der Waals force caused by the methyl groups, the interactions between the resonance systems of the solvent and that of the solute, and a favorable molecular structure of tetramethylurea." Amino acids, including the aromatic ones, are typically of low solubility in TMU. These low solubilities probably are related to the structure of these compounds since the solubilities increase as zwitterionic character decreases.[1] *p*-Aminobenzoic acid which apparently

TABLE IV

SOLUBILITIES OF MACROMOLECULAR SUBSTANCES[a]

Polymer	Temp. (°C)	Solubility (g/100 g)
Low-pressure polyethylene	22	—
	75	0.06
High-pressure polyethylene	22	—
	75	0.11
Polypropylene	22	—
	75	0.14
Poly(vinyl chloride)	22	Swells and dissolves
Rhovyl (Rhodiaceta)	22	Soluble[b]
PCU (Rhodiaceta)	22	Soluble[b]
Poly(vinylidene dicyanide) (Goodrich)	22	Soluble[b]
Poly(acrylonitrile)	175	—
Polycaprolactam	175	—
Nylon (Rhodiaceta)	100	—[b]
Viscose (American Enka)	100	—[b]
Viscose, crosslinked (Courtaulds)	100	—[b]
Cuprammonium rayon (Cupresa, Bayer)	100	—[c]
Cellulose 2,5-acetate (Rhodiaceta)	22	Soluble[b]
Cellulose triacetate (Rhodiaceta)	22	Soluble[b]
Polymethacrylate	22	Swells and dissolves
Polyester (Tergal, Rhodiaceta)	100	—[b]
Polyester (Terylene)	125	5–10
Polycarbonate (Bayer)	22	Swells and dissolves[c]
Polyurethane (Bayer)	175	Swells and dissolves
Polystyrene	22	ca. 30
Novolak	22	ca. 15
Bakelite	175	—

[a] Reproduced from A. Lüttringhaus and H. W. Dirksen, *Angew. Chem., Int. Ed. Engl.* **3**, 264 (1964) with permission of the copyright holder, Verlag Chemie.

[b] Data kindly furnished by Dr. *Weigand*, Deutsche Rhodiaceta Ag., Freiburg/Breisgau, Germany.

[c] Data kindly furnished by Drs. *Reichle* and *Prietzsch*, Farbenfabriken Bayer Ag., Dormagen, Germany.

does not form the zwitterion shows a high solubility as contrasted with the 2,4-diaminobenzenesulfonic acid which is not very soluble but is zwitterionic.[1] It has been suggested that the large differences in solubility between amino acids and their derivatives might lead to the use of tetramethylurea in the analysis of proteins.[1]

In Table IV are listed the solubilities of some selected polymers. Because of the good solubilities of a number of polymers TMU was marketed by Ott Chemical Company for a number of years as a polymer solvent. Further information on solubility behavior is available from Lüttringhaus and Dirksen.[1]

V. Chemical Reactions

Zaugg and co-workers[3–5] have studied reaction rate increases caused by *N*,*N*-dialkylcarboxamides on alkylation of alkali enolates. They find that adding small amounts ($<10\%$) of TMU considerably increases the rate of enolate alkylation in benzene solutions probably because TMU dissociates the ion-pair which exists in benzene. Lüttringhaus and Dirksen suggest that ". . . the high π electron density in the amide group, particularly in the π orbital of the carbonyl oxygen, accounts for the remarkable solvation of alkali metal ions." Additional discussion by these authors contrast dimethylformamide and TMU as rate enhancing solvents.

TMU has been used successfully as a reaction medium for a number of other reactions as discussed by Lüttringhaus and Dirksen,[1] for example:

1. The alkylation of tertiary heterocyclic amines.
2. The preparation of phosphonium salts from tertiary phosphines.
3. The use of carbonyl chloride–TMU mixtures as acylation media.
4. The use of TMU for increasing the rate of base-catalyzed prototropic double-bond shifts.
5. The use of TMU with free alkali metals in direct metalations.

VI. Applied Studies

A. Acid–Base Titrations

Studies such as those in dimethyl sulfoxide by Ritchie and co-workers[6] and in methyl isobutyl ketone by Bruss and Wyld[7] have clearly shown that dipolar aprotic solvents are powerful media for studying acid–base titration

TABLE V

TITRATION OF ACIDS AND BASES IN TETRAMETHYLUREA[a]

Acid/base	Recovery (%)	Acid	Recovery[c] (%)
Perchloric acid	100.00	Phenobarbital	97.49 ± 0.49
Oxalic acid	99.58	Amobarbital	98.74 ± 0.02
Salicylic acid	99.97	Secobarbital	98.96 ± 0.43
Phenol	100.39	Sulfamerazine	98.86 ± 1.41
p-Nitrophenol	99.66	Sulfapyridine	101.4 ± 0.4
m-Nitrophenol	99.48	Sulfathiazole	99.99 ± 0.50
Tri-*n*-butylamine	99.85	—	—
1,3-Diphenylguanidine	Standard	—	—

[a] Titrant, tetrabutylammonium hydroxide; electrodes, modified calomel–glass.
[b] Data summarized from references 8–10.
[c] Standard deviation based on three determinations.

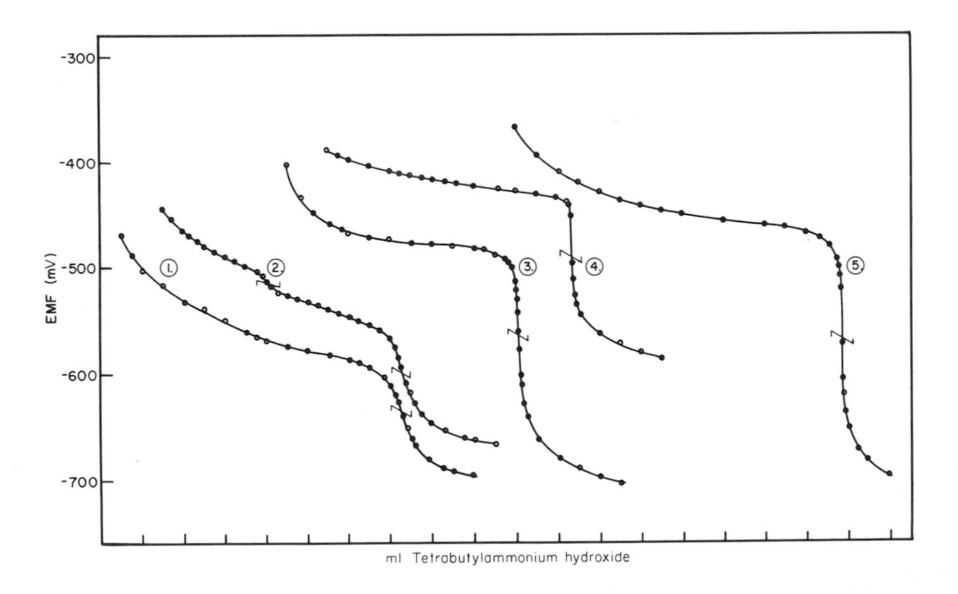

FIG. 1. Titration curves of (1) phenol, (2) benzoic acid and phenol, (3) 2-hydroxy-1-nitrosonaphthalene, (4) *p*-nitrophenol, and (5) benzoic acid in TMU. Reproduced from S. L. Culp and J. A. Caruso, *Anal. Chem.* **41**, 1331 (1969) with permission of the copyright holder, The American Chemical Society.

behavior. The weakly basic character as well as the limited autoprotolysis of these types of solvents contributes to such behavior.

As indicated previously, TMU is a liquid at room temperature, which property is unusual for ureas. During the past five years a number of acid–base titration studies have been completed in this solvent[8–10]. These studies have included both potentiometric[8,10] and indicator[9] titrations and have involved phenols and carboxylic acids[8,9] as well as barbiturates and sulfa drugs.[10] Since TMU has a good dissolving power for acids and bases, gives large potential breaks, and affords good recoveries as well as being inexpensive, commercially available, and easily purified it is considered an excellent titration solvent. Typical titration results are indicated in Table V.

Figure 1 shows titration curves of selected acids in TMU. Interestingly, while attempts to resolve a mixture of benzoic acid and phenol were unsuccessful the total recovery of 100.17% was excellent. Culp and Caruso[9] found that water levels as great as 5% could be tolerated without markedly altering the recoveries, although above a 1% water level the shoulders of the titration curves tended to become increasingly rounded. Also of particular interest was the observation of an additional inflection at half-neutralization in the titration of *m*-nitrophenol. This same behavior has been noted for *m*-nitrophenol in other dipolar aprotic solvents.[11] These observations suggest that TMU might promote an association **(I)** such as shown below which provides an additional species for titration.

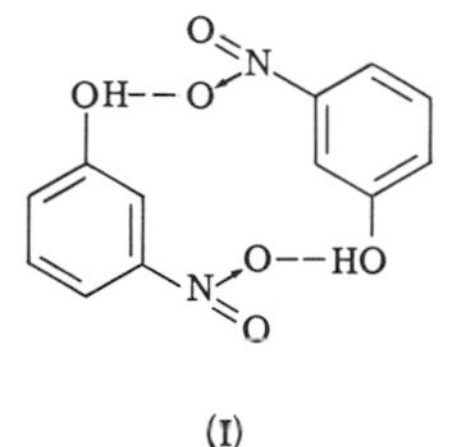

(I)

An additional inflection at 70% neutralization for perchloric acid was noted. The speculation was that the strong acid might protonate the nitrogen as well as the carbonyl oxygen, thereby leading to an additional species for titration.

Culp and Caruso also evaluated the effect of several indicators for acid–base titrations in TMU.[9] Thymol blue was found to be useful for the titration of carboxylic acids, as illustrated in Fig. 2. Bromophenol blue was found to be a good indicator for the titration of bases. Azc violet has the color transition in the proper potential range for phenols, but it is difficult to determine when the red fades from violet to blue.

Greenberg, Barker, and Caruso[10] successfully titrated a number of barbiturates and sulfa drugs in TMU. The experimental results are summarized in Table V. Indications are that TMU may be quite useful for drug titrations although the recovery data are not as encouraging as those given for other organic compounds. TMU is a differentiating solvent for barbiturates, but a leveling solvent for sulfa drugs.

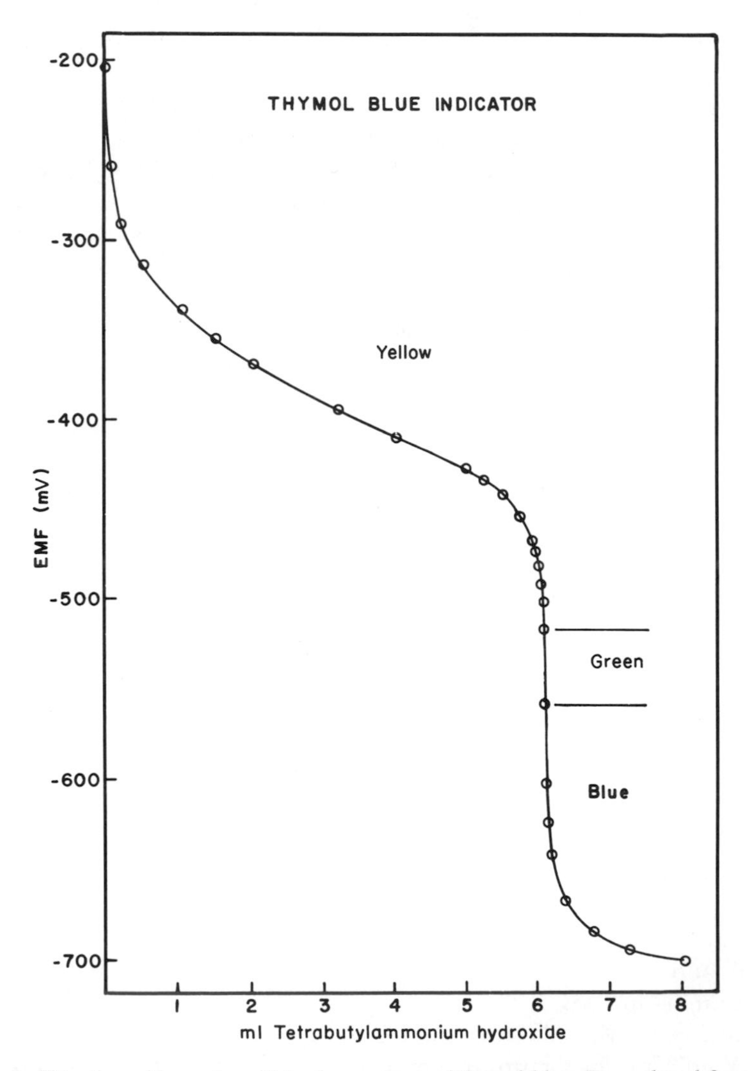

FIG. 2. Titration of benzoic acid in the presence of thymol blue. Reproduced from S. L. Culp and J. A. Caruso, *Anal.* Chem. **41**, 1877 (1969) with permission of the copyright holder, The American Chemical Society.

B. Polarography

Tetramethylurea has been used as a solvent for the polarographic reduction of both organic[11] and inorganic[12] compounds. In these studies the resistances of TMU solutions were found to be exceedingly high. Conventional plots of E vs. $\log i/(i_d - i)$ gave stright lines with slopes much greater than predicted (e.g., the slope of a log plot for 10^{-3} M Cd^{2+} in 0.1 M $NaNO_3$ was 0.072, whereas 0.030 was the expected slope). The broad plateaus extending from +0.3 to −0.1 V in all electrocapillary curves of TMU indicated surface

TABLE VI

POLAROGRAPHIC DATA IN TETRAMETHYLUREA[a]

Ion	$E_{1/2}$ (V)	Reference electrode	i_d (μA)
Cd^{2+}	−0.597	SCE	0.87
	−0.390	Hg pool	5.4
Cu^{2+}	+0.180	SCE	4.0
	−0.520	Hg pool	2.0
Zn^{2+}	−0.280	Hg pool	5.7
$HClO_4$	−0.710	Hg pool	3.6
HCl	−0.220	Hg pool	3.4
H_2SO_4	−0.880	Hg pool	2.8

[a] Data summarized from reference 12.

activity which was due to the solvent.[12] Table VI presents relative trends of diffusion currents and half-wave potentials for several inorganic substances in TMU. That HCl is more easily reduced than $HClO_4$ or H_2SO_4 is attributed to anionic solvation phenomenon.[12]

VII. FUNDAMENTAL INVESTIGATIONS

A. Conductance Studies

Several fundamental electrochemical investigations have elucidated the behavior of alkali metal and quaternary ammonium salts in tetramethylurea.[2,13,14] Values of Λ_0, K_A, and a_J for these salts were determined from the Fuoss–Onsager equation.[15]

$$\Lambda = \Lambda_0 - S(C\gamma)^{1/2} + [E \log C\gamma + (J - F\Lambda_0) - K_A \Lambda f^2] C\gamma \qquad (1)$$

The symbols in the equation have their usual meaning: $S = \alpha\Lambda_0 + \beta$ and $E = E_1\Lambda_0 - E_2$. The physical properties of TMU at 25°C lead to values of 1.403, 70.42, 19.82, and 145.7 for the coefficients α, β, E_1, and E_2, respectively. For unassociated electrolytes $\gamma = 1$ and $K_A = 0$; for associated electrolytes $\gamma < 1$ and $K_A > 0$. Initial Λ_0 values used in the Fuoss–Onsager evaluation were those obtained from the Shedlovsky[16] $y-x$ least-squares analysis of original data. Table VII summarizes the average values of the conductance parameters for all salts studied in tetramethylurea. For a graphical repre-

TABLE VII

AVERAGE VALUES OF FUOSS–ONSAGER CONDUCTANCE PARAMETERS FOR SALTS IN TETRAMETHYLUREA[a, b]

Salt	$C \times 10^4$	$\sigma\Lambda$	Λ_0	K_A	a_J
LiBr	1.260 –29.64	0.016	44.44	18.0	4.46
$LiNO_3$	2.303 –35.80	0.008	47.00	167	4.14
$NaBPh_4$	0.1916–11.92	0.006	29.39	—	6.31
$NaClO_4$	1.103 –30.10	0.005	44.52	—	4.65
NaI	0.4319–41.02	0.028	44.81	—	3.97
NaSCN	1.763 –30.98	0.021	49.12	67.2	3.79
NaBr	0.4822–29.69	0.007	46.20	316	4.67
$NaNO_3$	1.870 –14.28	0.018	48.46	1120	7.31
$KBPh_4$	0.2185–15.72	0.005	28.99	—	6.17
$KClO_4$	0.8643–23.26	0.010	43.96	—	4.57
KSCN	4.575 –38.50	0.005	48.77	—	3.77
$RbBPh_4$	0.2817–23.69	0.006	29.27	—	6.01
$RbClO_4$	0.8163–30.98	0.006	44.42	7.2	4.57
RbI	1.292 –29.93	0.021	44.81	9.7	4.60
$CsBPh_4$	0.2236–14.32	0.009	30.24	—	6.07
$CsClO_4$	0.8544–18.63	0.003	45.24	10.2	4.49
NH_4BPh_4	1.269 –43.43	0.016	31.29	—	5.26
Me_4NClO_4	1.428 –25.53	0.003	50.93	42.2	4.38
Et_4NClO_4	1.326 –40.05	0.026	50.32	19.7	3.88
Pr_4NClO_4	0.4078–41.36	0.014	45.31	17.6	3.89
Bu_4NClO_4	0.4994–32.53	0.011	43.71	21.8	4.04
Pen_4NBr	0.9255–32.88	0.007	44.13	186	3.60
Hex_4NBr	2.839 –33.40	0.004	41.89	132	3.27
$Hept_4NBr$	2.468 –37.17	0.004	41.64	126	3.08
$TABBPh_4$	0.3518–13.75	0.007	27.45	—	6.33
$TABClO_4$	0.1464–12.47	0.004	42.43	31.6	4.86

[a] No viscosity correction applied. Data obtained at 25°C.

[b] Me, Methyl; Et, ethyl; Pr, *n*-propyl; Bu, *n*-butyl; Pen, *n*-pentyl; Hex, *n*-hexyl; Hept, *n*-heptyl; TAB, triisoamylbutylammonium; BPh_4, tetraphenylborate.

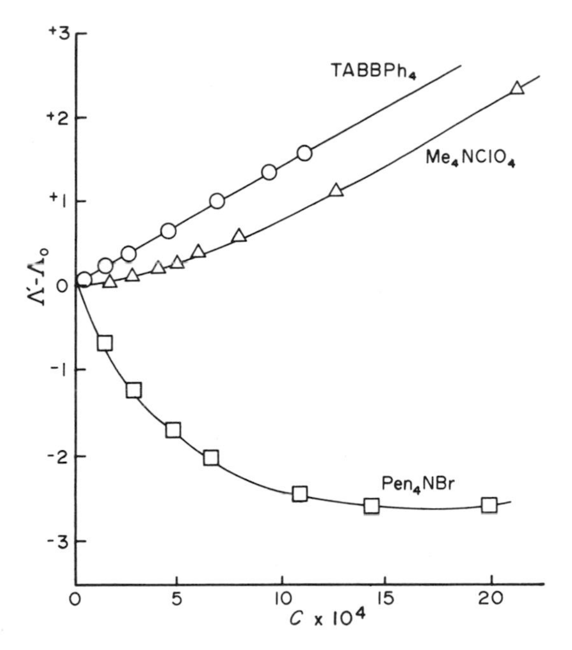

FIG. 3. Fuoss–Onsager $\Lambda' - \Lambda_0$ vs. C plots for quaternary ammonium salts in tetramethylurea. Reproduced from B. J. Barker and J. A. Caruso, *J. Phys. Chem.* **77**, 1884 (1973) with permission of the copyright holder, The American Chemical Society.

sentation of conductance data the Fuoss–Onsager equation is rearranged to the form[17]

$$\Lambda' \equiv \Lambda + SC^{1/2} - EC \log C = \Lambda_0 + (J - F\Lambda_0)C \qquad (2)$$

As illustrated in Fig. 3, increasing curvature in the Fuoss–Onsager plots indicates increasing association within the salt series.

Correlations between the properties of an electrolyte and solvent and the extent of association are limited. Generally in TMU, as the crystallographic radii of the cations increase, there is a slight increase in the association of alkali metal salts and as the crystallographic radii of the anions increase there is a slight decrease in the association of these salts. Association within the tetraalkylammonium salt series decreases as the crystallographic size of the cations increases.

Ionic limiting equivalent conductances were obtained indirectly by assuming that the limiting equivalent conductance of the triisoamylbutylammonium ion, shown to be equal to that of the tetraphenylborate ion in methanol,[18] is the same as that of the tetraphenylborate ion in tetramethylurea. Assuming $\Lambda_0(TAB^+) = \Lambda_0(BPh_4^-)$ in TMU, the set of ionic limiting equivalent conductances presented in Table VIII was obtained.

TABLE VIII

LIMITING EQUIVALENT CONDUCTANCES[a] AND CRYSTALLOGRAPHIC RADII (r_X) OF IONS IN TETRAMETHYLUREA

Cation	λ_0^+	r_X	Anion	λ_0^-	r_X
Li^+	14.13	0.60[b]	SCN^-	33.44	2.27[b]
Na^+	15.74	0.95[b]	NO_3^-	32.80	2.64[e]
K^+	15.26	1.33[b]	Br^-	30.38	1.95[b]
Rb^+	15.63	1.48[b]	I^-	29.12	2.16[b]
Cs^+	16.52	1.69[b]	ClO_4^-	28.71	2.92[e], 2.40[f]
NH_4^+	17.57	1.48[b]	BPH_4^-	13.72	4.94[d]
Me_4N^+	22.22	3.47[c]			
Et_4N^+	21.61	4.00[c]			
Pr_4N^+	16.60	4.52[c]			
Bu_4N^+	15.00	4.94[c]			
Pen_4N^+	13.75	5.29[c]			
Hex_4N^+	11.51	5.61[c]			
$Hept_4N^+$	11.26	5.89[c]			
TAB^+	13.72	4.94[d]			

[a] Determined at 25°C with $TABBPh_4$ as a reference electrolyte.
[b] L. Pauling, "Nature of the Chemical Bond," 2nd ed. Cornell Univ. Press, Ithaca, New York, 1948.
[c] R. A. Robinson and R. H. Stokes, "Electrolyte Solutions," 2nd ed. Butterworth, London, 1959.
[d] D. F. T. Tuan and R. M. Fuoss, *J. Phys. Chem.* **67**, 1343 (1963).
[e] E. R. Nightingale, Jr., *J. Phys. Chem.* **63**, 1381 (1959).
[f] J. E. Prue and P. J. Sherrington, *Trans. Faraday Soc.* **57**, 1795 (1961).

As expected, the limiting equivalent conductances of the alkali metal ions increase and the limiting equivalent conductances of the tetraalkylammonium ions decrease as the crystallographic radii increase. As the crystallographic radii of alkali metal ions increase the effective size of these ions decreases because of decreasing solvation. The order of decreasing relative cationic limiting equivalent conductances which is observed in tetramethylurea, $Me_4N^+ > Et_4N^+ > NH_4^+ > Pr_4N^+ > K^+ > Bu_4N^+$, indicates that the "effective" size of the solvated alkali metal ions is comparable to that of the large tetraalkylammonium ions in this solvent. The order of relative anionic limiting equivalent conductances which is expected[19] in tetramethylurea is the same as the order found in many other[20–22] polar nonaqueous solvents.

The high limiting equivalent conductance of the sodium ion in TMU is to be noted. The same anomalous behavior of the sodium ion has been observed in conductance studies in dimethylacetamide ($D = 37.78$),[20] dimethylpropionamide ($D = 33.1$),[23] and dimethylbutyramide ($D = 28.0$).[24]

In these solvents the sodium salts had Λ_0 values about 0.5 unit higher than corresponding potassium salts. This same difference has been found between sodium and potassium salts in TMU. The assumption was made[24] that because of the size and structure of the solvent molecules, steric factors may contribute significantly to the anomalous solvation behavior of the sodium ion in the three substituted amides. If a general structure $R\text{–}C(=O)\text{–}N(CH_3)_2$ is used to represent the amides and tetramethylurea, it is seen that when $R = CH_3$, C_3H_7, C_4H_9, or $N(CH_3)_2$ the unusual solvation behavior of the sodium ion is observed. In these media solvation probably occurs through the oxygen atom, although nitrogen may participate to a slight degree. Possibly, because of the crystallographic size and the charge density of the sodium ion, solvation through nitrogen occurs to a greater extent for this ion than for other alkali metal ions. Sodium ion solvated through nitrogen may be of smaller effective size and therefore more highly conducting than sodium ion solvated through oxygen. Extending the amide series, dimethylformamide ($R = H$) can form a larger solvation sphere around the sodium ion since the oxygen atoms of the solvent molecules are more accessible to the sodium ions. In DMF ($D = 36.71$) there is no anomalous behavior of the sodium ion; potassium salts are more conducting than corresponding sodium salts.[21]

For all tetraalkylammonium salts and for the larger alkali metal salts the sum of the crystallographic radii is much greater than the Fuoss–Onsager ion-size parameter a_J. Values of a_J which are small compared to the sum of cationic and anionic crystallographic radii frequently have been obtained[25, 26] for electrolytes in nonaqueous solvents. Small a_J values are to be expected[25] since the Fuoss–Onsager equations predict ionic separations which are too small to be physically realistic. In many nonaqueous solvents a_J values do not correlate with crystallographic radii; therefore, there is difficulty in interpreting a_J as an ion-size parameter. [27] Ion-size parameter in solution also can be determined from the equation

$$r_s = 0.820z/\eta_0 \lambda_0 \qquad (3)$$

in which r_s is the Stokes ionic radius, z is the absolute magnitude of the charge of the ion, η_0 is the solvent viscosity in poise, and λ_0 is the ionic limiting equivalent conductance. As Table VIII indicates, the Stokes radii of the alkali metal ions decrease and the Stokes radii of the tetraalkylammonium ions increase as the crystallographic size of these ions increases.[28] Figure 4 is a plot of Stokes radii of cations in TMU as a function of the crystallographic radii of these ions. As observed in other nonaqueous solvents, the sum of the Stokes cationic and anionic radii for any alkali metal salt is much greater than the sum of its crystallographic radii; the sum of the Stokes ionic radii for any tetraalkylammonium salt in TMU is much less than the sum of its crystallographic radii.

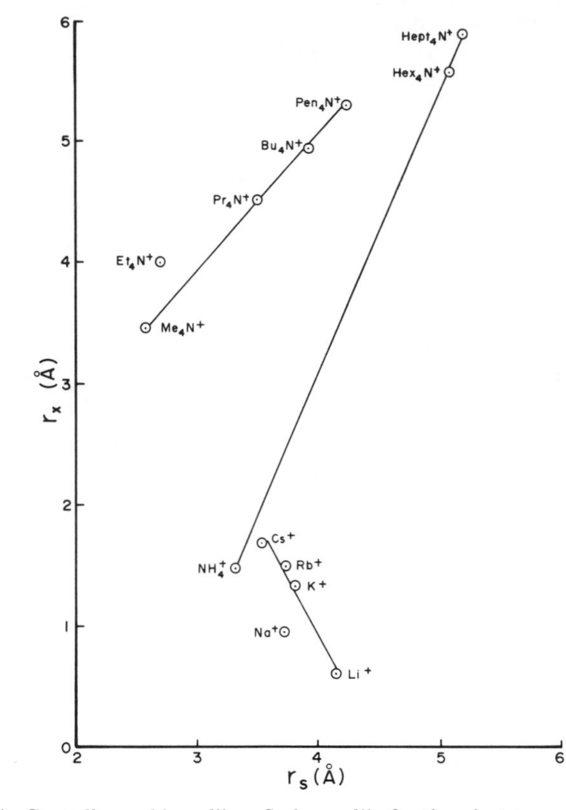

FIG. 4. Crystallographic radii vs. Stokes radii of cations in tetramethylurea.

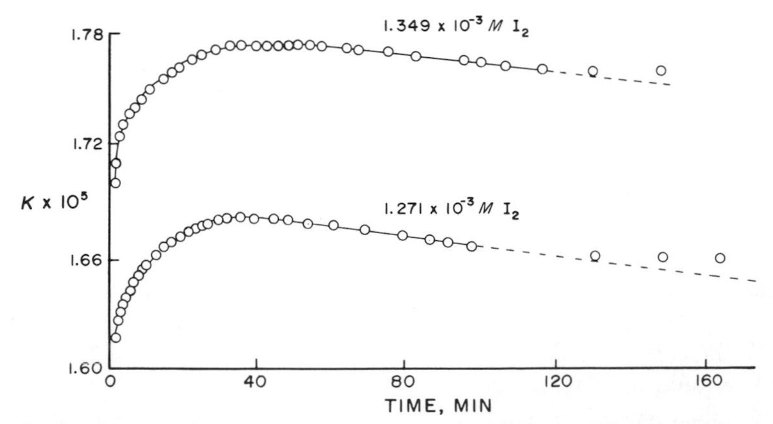

FIG. 5. Conductance–time plots for iodine in tetramethylurea. Reproduced from B. J. Barker and J. A. Caruso, *Electrochim. Acta* **18**, 315 (1973) with permission of the copyright holder, Pergamon Press.

The conductance of freshly prepared iodide–TMU and iodine–TMU solutions continually drifts with time as the solutions age.[13] The change in conductance is accompanied by a solution color change from colorless through light yellow to an intense dark yellow. The shape of the conductance–time plots (Fig. 5) is similar to that of kinetic plots for an intermediate species B in consecutive first-order reactions of the type A → B → C.

Reaction mechanisms of the following types have been considered in discussion of the conductance–time effect of hydrogen halides (HX)[29] and iodine monohalides (IX)[30] in acetonitrile.

$$\underset{}{CH_3CN + HX} \rightleftharpoons \underset{(II)}{CH_3CN \cdot HX} \rightleftharpoons \underset{(III)}{CH_3CNH^+X^-} \tag{4}$$

$$CH_3CNH^+X^- \rightleftharpoons CH_3CNH^+ + X^-_{\text{solvated}}$$

$$CH_3CN + HX \rightleftharpoons CH_3CNH^+X^- \rightleftharpoons CH_3C^+NH \cdot X^- \tag{5}$$

$$CH_3C^+NH \cdot X^- + HX \rightleftharpoons \underset{\substack{| \\ X}}{CH_3CNH_2{}^+X^-}$$

$$\underset{\substack{| \\ X}}{\overset{(IV)}{CH_3CNH_2{}^+X^-}} \rightleftharpoons \underset{\substack{| \\ X}}{\overset{(V)}{CH_3CNH_2{}^+}} + X^-_{\text{solvated}}$$

In the charge-transfer reaction (4) the outer orbital complex (**II**) slowly rearranges to form the inner orbital complex (**III**) which then ionizes. In reaction (5) the conductance–time effect is due to the finite rate of formation of the imino-type compound (**IV**) and its ionization to the imino hydrohalide (**V**).

Reaction schemes similar to (4) and (5) could be used to explain the shape of the k–t plots in tetramethylurea. The conductance of the solution increases with increasing formation of a charged iodine–TMU "intermediate" species. The solution conductance would decrease if the sum of the mobilities of the cation and the iodide ion which are formed from ionization of the "intermediate" species is less than the mobility of the "intermediate" species itself.

B. Viscosity Studies

The effect of several tetraalkylammonium salts on the viscosity of TMU was determined[14] by experimentally evaluating viscosity B coefficients of the Jones–Dole equation[31]

$$\eta/\eta_0 = 1 + AC^{1/2} + BC \tag{6}$$

in which η and η_0 are solution and solvent viscosities, respectively, and C

is the molar concentration of the solution. Since $[(\eta/\eta_0)-1]/C^{1/2}$ is a linear function of $C^{1/2}$ for concentrations up to approximately 0.1 M, the B coefficient, which represents ion–solvent interactions, can be obtained as the slope from a least-squares analysis of viscosity data. (The Falkenhagen coefficient A,[32] which represents ion–ion interactions, either can be obtained as the intercept from a least-square analysis of data or can be calculated[33] from electrolyte limiting equivalent conductances and solvent physical properties.)

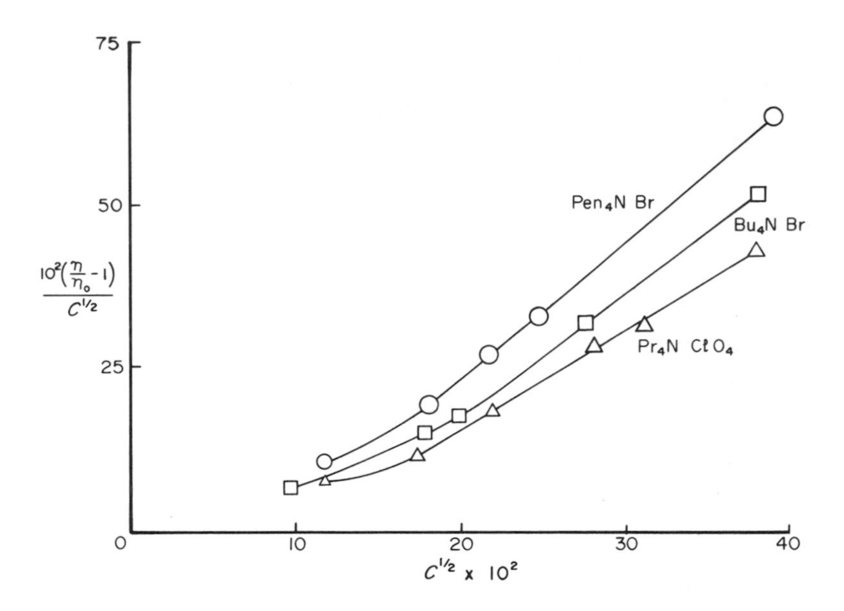

Fig. 6. Plot of the Jones–Dole equation for tetraalkylammonium salts in tetramethylurea. Reproduced from B. J. Barker and J. A. Caruso, *J. Phys. Chem.* **77**, 1884 (1973) with permission of the copyright holder, The American Chemical Society.

Viscosity B coefficients of 1.50 ± 0.04, 1.91 ± 0.05, and 2.12 ± 0.01 were obtained for Pr_4NClO_4, Bu_4NBr, and Pen_4NBr, respectively. Figure 6 is a plot of the Jones–Dole equation for these electrolytes in TMU.

The curvature below 0.03 M concentration indicates that, as expected[33,34] in a Jones–Dole plot for associated electrolytes, the linear region at high concentrations cannot be extrapolated to zero concentration to obtain the Falkenhagen coefficient. Extrapolation of viscosity data for tetraalkylammonium salts in TMU led to negative values of the Falkenhagen coefficient, similar to those obtained for salts in other nonaqueous solvents such as dimethyl sulfoxide[35] and aliphatic alcohols.[36]

VIII. Concluding Statements

In this chapter the use of tetramethylurea as a solvent for fundamental and applied chemical investigations has been reviewed. Since many electrolytes are readily soluble in tetramethylurea and since a wide range of solvation behavior has been found to exist in this medium, TMU will be a convenient, useful, and interesting solvent for future chemical studies.

References

1. A. Lüttringhaus and H. W. Dirksen, *Angew. Chem., Int. Ed. Engl.* **3**, 260 (1964).
2. B. J. Barker and J. A. Caruso, *J. Amer. Chem. Soc.* **93**, 1341 (1971).
3. H. E. Zaugg, B. W. Horrom, and S. Borgwardt, *J. Amer. Chem. Soc.* **82**, 2895 (1960).
4. H. E. Zaugg, *J. Amer. Chem. Soc.* **82**, 2903 (1960); **83**, 837 (1961).
5. H. E. Zaugg, D. A. Dunningan, R. J. Michaelis, L. R. Swett, T. S. Wang, A. H. Sommers, and R. W. Denet, *J. Org. Chem.* **26**, 644 (1961).
6. C. D. Ritchie and R. D. Usehold, *J. Amer. Chem. Soc.* **89**, 1721 (1967); **90**, 2821 (1968).
7. D. B. Bruss and G. E. A. Wyld, *Anal. Chem.* **29**, 232 (1957).
8. S. L. Culp and J. A. Caruso, *Anal. Chem.* **41**, 1329 (1969).
9. S. L. Culp and J. A. Caruso, *Anal. Chem.* **41**, 1876 (1969).
10. M. S. Greenberg, B. J. Barker, and J. A. Caruso, *Anal. Chim. Acta* **54**, 159 (1971).
11. S. Wawzonek and R. C. Duty, *Rev. Polarogr.* **11**, 1 (1963); *Chem. Abstr.* **62**, 1328 (1965).
12. W. E. Bull and R. H. Stonestreet, *J. Electroanal. Chem.* **12**, 166 (1966).
13. B. J. Barker and J. A. Caruso, *Electrochim. Acta* **18**, 315 (1973).
14. B. J. Barker and J. A. Caruso, *J. Phys. Chem.* **77**, 1884 (1973).
15. R. M. Fuoss and F. Accascina, "Electrolytic Conductance." Wiley (Interscience), New York, N.Y., 1959; R. M. Fuoss and L. Onsager, *J. Phys. Chem.* **61**, 668 (1957).
16. T. Shedlovsky, *J. Franklin Inst.* **225**, 739 (1938); R. M. Fuoss and T. Shedlovsky, *J. Amer. Chem. Soc.* **71**, 1496 (1949).
17. R. M. Fuoss and E. Hirsch, *J. Amer. Chem. Soc.* **82**, 1013 (1960).
18. M. A. Coplan and R. M. Fuoss, *J. Phys. Chem.* **60**, 1177 (1964).
19. A. J. Parker, *Quart. Rev., Chem. Soc.* **16**, 163 (1962).
20. G. A. Lester, T. A. Gover, and P. G. Sears, *J. Phys. Chem.* **60**, 1076 (1956).
21. J. E. Prue and P. J. Sherrington, *Trans. Faraday Soc.* **57**, 1795 (1961); P. G. Sears, E. D. Wilhoit, and L. R. Dawson, *J. Phys. Chem.* **59**, 373 (1955); D. P. Ames and P. G. Sears, *ibid.* p. 16.
22. R. Fernández-Prini and J. E. Prue, *Trans. Faraday Soc.* **62**, 1257 (1966); M. Della Monica, U. Lamanna, and L. Janelli, *Gazz. Chim. Ital.* **97**, 367 (1967); M. Della Monica, U. Lamanna, and L. Senatore, *J. Phys. Chem.* **72**, 2124 (1968); M. B. Reynolds and C. A. Kraus, *J. Amer. Chem. Soc.* **70**, 1709 (1948); C. R. Witschonke and C. A. Kraus, *ibid.* **69**, 2472 (1947); E. G. Taylor and C. A. Kraus, *ibid.* p. 1731; C. P. Wright, D. M. Murray-Rust, and H. Hartley, *J. Chem. Soc., London* p. 199 (1931).
23. E. D. Wilhoit and P. G. Sears, *Trans. Ky. Acad. Sci.* **17**, 123 (1956).
24. H. M. Smiley and P. G. Sears, *Trans. Ky. Acad. Sci.* **19**, 62 (1958).
25. W. A. Adams and K. J. Laidler, *Can. J. Chem.* **46**, 2005 (1968).

26. L. M. Mukherjee and D. P. Boden, *J. Phys. Chem.* **73**, 3965 (1969); C. Treiner and R. M. Fuoss, *Z. Phys. Chem.* (*Leipzig*) **228**, 343 (1965); R. L. Kay, C. Zawoyski, and D. F. Evans, *J. Phys. Chem.* **69**, 4208 (1965); P. Bruno and M. Della Monica, *ibid.* **76**, 1049 (1972).
27. M. A. Matesich, J. A. Nadas, and D. F. Evans, *J. Phys. Chem.* **74**, 4568 (1970).
28. R. A. Robinson and R. H. Stokes, "Electrolyte Solutions," 2nd ed. Butterworth, London, 1959.
29. G. J. Janz and S. S. Danyluk, *J. Amer. Chem. Soc.* **81**, 3846 (1959).
30. A. I. Popov and W. A. Deskin, *J. Amer. Chem. Soc.* **80**, 2976 (1958).
31. G. Jones and M. Dole, *J. Amer. Chem. Soc.* **51**, 2950 (1929).
32. H. Falkenhagen and M. Dole, *Phys. Z.* **30**, 611 (1929); H. Falkenhagen and E. L. Vernon, *ibid.* **33**, 140 (1932).
33. H. S. Harned and B. B. Owen, "The Physical Chemistry of Electrolytic Solutions," 3rd ed. Van Nostrand-Reinhold, Princeton, New Jersey, 1958.
34. M. Kaminsky, *Discuss. Faraday Soc.* **24**, 175 (1957).
35. N. P. Yao and D. N. Bennion, *J. Phys. Chem.* **75**, 1727 (1971).
36. J. P. Bare and J. F. Skinner, *J. Phys. Chem.* **76**, 434 (1972).

~5~

Inorganic Acid Chlorides of High Dielectric Constant

(with Special Reference to Antimony Trichloride)

E. C. BAUGHAN

Department of Chemistry and Metallurgy
Royal Military College of Science
Shrivenham, near Swindon, Wilts., England

I. Introduction

Compounds may dissolve in solvents without any specific interaction beyond the general van der Waals energies; or, like salt in water, an ionic compound may dissolve to give free ions; or, like hydrogen chloride in water, a covalent compound may react with the solvent to give free ions (Eq. 1). Other un-

$$HCl + H_2O \rightarrow OH_3^+ + Cl^- \tag{1}$$

predicted reactions may also occur such as the dissolution of alkali metals in liquid ammonia to give solvated electrons. The investigation of solutions in nonaqueous solvents by as many methods as feasible has therefore given (and continues to give) results of considerable chemical interest. The present review considers some inorganic halide solvents with particular reference to antimony trichloride.

Of the physical properties of solvents, the *dielectric constant* is of primary importance. A sphere of radius r and charge q electrostatic units transferred from a vacuum to a medium of dielectric constant D should[1] lose free energy ΔG (Eq. 2) if the dielectric constant remains constant right up to the surface

$$-\Delta G = \frac{q^2}{2r}\left(1 - \frac{1}{D}\right) \tag{2}$$

of the sphere and if no specific effects are involved. This expression is a rough approximation ($\pm 20\%$) to the true solvation energies of salts in water although about one-half the total solvation energy must come from the first neighbors of an ion. Some immediate qualitative consequences follow.

First, as the free energies of transfer of two (or more) ions from one solvent to another are often about equal, a class name is useful. We shall call such classes of ions "iso-solvate classes" (they have equal solvation energy differences between one solvent and another). Born's equation then implies that classes of ions can only be "iso-solvate" if they have the same charge and the same radius. Other conditions may also be involved, as Born's equation is not exact. This is to be determined experimentally.

Second, the Born solvation energy is very large; it is -110 kcal/mole for an ion of radius 1.5 Å transferred from vacuum to water at 25°C. A small lowering of dielectric constant can therefore cause a large fall in solubility; NaCl is almost insoluble[2] in dry acetone ($D = 23$ at 25°C).

Third, there is another important advantage to having a high dielectric constant: the concentration of ions, free to move under a low voltage, can be simply and unequivocally measured by electrical conductance. In solvents having a low dielectric constant ions may be electrostatically held together as ion-pairs or ion-triplets; in such cases conductance, which is always easy to measure, becomes very difficult to interpret. (See Chapter 1 in this volume.)

Other physical properties of solvents are less important. It is convenient to have a *freezing point* near room temperature so that the number of solute particles can be determined by cryoscopy. On the other hand, a low *boiling point* is an advantage since products can then be easily obtained by evaporating the solvent. Moreover, for normal liquids the boiling point is proportional to the heat of vaporization which is a major factor[3] in determining miscibility with other liquids and hence the possibility of obtaining products by solvent extraction. The *density* and *viscosity* of a solvent are of secondary importance, but need to be known. Most liquids are *transparent* in the visible and near-ultraviolet, but the halides of the heavy elements are also transparent in the infrared region where organic bonds absorb.

The *self-conductance* of a solvent needs special consideration, for if it is too high (e.g., H_2SO_4) the ionization caused by solutes cannot be measured by conductance. A perfect covalent liquid should contain no ions at all and its self-conductance should be zero. In fact, self-conductance arises from two main chemical causes: (a) self-ionization of the solvent itself, e.g.,

$$2H_2SO_4 \rightarrow H_3SO_4^+ + HSO_4 \quad (3)$$

$$2H_2O \rightarrow H_3O^+ + OH^- \quad (4)$$

or

$$2SbCl_3 \rightarrow SbCl_2^+ + SbCl_4^- \quad (5)$$

and (b) the production of ions from impurities or reactions of impurities with the solvent, e.g., in the dissolution of alkali from glass

$$CO_2 + H_2O \rightarrow H_3O^+ + HCO_3^- \qquad \text{(in water)} \quad (6)$$

$$H_2O + HF \rightarrow H_3O^+ + F^- \qquad \text{(in HF)} \quad (7)$$

Self-conductance is therefore essentially a chemical phenomenon, and much confusion has arisen in the literature by attributing to self-ionization conductances which were really caused by impurities. All these points will be relevant in the present review.

Jander[4] has classified as "waterlike" (Wasserähnlich) liquids which are essentially covalent (therefore volatile and of low self-conductance) yet capable of dissolving ions because of their high dielectric constant and of producing ions through their chemical reactions.

Some metallic chlorides and oxychlorides ($AsCl_3$, $SbCl_3$, $SeOCl_2$, for example) are "waterlike" in this sense and they have been studied to investigate interionic effects and also to find what ions are produced. One solvent of this type is antimony trichloride which has been investigated since 1899 by four schools of workers in Lwów[5–12] (Poland), Shrivenham[13–26] (England), Berlin,[27–29] and Paris.[30–38] Arsenic trichloride has also been investigated, but it has a much lower dielectric constant ($D = 12.8$ at 25°C), so the phenomena are less clear, and cryoscopy has not been attempted. Of the other such halides, selenium oxychloride $SeOCl_2$ ($D = 46$ at 20°C) has perhaps shown the clearest phenomena (see Section XIV).

The history of work on special solvents may be divided into three main periods. First, following the original discovery of ionization in water, many solvents were investigated and electrical conductances (and thermodynamic properties such as freezing point lowering) determined. Much information was obtained, but interionic interactions were not then understood, and many of the results are suspect because the solvents were inadequately purified. But extensive work was done, much of which remains valid and has led to the discovery of new modes of ionization. The standard reference is Walden's treatise[39] of 1924 which is still essential reading. In the 1920's and 1930's interest shifted from modes of ionization to the study of interionic attraction (Debye–Hückel theory). Recently, however, interest has returned to the study of new types of reaction. Theoretically, this has been stimulated particularly by G. N. Lewis' extension[40,41] of the concepts of acids and bases and by Mulliken's charge–complex theories. Experimentally, to the classic techniques of conductance and cryoscopy have now been added new spectroscopic techniques which study essentially the ground state of solute species: NMR, ESR, and infrared. Modern electronic techniques for electrochemistry are also contributing to the subject.

The study of absorption spectroscopy in the visible and ultraviolet has always been important. Visible color changes of indicators were used to define acids by Robert Boyle in the seventeenth century, and provided one of the chief arguments of G. N. Lewis[41] in 1938. But some care must be exercised for the following reasons.

First, the eye is sensitive to radiations between 6000 and 3000 Å (48 to 96 kcal/mole). Such high energies correspond to transitions from the ground state to excited states. One has little other experimental evidence about the excited states so that unless a given spectrum has already been observed when its ground state is otherwise *known*, the interpretation is doubly uncertain. An interaction may indeed, as in the so-called "contact charge-transfer complex," occur significantly only in the excited state.

Second, one of the very foundations of G. N. Lewis' theory was the close analogy of color reaction[41] shown by the usual acid–base indicators with Brønsted acids in water and by the same indicators with Lewis' enlarged class of acids. The colors are very similar, therefore the reactions are *analogous*, and this is a point of the greatest theoretical importance. But the reactions are not *identical* and one must find out whether a Lewis acid or a Brønsted acid is involved.

In simple language, color changes show that something important has happened, but can seldom (without other evidence) identify the process occurring.

As far as interionic attraction is concerned, antimony trichloride is an exceptionally simple solvent, as we shall show. Its chemistry is complicated, but the simple inorganic reactions are now fairly well understood. Recent work has been directed to the ionization of organic solutes, with interesting consequences for acid–base theory.

"Waterlike" inorganic chlorides undergo the following important reactions.

(1) They are Friedel–Crafts catalysts[42]. Such reactions involve chloride ion transfers, e.g.,

$$CH_3COCl + AlCl_3 \rightarrow CH_3CO^+ + AlCl_4^- \qquad (8)$$

$$CH_3CO^+ + C_6H_6 + AlCl_4^- \rightarrow CH_3COC_6H_5 + HCl + AlCl_3 \qquad (9)$$

The existence of the $AlCl_4$ intermediate was proved[43] by the radioactive indicator technique. Some Friedel–Crafts reactions have been proved to require the presence of water (or alcohols), e.g.,

$$BF_3 + 2H_2O \rightarrow OH_3^+ + BF_3OH^- \qquad (10)$$

in many others this is extremely probable, particularly for olefin polymerization. Because of such halide ion transfers, these inorganic halides can increase the strength of the appropriate Brønsted acids, e.g.,

$$HF + SbF_5 \rightarrow H^+(\text{solvated}) + SbF_6^- \qquad (11)$$

From this arises the whole modern study of "superacids."[44] The replacement of halogens by hydroxyl groups to give oxyacids also classifies these halides as *acid halides* just like acetyl chloride.

(2) Acid–base terminology may also be applied in the "solvent-system" sense. An acid (in this sense) is any solute which increases the concentration of the solvent cation (e.g., H_3O^+ in H_2O, NH_4^+ in NH_3, $COCl^+$ in phosgene, $SbCl_2^+$ in $SbCl_3$), a base any solute which increases the concentration of the solvent anion (e.g., OH^-, NH_2^-, Cl^- as $COCl_3^-$ or $SbCl_4^-$).

(3) G. N. Lewis[40–41,45] defined a base as a substance capable of donating two electrons to a corresponding acid thereby making a new chemical bond. The formation of the bond is an essential part of the process, for the reaction

$$Sn^{2+} + Pb^{4+} \rightarrow Sn^{4+} + Pb^{2+} \qquad (12)$$

would normally be classified as a two-electron oxidation–reduction reaction. It is, of course, possible to draw up a definition of acid–base behavior so as to include oxidation–reduction. One could indeed class together all chemical changes whatever in terms of a sufficiently vague general "electron shift." The real advantage of the Lewis definition is that it expresses the fact that compounds (amines, for example) which show basicity toward the proton, usually also show basicity toward such "acids" as Ag^+ in water, or $SbCl_2^+$ in $SbCl_3$. At the same time it does not include all oxidation–reduction reactions, which (in general) require different reagents.

But difficulties remain. What, precisely, is meant by a chemical bond? The addition of a proton to amines or aromatic hydrocarbons produces typical covalent bonds.[46] Yet the addition of Ag^+ to amines makes bonds which are much shorter and stronger than from the addition of Ag^+ yielding stoichiometric complexes to aromatic hydrocarbons.[17] Antimony trichloride itself forms stable stoichiometric solid phases with aromatic hydrocarbons[47–49] without chloride ion transfer, but in liquid antimony trichloride ionization may occur.[16,17]

Second, the Lewis extension of the acid–base concept requires that "the relative strength of the acids and bases depends not only on the chosen solvent, but also on the particular acid or base used for reference"[45] whereas the less general proton-transfer (Brønsted) concept explains and predicts simple quantitative relationships. But recent work has weakened this antithesis. On the one hand, quantitative relationships (though not of universal validity) are being discovered between various Lewis acid–base reactions.[50] On the other hand, complications have appeared in proton-transfer processes. For example, primary, secondary, and tertiary amines form three "iso-solvate classes," not one, when transferred from water to meta-cresol.[51] Finally, the newer distinction between hard and soft acids and bases has introduced some order into a wide range of chemistry.[52]

Many organic compounds with π electrons act as bases with "waterlike" inorganic halides (see a recent review[53]), but in few cases has conductance been measured and then mainly in systems whose low dielectric constants cause extensive ion association. Similarly, most of the few structure determinations of such solid complexes (and of complexes with iodine) show that halide ion transfer has not occurred. Yet the solid complex of pyridine (Py) with iodine is a *salt*[54]: $[Py–I–Py]^+I_3^-$. It is therefore essential to study Lewis acid–base reactions in solution using conductivity as well as modern spectroscopic techniques, for only by conductivity can the concentration of free ions be directly measured.

Such considerations also emphasize the precautions needed when studying solvents of this type. Because they are Lewis acids, they can also react as Brønsted acids in the presence of suitable hydrides (HCl or water, for example). Hence their purification presents special problems. These problems are increased by effects due to molecular oxygen, which can produce free radical cations from organic compounds in antimony trichloride, as in some other solvents.[55] These free radicals are usually intensely colored, but this color could hardly have been identified but for the discovery of electron spin resonance, which is diagnostic for unpaired spins as conductance is diagnostic for ions. This reaction was clearly recognized only fifteen years ago[55]; it is curious that oxygen, which itself has an unpaired spin and adds on to many free radicals (triphenylmethyl, nitric oxide, etc.) should itself generate free radicals from other compounds. This reaction is not caused by pure antimony trichloride itself, but is caused by antimony pentachloride present as an impurity.[15] Both are Lewis acids. Antimony pentachloride is an oxidizing agent as well, and this is an oxidation reaction. The two types of reaction must not be confused.

In practice, therefore, such investigations must include many independent older and modern techniques on solvents whose purity requires special consideration. The most useful coordinating theory has been the Lewis extension of the acid–base concept; this dates from 1923[40] but only more recently has its importance been generally recognized.

It is therefore not surprising that the earlier literature is somewhat confused. The present review therefore follows this plan: (a) It considers first the work done on antimony trichloride. For this solvent the interionic attractions are the simplest, and the work on organic solutes is the most extensive. Work in other solvents is quoted where relevant. The discussion is general since the facts have recently been summarized by Johnson.[25] (b) The review concentrates more on organic rather than inorganic reactions since these have recently been reviewed by Payne[56] who discussed other similar solvents. (c) Finally, it considers briefly some important recent work on other similar solvents.

II. Purification and Physical Properties

All methods consist of further purification of the best analytical grade reagents, which usually contain water, often HCl, and sometimes free chlorine. The purification is usually measured by using self-conductance.

Preliminary drying in desiccators is followed by distillation[25] in pure dry N_2 or CO_2. The presence of the purest metallic antimony to reduce chlorine ($SbCl_5$) is recommended[14].

This distillation is then followed by vacuum sublimation.[16] Samples having a self-conductance κ_0 of about $2\text{–}4 \times 10^{-6}$ ohms^{-1} cm^{-1} at 75°C have been regularly obtained (corresponding to $3\text{–}6 \times 10^{-6}$ at 99°C). If the solvent is sublimed directly into the working cell, the use of dry boxes and interchangeable glass grinds permits reproducible self-conductance measurements of this order, ranging somewhat from experiment to experiment. For details, see Atkinson *et al.*[16]

TABLE I

PHYSICAL PROPERTIES OF ANTIMONY TRICHLORIDE

Property	Temperature (°C)	Value	Ref.
Dielectric constant	75	33.2	57
	75	36.3	58
	99	34.0	58
Boiling point	—	222.6; 221.9	59, 60
Vapor pressure[a]	75	3.9 mm; 3.5 mm	59, 60
	99	12.3 mm; 12.3 mm	59, 60
ΔH_{vap} (liquid)[b]	—	12.2 kcal mole^{-1}	59
Density	75	2.679	61
	99	2.622	
Melting point	—	73.17°	13
ΔH_{fusion} (solid)[c]	—	3.61 kcal mole^{-1}	—
Cryoscopic constant[c]	—	15.1 °K mole^{-1} kg	—
Specific heat			
Liquid	—	33.5; 35.2	13, 62
Solid	—	25.8 cal °K^{-1} mole^{-1}	13
Viscosity	75	0.0258 poises	9
	99	0.0184 poises	9
Solubility parameter, δ	—	11.6 (cal$^{1/2}$ cm$^{-3/2}$)	—

[a] Extrapolated.
[b] At mean temperature of 140°C.
[c] See Section IV.

In an earlier study[10] freezing *in vacuo* had been recommended and the lowest value $\kappa_0 = 0.85 \times 10^{-6}$ at 99°C was obtained, which soon rose to $2\text{–}4 \times 10^{-6}$ at 99°C. The residual impurities are unknown, but the temperature coefficient of self-conductance[16] suggests strong electrolyte chlorides.

Table I lists the more important physical properties of antimony trichloride.

Throughout this paper concentrations in moles/1000 g of solvent will be called *molal* concentrations and given the symbol m; concentrations in g/molecules/liter will be given the symbol c. In most work the solvent has been directly weighed, and c is therefore obtained from the weights and the solvent density (Table I).[9,13,57–62]

III. Solvent Properties

Antimony trichloride is reactive and not very volatile (Table I). Analysis of dilute solutions in this solvent is therefore difficult; hence most observations on solubility are only qualitative. The general chemistry (up to the late 1940's) is thoroughly summarized in Gmelin (for later reviews, see Kolditz[63,64]).

Solubility

As would be expected from its "solubility parameter"[3] δ (see Table I), $SbCl_3$ is soluble in common organic solvents (and in CS_2, CCl_4, and most organic compounds are soluble in $SbCl_3$). It has therefore proved impracticable to isolate labile $SbCl_3$–organic compounds by liquid–liquid extraction, since it is also very soluble in (and reacts with) water. Addition compounds which are salts should be insoluble in solvents having a low dielectric constant. A few have been isolated by solidifying the melt and extracting $SbCl_3$[13,27–29] with CS_2 or $CHCl_3$. Many organic compounds with mobile electrons show marked colors (see later). (For polymers, see Section XIII.) The solubility of elements and inorganic compounds are given in Table II and may be summarized as follows: (a) Molecular solids are soluble [e.g., I_2, S_8, the polyvalent chlorides quoted, and SbOCl (which is also soluble in CS_2)]. This also would be expected. (b) True ionic solids are not soluble unless the ions are monovalent and large, and their lattice energies are small. (c) The reaction of nitrates and carbonates also shows that many simple anions react with this solvent.

TABLE II

SOLUBILITIES IN $SbCl_3$[a]

Easily soluble (sometimes with evolution of heat)
- Elements: chlorine, iodine, sulfur
- Monovalent halides: chlorides and bromides of K, Rb, Tl, NH_4, and substituted ammonium ions; KI (with color formation)[b]
- Polyvalent halides: $HgCl_2$, $HgBr_2$, $AlCl_3$, SbF_3, $SeCl_4$, $TeCl_4$
- Antimony oxychloride: see below
- Oxides and sulfides: only Sb_2O_3, Sb_2S_3, As_2O_3 (see below)
- Oxyacid salts: sulfate and perchlorate of $N(CH_3)_4$; acetates of K, Hg(II), Sb(III)

Slightly soluble: LiCl, NaCl, $FeCl_3$, $BiCl_3$ (for HCl, see below)

Insoluble: KF; CuCl, AgCl
- Divalent chlorides: Ca, Sr, Ba, Zn, Cd, Pb, Mn, Co, Ni, Hg_2Cl_2 (and $CrCl_3$)
- Divalent oxides: Mg, Ca, Zn, Mn, Pb (and Al_2O_3)
- Divalent sulfides: Zn, Cu, Hg, Pb
- Oxyacid salts: Sulfates of Na, K, NH_4, Mg, Ba, Zn, Sb(III); $KClO_4$; $AgClO_4$; K_2CrO_4; $KMnO_4$
- KCN

Soluble with evolution of gas: KNO_3 (nitrous gas); $K_2CO_3(CO_2)$

Slightly soluble or insoluble, with gas evolution
- Nitrates: Na, Ag
- Carbonates: Li, Na, Mg, Ba, Zn, Mn, Pb

[a] This table follows the summary of Jander and Swart.[27]
[b] From Klemensiewicz and Balowna.[11]

IV. Cryoscopy: "Normal Solutes" and Inorganic Salts

Cryoscopy is the simplest way of finding the number of molecules produced per molecule of solute. Chemically reactive solvents such as $SbCl_3$, however, cause several difficulties. (a) First, analysis is impracticable; so the inaccurate Beckmann technique must be used. (b) Second, it is not easy to find non-reacting "normal solutes." (c) Third, it is advisable to measure directly the heat of fusion to check the cryoscopic constant deduced from supposedly "normal solutes."

These points are all familiar in the study of cryoscopy in anhydrous H_2SO_4 and are important for cryoscopy in $SbCl_3$.

The first thorough investigation was made by Tolloczko[5] in 1899, who quoted depression of freezing point Δt for weights of solute in g grams of $SbCl_3$ and for each concentration calculated* a cryoscopic constant E (which is in units ten times greater than the modern units "degrees mole^{-1} kg"). He realized the importance of drying the air in the cryoscope and of surrounding it by an air gap itself surrounded by a large well-stirred tank of liquid kept within a degree or so of the solvent's freezing point. The solvent was, however, only distilled once, though freshly for each run.

The "normal solutes" used were xylene, anthracene, diphenylmethane, acetophenone, and benzophenone. Later conductance work[25] has shown that some of these should be slightly ionized at such concentrations, but this would have little effect on his cryoscopic constant. But the color reactions as shown below

Mesitylene: intensive violet
Anthracene: deep green
Phenanthrene: pale blue
Pyridine: bright yellow

naturally cause some doubt as to whether these are really "normal solutes," although they gave a consistent cryoscopic constant k_f.

$$k_f = 18.4 \text{ degrees mole}^{-1} \text{ kg} \tag{13}$$

Tolloczko also published results for KCl, which he showed to be extensively dissociated, and for KBr which showed a somewhat greater depression.

In 1958, almost 60 years later, Porter and Baughan[13] published data using the Beckmann technique on four hydrocarbons, fluorene, dibenzyl, anthracene, and stilbene, on benzophenone, and also on the salts, NMe_4Br, NMe_4Cl, CsCl, KCl, and the organic halide triphenylmethyl chloride

* There are some arithmetic errors in the calculation of his concentrations, which do not affect the calculation of the cryoscopic constant.

CPh_3Cl. The solvent was more carefully purified, the new technique of interchangeable glass grinds was available, yet the results for KCl are practically identical with those of Tolloczko.[5] This confirms the applicability of the Beckman technique and the accuracy of Tolloczko's old results.[5]

The cryoscopic constant recommended by Porter and Baughan[13] for the hydrocarbons and benzophenone was, however, considerably lower.

$$k_f = 15.6 \pm 0.2°\text{K mole}^{-1}\text{ kg} \tag{14}$$

This difference of almost 20% does not however imply a 20% inaccuracy in Δt, the depression of freezing point. Tolloczko's solutions were much more concentrated, and Porter and Baughan's values of Δt vs. concentration showed a significant upward trend. The solutions showed marked deviations from Raoult's Law and were clearly colored.

Anthracene: deep green
Fluorene: deep blue
Benzophenone: green
Dibenzyl: pale green
Stilbene: pale green

Stilbene is exactly dimerized, and the dimers have been identified.[23]

In a later paper[18] the cryoscopy of naphthalene, perylene, sulfur (S_8), and hexachloroethane were investigated, as these were known to be little ionized. From those measurements done in more dilute solution (to minimize nonideality), a value

$$k_f = 14.7 \pm 0.4°\text{K mole}^{-1}\text{ kg} \tag{15}$$

was recommended. These authors sent a large sample (20 cm^3) of their specially purified $SbCl_3$ to the National Physical Laboratory at Teddington, England where Dr. J. E. Martin[65] measured the heat of fusion, obtaining

$$k_f = 15.1°\text{K mole}^{-1}\text{ kg} \qquad \text{(recommended value)} \tag{16}$$

The latent heat of fusion has since been determined elsewhere[66] on a microscale, giving $\Delta H_f = 3{,}370 \pm 40$ kcal/mole, corresponding to

$$k_f = 16.1 \pm 0.2°\text{K mole}^{-1}\text{ kg} \qquad \text{(in rough agreement)} \tag{17}$$

The behavior of "normal solutes" is therefore established, and cryoscopy can be used as a standard method. Results for special solutes are quoted later.

Porter and Baughan[13] also considered the salts NMe_4Cl, KCl, and CsCl and showed that the organic covalent compound CPh_3Cl behaved in the same way. All these compounds were strong 1:1 electrolytes, giving limiting *i* factors of 2, and osmotic coefficients in agreement with Debye–Hückel theory in very dilute solution. Differing values for the "radius" *a* represent well the data in less dilute solutions (for details, see the original paper[13]).

Electrolyte theory is beginning to explain these *a* parameters; the results of Porter and Baughan[13] should provide a useful test for such improved theories.

The cryoscopy of bromide salts has proved instructive. The very close agreement (see Section V) between the conductance of some bromides and their corresponding chlorides suggests a common ionic mechanism. If Br^- converts into Cl^-, additional covalent solutes must also be produced, e.g.,

$$6Br^- + 2SbCl_3 \rightarrow 2SbBr_3 + 6Cl^- \quad (18)$$

$$6Br^- + 3SbCl_3 \rightarrow 3SbBr_2Cl + 6Cl^- \quad (19)$$

$$6Br^- + 6SbCl_3 \rightarrow 6SbBrCl_2 + 6Cl^- \quad (20)$$

with the production of an additional 1:3, 1:2, 1:1 moles, respectively, of solute per mole of bromide ion. Cryoscopy shows the first process—formation of $SbBr_3$—to be of predominant importance (see Table III).

TABLE III

FREEZING POINT DEPRESSIONS Δt OF BROMIDES AND CORRESPONDING CHLORIDES

			Δt (°C)		$\Delta\Delta t$	
Cation	Ref.	100 m[a]	Bromide[b]	Chloride (interpolated)	Observed	Calculated[c]
K^+	5	1.77	0.56	0.45	0.11	0.09
		4.90	1.45	1.15	0.30	0.25
		7.80	2.20	1.75	0.45	0.39
NMe_4^+	13	0.799	0.27	0.20	0.07	0.04
		1.585	0.55	0.45	0.10	0.08
		2.744	0.96	0.81	0.15	0.14
		4.195	1.48	1.25	0.23	0.21
		5.698	2.01	1.75	0.26	0.29
		7.322	2.64	2.28	0.36	0.37

[a] m is concentration in g/moles/1000 g.

[b] If the bromide XBr is entirely converted to XCl, then Δt for bromide should be greater than that of the chloride at the same concentration because of the formation of an nonionized solute, whose effect $\Delta\Delta t$ should be simply additive.

[c] $\Delta\Delta t(\text{calc}) = \frac{1}{3}mk_f$; $k_f = 15.1$.

In 1959 Jander and Swart[27–29] also published some cryoscopic results. Individual points are not given, and they interpreted their results using the out-of-date value[5] $k_f = 18.4$°K mole^{-1} kg. Their results also confirm Tolloczko's[5] experiments on KCl and show that CPh_3Cl and $N(CH_3)_4Cl$

(also anilinium chloride) are strong 1:1 electrolytes, but are not precise enough for useful calculation of osmotic coefficients, though in general agreement with Porter and Baughan.[13] $TeCl_4$ and $SeCl_4$ appear undissociated; their results for Sb_2O_3, SbOCl, we will discuss later.

V. Conductance of Simple Salts

Antimony trichloride is difficult to work with and to purify except on a small scale. Conductance data are therefore only accurate to 1 or 2%. The results for simple salts (chlorides, bromides of the alkalies, ammonium, monovalent thallium, and divalent mercury) are mainly due to the careful work of Z. Klemensiewicz and co-workers[7–12] between 1908 and 1934. Most of their results were obtained at 99°C, but a few cover the range between 79°C and 202°C.

The alkali and ammonium bromides and chlorides (cf. the freezing point depressions) are typical strong 1:1 electrolytes; indeed successive improvements in conductance theory have shown improved agreement with these data. In 1924 Klemensiewicz[9] explained the earlier data by the partially incorrect theory of Ghosh, which he applied to his later work (at 99°C) with Balowna in 1930[10] and 1931[11]. These two papers, together with a third paper by Klemensiewicz and Zebrowska[12] in 1934 (see Section VI), are still the most accurate conductance work, and have been confirmed by later workers.

In 1959 Porter and Baughan[13] showed that these results in dilute solution obeyed the Debye–Hückel–Onsager equation

$$\Lambda_c = \Lambda_0 - Sc^{1/2} \tag{21}$$

(Λ equals the equivalent conductance at concentration c, Λ_0 at infinite dilution, and S a theoretically derivable coefficient). The later theory of Fuoss and Onsager[67] adds additional terms

$$\Lambda = \Lambda_0 - Sc^{1/2} + Ec \log c + Jc \tag{22}$$

where the value of E also is known from Λ_0, viscosity, and dielectric constant; the value of J also depends on ionic size. Fuoss and Accascina[68] have shown the applicability of this equation. For KCl in water at 25°C between 0.0005 and 0.0177 M

$$\Lambda = 149.96 - 94.56c^{1/2} + 58.75c \log c + 193c \tag{23}$$

(with the theoretical values of S and E); the data is reproduced with five-figure accuracy. Other refined conductance theories* lead to similar formulas,

* See Chapter 1 of this volume.

TABLE IV

EQUIVALENT CONDUCTANCE Λ OF SIMPLE INORGANIC HALIDE SALTS IN $SbCl_3$[a]

	KCl		KBr	NH_4Cl	NH_4Br	RbCl	TlCl	TlBr
c (g/mole/liter)	Calculated (Eq. 23)	Observed						
0.0001	155	155	153	149	150	155	147	148
0.0002	154	154	151	148	148	154	145	145
0.0005	153	151	149	146	146	151	140	140
0.001	151	148	147	144	144	149	135	136
0.002	148	145	144	141	141	146	130	131
0.005	143	140	139	136	137	142	122	119
0.01	137	135	134	131	132	137	114	114
0.02	128	129	129	126	127	133	104	106
0.05	112	119	119	117	118	123	89.5	93.5
0.1	101	110	111	109	110	115	77.6	84.1
0.2	100	99.1	101	99.3	100	107	65.9	74.1
0.5	—	82.8	85.1	84.3	85.1	89.1	—	61.0
1	—	64.9	66.5	70.8	67.9	67.8	—	—
2	—	42.7	40.7	44.5	—	39.8	—	—
3	—	29.0	—	—	—	—	—	—

[a] At 99°C. Interpolated from Klemensiewicz and Balowna.[11]

but with different interpretations[69] of the coefficient J. It is therefore remarkable that equations of this form may be applied to the more concentrated solutions in $SbCl_3$.

Table IV is based on Table 8 of the Klemensiewicz–Balowna paper of 1931.[11] The results for chlorides and corresponding bromides are practically identical. This has been explained by the reaction

$$3Br^- + SbCl_3 \rightarrow SbBr_3 + 3Cl^-$$

($SbBr_3$ by itself does not conduct in $SbCl_3$.[17])

The results for KCl/KBr, NH_4Cl/NH_4Br, and RbCl are represented by equations of the Fuoss–Onsager type. For KCl/KBr

$$\Lambda = 156 - 130c^{1/2} + 500c \log c + 360c \tag{24}$$

covers the data up to $c = 0.02$ as can be seen from Table IV (S and E have the theoretical values). The close similarity of results for the more concentrated solutions suggests that the phenomena are particularly simple. These results, like those of Porter and Baughan[13] on freezing points, should provide a useful check on future theories for more concentrated electrolyte solutions.

The results depend little on the nature of the cation, and Λ_0 for KCl at 99°C is about equal to Λ_0 for KCl in H_2O at 25°C although the viscosity of $SbCl_3$ is 2.2 times greater. These facts suggest that the chloride anion is abnormally mobile. This has been checked: (a) by a direct determination[8] of transport numbers for KCl and NH_4Cl by Hittorf's method (the transport number of the chloride ion at 99°C is about 0.9); and (b) indirectly from Walden's rule and the results[14] for NMe_4Cl at 75°C.

These results and the freezing point data lead to the simple conclusions which are basic to later work: that interionic effects are as well understood as they are in water and that the chloride ion is particularly stable (cf. reaction of bromide) and particularly fast. Hence its presence may be simply recognized.

In this solvent the Bjerrum critical distance y for ion-pair formation is (at 99°C)

$$y = e^2/2DkT = 7.2\ \text{Å} \tag{25}$$

If the chloride ion is really the large ion $SbCl_4^-$, ion-pairing should be unimportant; but recent Raman spectrum work[70] on $KCl/SbCl_3$ shows no evidence for $SbCl_4^-$.

Klemensiewicz' results at 99°C were completely confirmed by the less accurate work of Jander and Swart[27–29] (including the lower Λ values for TlCl) who also showed that NMe_4Cl, CPh_3Cl, and anilinium chloride were strong 1:1 electrolytes. Only in the case of SbOCl (see Section VI) is there any significant disagreement.

VI. Conductance of Other Inorganic Halides

The results of Jander and Swart[27–29] (presented graphically) show that Λ for aluminum chloride is about one-third of the Λ values for KCl at equal molecular concentrations (see Table V). Texier[33] also showed that the Debye–Hückel–Onsager equation was obeyed by solutions in dry $SbCl_3$ (with c the molecular concentration) and with $\Lambda_0 \simeq 48$.

These results show that $AlCl_3$ is a 1:1 electrolyte but with a different *normally* mobile anion, presumably

$$AlCl_3 + SbCl_3 \rightarrow SbCl_2^+ + AlCl_4^- \tag{26}$$

Suppose that $SbCl_2^+$, K^+, and $AlCl_4^-$ all had the same normal mobility, then $SbCl_4^-$ would have five times this mobility.

This interpretation is supported by the results for silver perchlorate. Jander and Swart[29] showed that this precipitated AgCl, the solution then having about the same conductance as $AlCl_3$. The obvious reaction is

$$AgClO_4 + SbCl_3 \rightarrow AgCl + SbCl_2^+ + ClO_4^- \tag{27}$$

TABLE V

CONDUCTANCES OF KCl AND $AlCl_3$[a] AT 99°C

c (moles/liter)	Λ KCl (Table IV)	Λ $AlCl_3$[b]	Λ $AlCl_3$/Λ KCl
0.25	99	35	0.35
0.5	83	28	0.34
0.75	70	25	0.35
1.0	65	22.5	0.35

[a] Regarding the latter as a 1:1 electrolyte.
[b] From Jander and Swart.

$FeCl_3$, $SeCl_4$, $TeCl_4$, and the halides of divalent mercury show smaller conductances whose cause is still uncertain. Some early cryoscopic work[6] showed abnormally large freezing point depressions for the solutes $AsBr_3$, AsI_3, and $BiCl_3$; these have not been reinvestigated.

Recently it has been reported[71] that $GaCl_3$ conducts in $SbCl_3$.

VII. Conductance of Probable Impurities

Dilute solutions of $SbCl_3$ in water may give many products including HCl, SbOCl, and even Sb_2O_3. Conversely, H_2O, HCl, SbOCl, Sb_2O_3, and $SbCl_5$ (from excess chlorine) are of particular interest in $SbCl_3$ (cf. Klemensiewicz and Zebrowska[12]). Conductance and cryoscopic data are now available for H_2O, HCl, and SbOCl. H_2O and HCl have also been investigated by NMR.[32] Cryoscopic data only is available for Sb_2O_3 and conductance data only for $SbCl_5$. The results are in good agreement (except for cryoscopy of SbOCl).

(1) *H_2O.* In dilute solutions water has a very small conductance value[17] ($\Lambda \simeq 0.01$, 10^{-4} of KCl). Cryoscopy shows it to be a normal solute,[18] and NMR detected no new species.[32] In concentrated solutions some reaction presumably occurs, but this is not yet known.

(2) *HCl.* Klemensiewicz and Zebrowska[12] quote Λ values for HCl varying irregularly between 0.1 and 0.5 at 99°C. Atkinson, Jones and Baughan[16] measured the freezing point depression of concentrated solutions of dry HCl gas at 1 atm and 75°C. This would for a nonionized solute imply a solubility of 0.06 m (ideal solubility theory[3,16] would predict about 0.09 m) and the conductances of such solutions gave Λ values of 0.04 and 0.4 (two experiments).[16] HCl shows no new species with NMR.[32]

H_2O does not therefore react significantly with the solvent; this is rather surprising. Dry HCl dissolves as a molecule. This would have been expected; the small and irregular conductances observed are probably due to reactive impurities. If H_2O were to react, it would presumably give $OH_3^+Cl^-$ which is (like NH_4Cl) a strong electrolyte as shown by the conductance[16] of a few drops of concentrated aqueous HCl in $SbCl_3$. One may perhaps speculate that the conductance of HCl is due to formation of OH_3Cl which (if complete) would imply that careful working can keep the water concentration down to 10^{-5} *M*.

(3) *SbOCl*. Klemensiewicz and Zebrowska[12] quote Λ values of about 0.03. Jander and Swart[27] give somewhat higher values, but they are still very low. Jander and Swart[28] however gave cryoscopic graphs suggesting extensive dissociation. This discrepancy was resolved by later work which gives a normal freezing point depression,[18] which is supported by Jander and Swart's data on Sb_2O_3.

(4) Sb_2O_3 (Sb_2S_3). Alone among polyvalent oxides and sulfides these are soluble in $SbCl_3$; evaporation leaves SbOCl and the hitherto unknown SbSCl. Cryoscopy by Jander and Swart gives i values between 2 and 3 for Sb_2O_3. As the formation of SbOCl (SbSCl) presumably takes place in solution (Eq. 28) an i factor of 3 would be expected if SbOCl were a normal

$$Sb_2O_3 + SbCl_3 \rightarrow 3SbOCl \qquad (28)$$

solute and an i factor greater than 3 if it were dissociated.

(5) $SbCl_5$. Jander and Swart[27–29] quote (at 80°C) a Λ value of 0.8 at $c = 0.5$ mole/liter. The curve suggests a weak 1:1 electrolyte, but the ions are yet unknown. One later determination[16] roughly supports these values.

(6) *The self-conductance of the solvent and its ionic product.* The lowest observed value[9] of κ_0 (0.85×10^{-6} ohms^{-1} cm at 99°C) would, if caused by chloride anion, imply a concentration of 5.6×10^{-6} moles/liter, and therefore an upper limit of 3×10^{-11} (moles^2/liter^{-2}) for the ionic product of the reaction

$$2SbCl_3 \rightarrow SbCl_2^+ + SbCl_4^-$$

Other workers[25] have regularly obtained κ_0 values varying between 6 to 8 and 10 to 15 times this minimal value. Such variations of themselves imply impurity; the temperature coefficient[16] of κ_0 is about that of Λ for KCl, and therefore the impurity is probably a strong chloride electrolyte. Correction of observed conductances of other solutes based on this hypothesis improves their self-consistency.[14,16]

Recently, however, EMF data have been quoted[33] as supporting a much higher value for the ionic product $\kappa_i = 10^{-7.8}$. Impurities can easily make the observed ionic concentration of a solvent sample much greater than corresponds to its self-ionization (this is usually the case with water), but can hardly make it less. These emf data are therefore presumably caused by some adventitious chloride, and this hypothesis explains how an apparent pK_i can be practically independent of temperature over the whole temperature range from 79° to 150°C.

The ionic product of this solvent is therefore still unknown, but cannot be greater than 3×10^{-11} at 99°C.

VIII. Organic Solutes: Introduction

Antimony trichloride (if sufficiently pure) is therefore a particularly simple ionizing solvent. The interionic effects exemplify the Debye–Hückel theory, cryoscopy is a proven technique, and the chloride anion is abnormally mobile and hence easily recognized (and its concentration measured) by conductance. Conversely, the conductance of chloride-containing solutions proves little about the number and types of cations beyond the rigid requirements of stoichiometry and electrical neutrality. Cryoscopy (or its thermodynamic equivalent), by which total solute concentrations may be measured, is therefore essential for the interpretation of the conductance results. Such information on types and concentrations of solute then provides the essential foundation for spectroscopic investigations (visible, infrared, Raman, NMR) and for such refined electrochemistry as the search for reversible electrodes.

Since 1958 papers have appeared from our laboratory on cryoscopy and the application of cryoscopy and conductance to organic solutes[13–26]; from 1963 onward[15] proton electron spin resonance has also been applied. These results have been used[22] to explain some complicated color phenomena. All solution measurements were made at 75°C, just above the melting point of 73.17°C, to minimize the effects of volatility and reactivity of the solutes.

We will consider first this work which has provided systematic data on Lewis basicity, then the electrochemical work from Paris since 1968, the independent work on spectroscopy, and finally other applications of this solvent.

IX. Organic Chloro Compounds

The conductance of two true salts KCl and NMe_4Cl was first measured at 75°C to check roughly the Debye–Hückel–Onsager (D.H.O.) equation and the abnormal mobility of the chloride ion (see Section IV). These and other results establish the D.H.O. equation for a chloride strong 1:1 electrolyte at 75°C as

$$\Lambda = \Lambda_0 - 79c^{1/2} \tag{29}$$

with $\Lambda_0 \simeq 88$ ($\pm 3\%$, perhaps); this coefficient of $c^{1/2}$ is based on the recent measurements of dielectric constant D (Table I). At 99°C

$$\Lambda = \Lambda_0 - 130c^{1/2}$$

with $\Lambda_0 \simeq 155$ ($\pm 2\%$, see Table IV).

Triphenylmethyl chloride is a strong 1:1 electrolyte both from freezing point[13,27] and conductance measurements[14,27,31] at 75° and 99°C. The solution is orange-yellow. Triphenylmethyl chloride is therefore a strong base in the solvent-system sense; in other solvents such as SO_2 it is very weakly ionized.

The idea that $SbCl_3$ should accept chloride ion from organic chlorides led, indeed, to the whole study of organic solutes. The work of Meerwein, Polanyi, Ingold, and others had long ago shown that some reactions of organic chlorides occurred by ionic mechanisms (S_N1 !) and rough calculations[72] suggested that such activation energies should be about equal to the (endothermic) equilibrium heats of ionization. If so, a few kilocalories per mole from a covalent bond on top of the electrostatic solvation energy (Born's equation) might cause detectable equilibrium ionization and it was hoped that the process

$$Cl^- + SbCl_3 \rightarrow SbCl_4^- \tag{30}$$

would do so. This process occurs in water–HCl solutions[73] of $SbCl_3$ if (perhaps)[70] not in $SbCl_3$ itself. The results with CPh_3Cl seemed encouraging; and later work, particularly by Olah,[74,75] has detected by NMR the carbonium ions formed by protonation of olefins by HF–SbF_5 and similar mixtures ("Magic-Acid").

Several other organic chloro compounds were therefore tested for conductance in $SbCl_3$. Some (e.g., chlorobenzene, C_2Cl_6) showed no effect. Others gave definite reproducible conductances. The facts are clear although their original structural *interpretation* may be incorrect. The case of bornyl chloride (chosen as nonvolatile and unlikely to react by an S_N2 mechanism) is instructive. Values of Λ over a wide concentration range are about one half the Λ values for triphenylmethyl chloride at the same concentration. Table VI shows Λ_B for bornyl chloride at concentration c_B divided by Λ_0^*, the Λ_0 value taken for triphenylmethyl chloride (89.4). The drift in the ratio Λ_B/Λ_0^* disappears almost entirely if we assume that two bornyl chloride molecules give one chloride anion, when Λ_B is compared with Λ_c^*, the value for triphenylmethyl chloride at a concentration c one-half c_B. This ratio is shown in the last line of Table VI, the mean value

$$\Lambda_B/\Lambda_c^* = 0.467 \pm 006 \tag{31}$$

or 94% of the simple ratio one-half. Later cryoscopic work[18] showed an i

TABLE VI

COMPARISON OF Λ_B FOR BORNYL CHLORIDE[a]

					Run 1					
c_B	2.37	4.20	7.76	9.95	12.63	16.21	19.54	24.59	31.81	40.53
Λ_B	41.4	40.4	39.1	38.7	38.4	38.2	38.0	37.7	37.9	38.2
$\Lambda_B/\Lambda_0{}^*$	0.463	0.452	0.457	0.434	0.430	0.427	0.425	0.422	0.424	0.427
$\Lambda_B/\Lambda_C{}^*$	0.479	0.472	0.462	0.461	0.461	0.461	0.462	0.463	0.464	0.472
					Run 2					
c_B	1.54	3.69	5.83	7.85	10.52	12.85	15.53	18.82	24.95	30.47
Λ_B	43.7	40.4	39.3	38.8	38.0	37.6	37.9	37.3	38.1	38.7
$\Lambda_B/\Lambda_0{}^*$	0.489	0.452	0.440	0.434	0.425	0.421	0.424	0.417	0.426	0.433
$\Lambda_B/\Lambda_C{}^*$	0.502	0.482	0.462	0.459	0.454	0.451	0.457	0.454	0.469	0.480

[a] At concentrations c_B (mmoles/liter) with Λ^* for triphenylmethyl chloride. From Davies and Baughan.[14]

factor of 1.24 (nearer 1 than 2) in this solvent which was reproducible independent of concentration and time. As pointed out in Section III many organic solutes give somewhat large *i* factors through nonideality in solution.

These observations (closely paralleled with *n*-decyl chloride) show from the number of chloride ions (conductance) and solute particles (cryoscopy) that the halides RCl react

$$2RCl \rightarrow [R_2Cl]^+ + Cl^- \tag{32}$$

Other organic chlorides (e.g., cyclohexyl, ethyl cyclohexyl, diphenylmethyl, cinnamyl, and benzyl) are not so fully ionized and (because the interionic effects are simple) equilibrium constants (Eqs. 33 and 34) can be evaluated

$$K' = [R^+][Cl^-]f^2/[RCl] \tag{33}$$

$$K = [R_2Cl^+][Cl^-]f^2/[RCl]^2 \tag{34}$$

(f is the mean ionic activity coefficient); for the detailed calculations, see Davies and Baughan.[14] The values of these equilibrium constants are shown in Table VII. The ratio of the two dissociation constants is almost constant.

Recent knowledge of carbonium ions (based mainly on NMR) argues against this view. Although alkyl fluorides produce them with SbF_5 this

$$RF + SbF_5 \rightarrow R^+ + SbF_6^- \tag{35}$$

reaction does not occur[74] with alkyl chlorides and the liquid Lewis acid halides $SnCl_4$, $TiCl_4$, and $SbCl_5$, whose dielectric constants are, however, much less than that of $SbCl_3$. Moreoever, if RCl formed R^+ and Cl^-, there should be extensive and quick exchange of chlorine between RCl and the

TABLE VII

EQUILIBRIUM CONSTANTS FOR ORGANIC CHLORIDES[a]

Chloride	K	K'
Triphenylmethyl	—	S.E. (> 0.2)[b,c]
Bornyl (*n*-decyl)	S.E.[b]	—
1-Ethylcyclohexyl	$2.2 \pm 0.1 \times 10^{-1}$	$2.5 \pm 0.6 \times 10^{-4}$
Cyclohexyl	$9.7 \pm 0.2 \times 10^{-2}$	$4.1 \pm 1.2 \times 10^{-4}$
Diphenylmethyl	$5.8 \pm 0.3 \times 10^{-2}$	$2.6 \pm 1.5 \times 10^{-5}$
Cinnamyl	$7.5 \pm 0.5 \times 10^{-3}$	$2.1 \pm 0.8 \times 10^{-5}$
Benzyl	$3.7 \pm 0.4 \times 10^{-5}$	$1.0 \pm 0.5 \times 10^{-7}$

[a] From Davies and Baughan.[14]
[b] S.E., Strong electrolyte.
[c] The estimate for K' for CPh_3Cl is based on the assumption that 10% of undissociated CPh_3Cl would be just detectable.

solvent. Radioactive tracer work[26] however shows only slow exchange with bornyl and *n*-decyl chlorides and the solvent.

It seems therefore probable that these conductance and cryoscopic results are to be interpreted as Lewis acid reactions of the $SbCl_2^+$ cation (see later). Thus,

$$\text{“}R^+\text{”} = RCl \rightarrow SbCl_2^+ \tag{36}$$

$$\text{“}R_2^+Cl\text{”} = \begin{matrix} RCl \searrow \\ \qquad SbCl_2^+ \\ RCl \nearrow \end{matrix} \tag{37}$$

The application of NMR should be decisive; preliminary work[76] in $AsCl_3/SbCl_3$ mixtures has not shown the shifts expected of carbonium ions.

X. Solutions of Aromatic Hydrocarbons

Aromatic hydrocarbons show a complicated picture. The more reactive hydrocarbons (e.g., tetracene, Table VIII, **VI**) gave intensely colored solutions conducting like strong 1 : 1 electrolyte chlorides, although the solutes contain no chlorine. Less reactive hydrocarbons (e.g., naphthalene, **II**) show no conductance or much weaker colors (e.g., anthracene, **IV**). Electron spin resonance[15] showed that solutions of the more reactive hydrocarbons contained their cation free radicals R^+, which had previously been detected in

TABLE VIII

AROMATIC HYDROCARBONS WHOSE CONDUCTANCE HAS BEEN MEASURED IN PURE $SbCl_3$

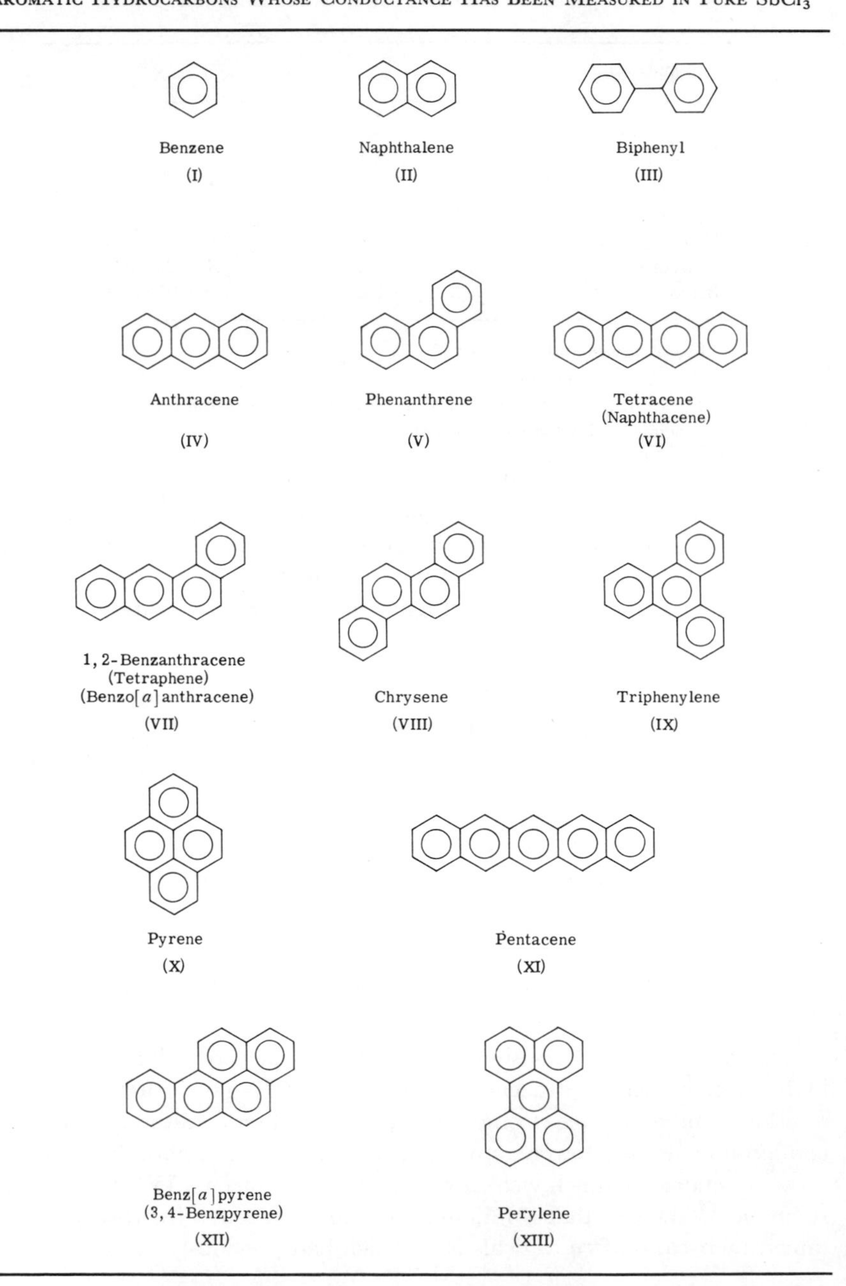

strongly oxidizing media.[77] But $SbCl_3$ is not an oxidizing agent and, in fact, the production of free radicals requires traces of oxygen or antimony pentachloride.[15]

It was therefore advisable to study a hydrocarbon of intermediate reactivity in the purest possible solvent and the effects of added oxidizing agents (molecular oxygen, $SbCl_5$, etc.) and impurities (HCl, H_2O) which could give Brønsted acids with the solvent.

A final vacuum sublimation stage was therefore added[16] to the "previous" distillations in high-purity nitrogen; this has slightly reduced the specific conductance κ_0 of the solvent, but not affected the conductance of less reactive hydrocarbons. For hydrocarbon solute *perylene* (Table VIII, **XIII**) was chosen because it gives the free-radical cation[55] with molecular oxygen in HF and CF_3COOH. This cation proved extremely stable—no changes could be observed in the ESR signal over a period of eighteen months at 80°C in $SbCl_3$.[16] Three reactions were observed.[16]

(1) *Perylene with molecular oxygen.* O_2 itself is hardly soluble in $SbCl_3$ and shows no conductance; but with perylene (Pn) the following reaction occurs

$$O_2 + 4Pn + 6SbCl_3 \rightarrow 4Pn^+ + 4SbCl_4^- + 2SbOCl \tag{38}$$

(*i*) The conductance is that of a strong 1:1 electrolyte chloride. (*ii*) The oxygen uptake is 1 mole per 4 moles perylene. (*iii*) The *i* factor is about right (2.30 ± 0.10; theoretical value, 2.30, allowing for interionic effects). (*iv*) The strong ESR signal suggests complete formation of Pn^+ in dilute solution. This has been checked from the paramagnetic susceptibility of dilute solutions; in more concentrated solutions the paramagnetism is less.[19] (*v*) The solutions are deep purple in color, and the absorption spectrum shows that the formation of this cation is almost quantitative in other systems.[19] This is a general reaction[78,79] for producing the free-radical cations of hydrocarbons, some of which (e.g., from pentacene, **XI**) had not been obtained otherwise. (*vi*) $SbCl_5$ also gives the ESR signal, the deep purple color and the reaction

$$2Pn + SbCl_5 + SbCl_3 \rightarrow 2Pn^+ + SbCl_4^- \tag{39}$$

and $FeCl_3$ also gives the same qualitative results. (*vii*) There is some evidence that these reactions are reversible.[16] The presence of excess $SbCl_3$ helps the reaction, but there is a similar reaction even in CCl_4 where 4,4′-dimethoxydiphenylamine gives with $SbCl_5$ a black precipitate whose paramagnetic susceptibility proved the free radical and whose analysis repeatedly gave one antimony to each four chlorines.[80]

(2) *Perylene with HCl.* The perylene also becomes a strong 1:1 electrolyte chloride, but the solution is green and gives no ESR signal. This is due to the

$$Pn + HCl + SbCl_3 \rightarrow PnH^+ + SbCl_4^- \tag{40}$$

hydrocarbon behaving as a Brønsted base. On addition of oxygen, the solution immediately becomes deep purple and the ESR signal appears strongly.

(3) *Perylene in pure solvent.* In the absence of oxygen or HCl, the perylene is a weak 1 : 1 electrolyte whose conductance is the same in vacuum-sublimed $SbCl_3$ as in the "previous" nitrogen-distilled solvent. The same reproducibility was also obtained with anthracene.[16]

More reactive hydrocarbons are sufficiently ionized to show the formation of one abnormally fast chloride anion per molecule. The reaction must therefore be

$$Pn + 2SbCl_3 \rightarrow Pn{\rightarrow}SbCl_2^+ + SbCl_4^- \tag{41}$$

and measures the Lewis basicity of the hydrocarbon. This process is not affected by small additions of water.

These results could lead in three directions: (1) to further study of organic oxidation–reduction reactions, as in the later work of the Paris School[30–38]; (2) to further study of protonation reactions (not yet done), and (3) to study of other Lewis base reactions, which we now consider.

Other Aromatic Hydrocarbons in Pure Solvent: Conductance[12]

In all 13 hydrocarbons have been tested for conductance in pure $SbCl_3$ (shown in Table VIII). The principal problem, again, is purity. On the whole, the bigger rings conduct better, so the problem is to obtain the lower hydrocarbons free from their higher homologs. Vacuum sublimation appeared particularly suitable and this was applied to samples which had already been refined by zone melting. Even so, in the case of chrysene (**VIII**) (which ionizes very little) adequately reproducible results could not be obtained because of its well-known[81] impurity tetracene (**VI**) which ionizes a lot.

Of the other 12 hydrocarbons, 4 gave no detectable conductance: benzene (**I**), naphthalene (**II**), biphenyl (**III**), and triphenylene (**IX**). For the remaining seven, conductances were obtained, showing them to be weak 1 : 1 electrolytes. For pentacene and tetracene the Λ results in dilute solution show that each molecule gives one abnormally mobile chloride anion, on which assumption all the other hydrocarbons also give satisfactory 1 : 1 dissociation constants (corrected for interionic effects and the self-conductance of the solvent).

The values of these thermodynamic dissociation constants (as logarithms) are given in Table IX.

For the reasons given, they are attributed to

$$R + 2SbCl_3 \rightarrow R{\rightarrow}SbCl_2^+ + SbCl_4^- \tag{42}$$

measuring, therefore, the Lewis basicity of the hydrocarbons R against the common sextet cation and Lewis acid $SbCl_2^+$ in the common solvent $SbCl_3$.

TABLE IX

THERMODYNAMIC DISSOCIATION CONSTANTS[a]

	Pentacene (XI)	Naphth-acene (VI)	Anthr-acene (IV)	Perylene (XIII)	1,2-Benz-anthr-acene (VII)	Pyrene (X)	Phenan-threne (V)
$-\log K_{th}$	2.66	3.33	4.30	4.43	4.61	5.09	5.83
m_m	0.220	0.295	0.414	0.347	0.452	0.445	0.605
N_r	0.80	1.03	1.26	1.33	1.35	1.51	1.79
$-\log K_{calc}$	2.73	3.47	4.21	4.43	4.50	5.01	5.91

[a] From Johnson and Baughan.[17]

These results may be compared with the molecular orbital (MO) theory of reactivity of the hydrocarbons R and their Brønsted basicity, measured in anhydrous HF at 0°C.

In MO theory both electrons come from the highest occupied π level of the molecule. Simple Hückel (HMO) theory gives these energies as a common integral β multiplied by coefficients[82] m_m given in Table IX, line 2. A plot of $\log K_{th}$ against m_m gives a straight line except with perylene and pyrene. The later refinements of Dewar[83] give reactivity numbers N_r (line 3, Table IX), and $-\log K_{calc}$ (line 4) shows good agreement with the equation

$$-\log K_{th} = 0.15 + 3.22N_r \tag{43}$$

for all except benz[a]pyrene (**XII**), with probable error ± 0.08.

In the same way the basicity constants of 12 aromatic hydrocarbons[84] in HF at 0°C are given by

$$-\log K = -20.5 + 12.9N_r \tag{44}$$

though with rather greater scatter. Therefore the free energy changes in Lewis basicity and Brønsted basicity are mutually linear in one another with slope $(3.2 \times 348)/(12.9 \times 273) = 0.32$ (for interpretation, see Johnson and Baughan[17]).

With the information from cryoscopy, conductance, and ESR the colors of these solutions can now be interpreted. Johnson[22] has investigated the UV–visible absorption spectra of 16 polycyclic aromatic hydrocarbons in pure (colorless) $SbCl_3$. All 16 hydrocarbons showed at least one clearly marked absorption band not observed in inert solvents. One band (his Band 1) appeared alone in hydrocarbons whose ionization is not measurable by conductance [benzene (**I**), naphthalene (**II**), biphenyl (**III**), triphenylene

(IX)]. Together with Band 1, a new Band 2 was observed for hydrocarbons [phenanthrene **(V)**, chrysene **(VIII)**, pyrene **(X)**, anthracene **(IV)**, and benz[*a*]pyrene **(XII)**] of intermediate ionization. For the three strongly ionizing hydrocarbons[17] [perylene **(XIII)**, naphthacene **(VI)**, pentacene **(XI)**] Band 1 disappeared, leaving Band 2 alone; and with these three reactive hydrocarbons molecular oxygen produced the intense new Band 3. The results are summarized in Table X, which also shows similar colors observed elsewhere.

Band 1 may be attributed to a charge-transfer complex between the hydrocarbon and $SbCl_3$ itself; Band 2 to the Lewis complex $R \rightarrow SbCl_2^+$, and this color (Eq. 42) may be reduced by adding KCl; Band 3 to the free radical cation. Supporting evidence for all these assignments has been given.[22]

This work on aromatic hydrocarbons provides strong evidence for the Lewis acid–base theory from (a) the linearity between free-energy charges for

$$\begin{aligned} R + SbCl_2^+ &\rightarrow R \rightarrow SbCl_2^+ \\ R + H^+ &\rightarrow RH^+ \end{aligned} \qquad (45)$$

(b) the remarkable similarity between their color changes (Table X) which involve ground states and excited states, and (c) the correlation with refined MO theory.

The principal reaction of the pure oxygen-free solvent with organic solutes seems due to the strong Lewis acid $SbCl_2^+$. This may, perhaps, explain the curious results with alkyl chlorides and is certainly supported in later sections.

XI. Amines

Other types of proton bases also conduct in $SbCl_3$. Johnson and Baughan[24] investigated 12 amines (8 primary, 2 secondary, and 2 tertiary). These solutions conduct well, and four of the amines are "leveled" as strong 1:1 electrolyte chlorides; for two others the results are complicated by micelle-formation. For the remaining six equilibrium constants K_{th} can be obtained for the reaction

$$Am + 2SbCl_3 \Rightarrow Am \rightarrow SbCl_2^+ + SbCl_4^- \qquad (46)$$

where

$$K_{th} = f^2(Am \rightarrow SbCl_2^+)(SbCl_4^-)[C - (Am \rightarrow SbCl_2^+)] \qquad (47)$$

C being the stoichiometric concentration of amine; the effect of κ_0 is negligible.

TABLE X

ABSORPTION SPECTRA OF AROMATIC HYDROCARBONS IN LIQUID ANTIMONY TRICHLORIDE[a]

Hydrocarbon	Band attributed to charge transfer: Band 1	Band attributed to $R \rightarrow SbCl_2^+$ ion: Band 2	Bands due to free radical positive ion (oxygen addition): Band(s) 3	Oxidation potential R/R^+ (Texier[36]) (V)	Proton or Lewis acid complex	Free radical positive ions
Pentacene	—	484, 582	432	—	475, 568[b]	382,[b] (429)[c]
Naphthacene	—	468	400, 753	—	444	388, 746, 847
Perylene	—	439	554	0.24	402, 604	539
Benz[*a*]pyrene	416.5	490	Unstable	0.48	—	543
Anthracene	410	420	*d*	0.52	410	312, 735
Fluoranthene	383	Not formed	*d*	—	—	—
Benzo[*a*]anthracene	376	—	*d*	0.60	460, 532	—
Pyrene	380	457	*d*	—	464, (378 weak)	—
Chrysene	380	456	*d*	—	—	—
Phenanthrene	370	416	*d*	—	410, (510 weak)	—
trans-Stilbene	357	*d*	*d*	—	—	—
Naphthalene	360	*d*	*d*	—	390, (410 shoulder)	—
Triphenylene	348	*d*	*d*	—	—	—
Biphenyl	348	*d*	*d*	—	286	—
Benzene	345	*d*	*d*	—	370, 260	—
Fluorene	333	*d*	*d*	—	375	—

[a] At 80°C. Data given in nanometers. From Johnson.[22]
[b] Measurements by author in 98% H_2SO_4 solution.
[c] Measurement in Carr–Price reagent.
[d] Not formed.

TABLE XI

EQUILIBRIUM CONSTANTS FOR AMINES[a]

Amine	Class	$-\log K_{th}$ [$=$pK]	pK_a
Amylamine	P	S.e.	10.44
Cyclohexylamine	P	S.e.	10.6
p-Toluidine	P	1.80	5.1
Aniline	P	1.67	4.6
p-Nitroaniline	P	2.29	1.0
m-Nitroaniline	P	2.47	2.50
Diphenylamine	S	2.64	0.78
Indole	S	2.23	−2.4
Acridine	T	S.e.	5.6
Tribenzylamine	T	S.e.	9.2
Octylamine	P	*b*	—
Octadecylamine	P	*b*	—

[a] From Johnson and Baughan.[24] P, Primary; S, secondary; T, teriary; S.e., strong electrolyte.
[b] Micelle formation.

In Table XI these results are compared with the pK_a values of the same compounds in water at 25°C.

From these results we note that: (a) Tertiary amines ("leveled" as strong electrolytes) are stronger Lewis bases toward $SbCl_2^+$ (by at least one power of ten in K_{th}) than primary amines of equal pK_a in water; this behavior is also found with Ag^+ in water itself.[85] Moreover, tertiary amines in other solvents also show much stronger proton basicity than primary amines having the same pK_a in water. It is therefore their proton basicity in water that requires special explanation.[86] (b) Primary amines are strong electrolytes toward $SbCl_2^+$ except for the weakly basic aniline derivatives. For these (Table XI) a linear relationship holds with a low slope, which is about the same as that

$$-\log K_{th} = -0.20\text{p}K_a + 2.7 \tag{48}$$

for the hydrocarbons (Section X). Indeed, a common straight line may perhaps lie between both sets of points; at present the chief uncertainty lies in the relation between the hydrocarbon dissociation constants[84] in HF and their (inferred) pK_a values in water.

Such variations in basicity may be attributed to electron densities at two common centers of about the same radius: the most reactive carbon in the hydrocarbons and the basic nitrogen in the primary amines. A common relationship between basicity to $SbCl_2^+$ and to proton would not therefore be surprising. These slopes (which are, of course, substituent constants σ in Hammett's sense) are less than unity because the $SbCl_2^+$ is much larger than the proton and cannot therefore get so close to the basic center (C or N in this case). A similar slope (0.25) was also obtained with amines[85] and Ag^+ (which is also large); Ag^+ also may coordinate with one or two molecules of amine, and this may be analogous to the reactions of alkyl chlorides in $SbCl_3$ (Section VIII). Finally, Ag^+ forms complexes with aromatic hydrocarbons in water solution.[87] The analogy between these two Lewis acid cations is therefore wide-reaching.

It is therefore surprising that the corresponding slope (substituent constant) for complexes between bases and *molecular* metallic chlorides ($SnCl_4$, etc.) seems generally[50] to be between 0.7 and 1.0. Perhaps, as with acid–base catalysis,[88] a general curve is involved which (over short sections) is roughly linear, but more work is needed.

Antimony trichloride has shown for these two types of base a close parallelism between Lewis basicity and proton (Brønsted) basicity. Work on different types of base has begun. Thus, the series NPh_3, PPh_3, $AsPh_3$, $BiPh_3$ showed[25] a maximum basicity with phosphorus triphenyl which gives a strong 1:1 electrolyte chloride; the nitrogen and arsenic compounds are medium-strength electrolytes, and conductance could not be detected with the antimony or bismuth compounds. $SbCl_2^+$ seems therefore to be a rather "soft" acid. Aromatic hydrocarbons add to it because they are rather "soft" bases. They add to Ag^+ even in water, but to the "hard" proton only in very acid media.[52]

Triphenylamine shows no measurable proton basicity in water, but with strongly oxidizing acids gives the deep blue free-radical cation. Similarly, triphenylamine and even diphenylamine[24] give such colors in $SbCl_3$ with $SbCl_5$ and molecular oxygen, just like perylene (Section X). Once more, "aromatic" and "lone-pair" electrons show very similar reactions.

XII. Voltammetry in $SbCl_3$

The study of this solvent has recently been extended to voltammetry.[33–38] Such techniques require that the ions present should sometimes be known to establish the behavior of the electrode system. Potassium chloride was used for solutions of high chloride ion concentration or low pCl where

$$pCl \equiv -\log c(Cl^-) \quad (49)$$

This is well-established (Section V). Aluminum trichloride was used for solutions of low chloride ion concentration (high pCl) because of the reaction

$$Cl^- + AlCl_3 \rightarrow AlCl_4^- \quad (50)$$

If suitable electrode systems can be found, oxidation–reduction processes can be studied electrochemically. In this solvent the hydrogen electrode cannot be used; the solvent is reduced. Oxidation–reduction reactions are too slow at a gold electrode, while the less noble metals react to give metallic antimony. Electrodes of antimony or vitreous carbon appear satisfactory.

The purification of the solvent during this work was less rigorous than in that of Klemensiewicz and Baughan and their co-workers; but it was found that H_2O, HCl, and SbOCl conducted very little (see Section VII). However, HCl (a nonconductor) produced a strong electrolyte from pyridine as it does from perylene (Section IX). The nonconductor water can react with pyridine to cause conductance; the pyridine presumably disturbs the hydrolysis equilibrium by reacting with HCl. Conversely water lowers the conductance of $AlCl_3$ by reacting

$$AlCl_3 + H_2O \rightarrow AlOCl + 2HCl \tag{51}$$

TABLE XII

Half-Wave Potentials of Some Aromatic Hydrocarbons in Molten Antimony Trichloride[a]

Compound	$E_{1/2}$ (V)	Compound	$E_{1/2}$ (V)
Tetracene	0.21	Anthracene	0.51
Perylene	0.24	Pyrene	0.59
9,10-Dimethylanthracene	0.31	1,2-Benzanthracene	0.63
3-Methylcholanthrene	0.34	Acenaphthene	0.64
9,10-Diphenylanthracene	0.39	Coronene	0.64
3,4-Benzpyrene	0.44	1,2-Benzpyrene	0.66

[a] From Bauer *et al.*[36]

These reactions, like the conductance of dry HCl, have been proposed for titrating the water content; the HCl conductance seems the simplest (like the conductance of salt in acetone[2]), but still needs calibrating with HCl–H_2O mixtures.

The work on the electrochemical oxidation of polynuclear aromatic hydrocarbons gave the half-wave potentials shown in Table XII.[36] These results are similar to those obtained in acetonitrile,[89] but they also correlate thermodynamically with our earlier results that some less reactive hydrocarbons (whose Lewis basicity is measurable) do not, however, give free radicals with molecular oxygen.

The possibilities of this curious oxidation reaction in $SbCl_3$ can therefore be assessed from well-known half-wave potentials in acetonitrile.

These methods have also been applied[37] to eight aromatic amines at 100°C. Their electrochemical oxidation was shown to proceed through a one-electron step, but the free-radical ions underwent further reactions. This work also confirmed that some amines were strong 1:1 electrolyte chlorides, but it shows a few unresolved differences in conductance with earlier work.[24]

XIII. Other Applications of This Solvent

A. Spectroscopic Applications

Many organic compounds give peculiar visible (and UV) colors in this solvent. Johnson's work on aromatics[22] showed two distinct color reactions to which must be added colors due to Brønsted acids and oxidizing agents in less pure solvent. Thus, even the well-known Carr–Price color test for vitamin A requires small quantities ($<0.1\%$) of antimony pentachloride.[90]

The phenomena in the infrared are simpler.[91–93] The pure solvent is transparent from 1 to 10 μm, when overtone bands appear from $SbCl_3$ molecules. Water and hydrolysis products cause absorption at 2.75 and 6 μm. Many complex organic molecules dissolve well, and the spectra of pyrimidines, purines, amino acids, and antibiotics have been published.

On the whole, these spectra show small differences particularly in C–H, O–H, and N–H stretching frequencies from more conventional solvents (although the antibiotic terramycin shows an intense new band). Extensive ionization must occur with the nitrogen atoms; its small effect on infrared can be understood if the ionization is due to inert-pair electrons.

Antimony trichloride dissolves many polymers (including polyimides[94]) and infrared spectra may be obtained by allowing such mixtures to solidify into discs on a surface of salt.[95]

The Raman spectroscopy of the solid is well established.[96] Recently, using laser excitation, the liquid has been investigated[70] and solutions of KCl (at 50 and 33 mole %) and $AlCl_3$ (50 mole %) give however no clear evidence of $SbCl_4^-$ or $AlCl_4^-$ anions. The conductivity evidence for the ionization of Lewis bases only requires that such bases give the same abnormally fast anion as given by alkali chlorides which could possibly be free chloride; the conductivity evidence (Section IV) on $AlCl_3$ solutions seems, however, to require a different (and slower) anion. But it is very difficult (if not impossible) to obtain clear Raman evidence for the well-established OH_3^+ in water.[45]

B. As a Solvent for Polymers

A large number of polymers are soluble[94] in antimony trichloride (of unspecified purity), arsenic trichloride (Section XIV), and their mixtures; solubility of wool and nylon has also been reported.

Conductance data on such solutes in pure solvent are not yet available.

C. For Energy Conversion

Antimony pentachloride dissociates as the temperature rises and the

$$SbCl_5 \rightarrow SbCl_3 + Cl_2 \quad (52)$$

chlorine can generate an emf. This process could therefore, in principle, be used to convert low-temperature heat into electricity. This reaction and

$$TeCl_4 \rightarrow TeCl_2 + Cl_2 \quad (53)$$

seem the most feasible of such processes.[97]

XIV. Other Similar Solvents

The results with antimony trichloride have been discussed in detail because (a) the data on conductance and freezing point are the most accurate, (b) the effects of impurity are best understood, (c) the results with organic solutes are the most extensive, (d) much of the work has been done in the last twenty years and new techniques are being applied. Work on other "waterlike" inorganic acid halides, mainly with inorganic solutes, was reviewed a few years ago.[56] We have therefore considered a few such solvents briefly with particular reference to (a) their similarities and differences with $SbCl_3$, (b) organic solutes, and (c) recent work. The three solvents considered are arsenic trichloride, antimony tribromide, and selenium oxychloride.

A. Arsenic Trichloride

The melting point of $AsCl_3$ is $-18°C$ and the dielectric constant[58] about 12 at 20°C. It is therefore not as good an ionizing solvent as $SbCl_3$, but it is liquid at room temperature. Solubility relations are similar to those in $SbCl_3$, but the alkali chlorides are not soluble.[98] Nothing has been published on its cryoscopy. Gutmann studied by conductance[99] several reactions of chloride ion transfer and also showed (1) that $SbCl_5$ is a weak 1:1 electrolyte; (2) that NMe_4I is a strong 1:1 electrolyte obeying the D.H.O. equation; (3) that pyridine is a weak 1:1 electrolyte (Walden[100] had previously shown this for quinoline); this conductance Gutmann interpreted as in Section XI above. Transport number measurements showed[101] that the chloride ion is abnormally mobile.

These results are analogous to those in $SbCl_3$. Also, Andersson and Lindqvist[102] used silver–silver chloride electrodes for potentiometric titration and obtained an ionic product $K_i = 10^{-15}$ moles^2/liters^{-2}.

Extensive infrared and proton NMR studies have been made in this solvent.[103] Water reacts very slowly and hydrogen chloride is almost insoluble in dry solvent, but in water-containing solvent it gives a new spectrum, the same as when some aqueous HCl is added. Infrared and Raman lines have been attributed to $OH_3{}^+$. These results using different techniques are very similar to those in antimony trichloride. Infrared spectra have also been published on carboxylic acids and organic phosphorus compounds. The carbonyl vibration frequency of acetone appears at 1696 cm^{-1}, not found in its vapor or pure liquid.[103]

Mixtures of arsenic and antimony trichloride show a simple eutectic close to pure $AsCl_3$; vapor pressure composition curves are also known.[104] Such mixtures should have a higher dielectric constant than arsenic trichloride.

B. Antimony Tribromide

The melting point (97°C) is higher and the dielectric constant[58] ($D = 21$ at 100°C) is lower than those of antimony trichloride. No work has been done on organic solutes or infrared spectroscopy. Three papers by Jander and Weis[105–107] discuss solubility (similar to Table II; the bromides of K, NH_4, Rb, and Tl are soluble in $SbBr_3$) some inorganic adducts, cryoscopy, and conductance.

Cryoscopic work by Tolloczko[6] with dibenzyl, diphenylmethane, and benzophenone as solutes gave a cryoscopic constant k_f

$$k_f = 26.5 \pm 0.6\ °\text{K mole}^{-1}\text{ kg} \tag{54}$$

From their own data, Jander and Weis[105] recommend

$$k_f = 26.7\ °\text{K mole}^{-1}\text{ kg} \tag{55}$$

since confirmed by the microcalorimeter determination[66] of the heat of fusion.

Conductivity and cryoscopic data[105–107] (presented only as graphs) are in many ways like the more accurate data for $SbCl_3$. Thus, $N(CH_3)_4Br$, NH_4Br, and KBr give similar strong electrolyte type Λ–concentration curves with $\Lambda_e \simeq 67$. The application of Walden's rule to $N(CH_3)_4{}^+$ shows that Br^- is abnormally mobile (the viscosity is 6.8 cP at 100°, $3\frac{1}{2}$ times that of $SbCl_3$ at

99°); for TlBr the Λ value is slightly lower. For $AlBr_3$ and $GaBr_3$ the Λ value is practically independent of concentration, but only about 10. A strong electrolyte with a normal mobility would give this, but the *i* factor would be about 2, whereas Jander and Weis[105] report a value unity. This discrepancy needs clarification. $TeBr_4$ also is a weak electrolyte. These papers[105–107] contain conductance (and cryoscopic) data on some interesting inorganic antimony compounds (Sb_2O_3, Sb_2S_3, Sb_2Se_3, Sb_2Te_3, SbN, etc.). No systematic study of likely impurities has been made.

On the whole, the results in this solvent agree with those in (rather more convenient) antimony trichloride. $SbBr_3$ melts are much more easily oxidized (to give bromine), e.g., by $AgClO_4$, $AgNO_3$, or NEt_4NO_3.

C. Selenium Oxychloride

The interesting chemical reactions of this liquid ($SeOCl_2$) were investigated by Lenher and co-workers[108]; its properties as a solvent were reviewed by G. B. L. Smith.[109] The freezing point (11°C) should be convenient for cryoscopy, and the dielectric constant is high (about $D = 50$ at 20°C). This early work of great chemical interest does not permit quantitative evaluation of equilibria, nor have the effects of impurity been thoroughly studied. Smith showed the importance of the chloride ion exchange reactions; the solvent ionized thus

$$SeOCl_2 \rightleftharpoons SeOCl^+ + Cl^- \tag{56}$$

and pyridine ionized by coordinating this solvent cation (just as in $SbCl_3$ or $AsCl_3$).

Recently this solvent has been reinvestigated[110,111] by voltammetric and potentiometric methods. $NEt_4{}^+Cl^-$ is a strong electrolyte and so is pyridine (confirming Smith's results). For the few bases B studied (acetonitrile, tetrahydrofuran, dimethylformamide) the equilibrium correlates with Gutmann's[112] "donor number," i.e., the reactivity of $SbCl_5$ solute with these bases as pure liquid solvents.

XV. Summary

These solvents show acidity in several ways: as chloride ion acceptors, as Lewis acids in two distinct ways ($SbCl_3$, $SbCl_2{}^+$, for example), and as Brønsted acids if such impurities as HCl and water are present. They can also generate free-radical cations with molecular oxygen. Cryoscopy, conductance, and electron spin resonance studies on antimony trichloride solutions, with particular emphasis on the effects of impurities, have shown the main reactions occurring with organic solutes. The reactions of common inorganic solutes have been known for some years. Arsenic trichloride, antimony tribromide, and selenium oxychloride seem to be analogous. Interionic effects in dilute solutions in antimony trichloride are well explained by the Debye–Hückel theory; some results in more concentrated solutions are available for testing modern theoretical refinements.

This work has laid the foundations necessary for investigations by NMR and refined electrochemical techniques. Such work has now been started, and the complicated color phenomena are beginning to be explained.

References

1. M. Born, *Z. Phys.* **1**, 45 (1920).
2. A. Lannung, *Z. Phys. Chem.* **161**, 255 (1932).
3. J. H. Hildebrand, J. M. Prausnitz, and R. L. Scott, "Regular and Related Solutions" (particularly Appendix 5). Van Nostrand-Reinhold, Princeton, New Jersey, 1970.
4. G. Jander, "Die Chemie in Wasserahnlichen Losungsmitteln." Springer-Verlag, Berlin and New York, 1949.
5. S. Tolloczko, *Z. Phys. Chem.* **30**, 705 (1899).
6. S. Tolloczko, *Bull. Int. Acad. Sci. Cracovie* p. 1 (1901).
7. Z. Klemensiewicz, *Bull. Int. Acad. Sci. Cracovie* p. 485 (1908).
8. K. Frycz and S. Tolloczko, *Festschr. Univ., Lwow* **1**, 1 (1912).
9. Z. Klemensiewicz, *Z. Phys. Chem.* **113**, 28 (1924).
10. Z. Klemensiewicz and Z. Balowna, *Rocz. Chem.* **10**, 481 (1930).
11. Z. Klemensiewicz and Z. Balowna, *Rocz. Chem.* **11**, 683 (1931).
12. Z. Klemensiewicz and A. Zebrowska, *Rocz. Chem.* **14**, 14 (1934).
13. G. B. Porter and E. C. Baughan, *J. Chem. Soc., London* p. 744 (1958).
14. A. G. Davies and E. C. Baughan, *J. Chem. Soc., London* p. 1711 (1961).
15. E. C. Baughan, T. P. Jones, and L. G. Stoodley, *Proc. Chem. Soc., London* p. 274 (1963).
16. J. R. Atkinson, T. P. Jones, and E. C. Baughan, *J. Chem. Soc., London* p. 5000 (1964).
17. P. V. Johnson and E. C. Baughan, *J. Chem. Soc., A* p. 2686 (1969).
18. J. R. Atkinson, E. C. Baughan, and B. Dacre, *J. Chem. Soc., A* p. 1377 (1970).
19. G. B. Porter, J. Simpson, and E. C. Baughan, *J. Chem. Soc., A* p. 2806 (1970).
20. E. C. Baughan, *Annu. Rep. Chem. Soc.* **65**, 105 (1970).
21. E. C. Baughan, *J. Electroanal. Chem.* **29**, 81 (1971).
22. P. V. Johnson, *J. Chem. Soc., A* p. 2856 (1971).
23. M. Hiscock and G. B. Porter, *J. Chem. Soc., Perkin Trans. 2* p. 79 (1972).
24. P. V. Johnson and E. C. Baughan, *J. Chem. Soc., Perkin Trans. 2* p. 1958 (1973).
25. P. V. Johnson, *in* "Nonaqueous Electrolytes Handbook" (G. J. Janz and R. P. T. Tomkins, eds.), Vol. 2, p. 784. Academic Press, New York, 1973.
26. B. Dacre, unpublished work in these laboratories.
27. G. Jander and K.-H. Swart, *Z. Anorg. Allg. Chem.* **299**, 252 (1959).
28. G. Jander and K.-H. Swart, *Z. Anorg. Allg. Chem.* **301**, 54 (1959).
29. G. Jander and K.-H. Swart, *Z. Anorg. Allg. Chem.* **301**, 80 (1959).

30. P. Texier and J. Desbarres, *C. R. Acad. Sci., Ser. C* **266**, 503 (1968).
31. D. Bauer and P. Texier, *C. R. Acad. Sci., Ser. C* **266**, 602 (1968).
32. D. Bauer, J.-P. Beck, and P. Texier, *C. R. Acad. Sci., Ser. C* **266**, 1335 (1968).
33. P. Texier, *Bull. Soc. Chim. Fr.* p. 4716 (1968).
34. J. Desbarres and P. Texier, *Bull. Soc. Chim. Fr.* p. 5061 (1968).
35. D. Bauer, J.-P. Beck, and P. Texier, *C. R. Acad. Sci., Ser. A* **269**, 822 (1969).
36. D. Bauer, J.-P. Beck, and P. Texier, *Collect. Czech. Chem. Commun.* **36**, 940 (1971).
37. D. Bauer and J.-P. Beck, *Collect. Czech. Chem. Commun.* **36**, 323 (1971).
38. D. Bauer, C. Colin, and M. Caude, *Bull. Soc. Chim. Fr.* p. 942 (1973).
39. P. Walden, "Elektrochemie Nicht-wässriger Losungen." Teubner, Leipzig, 1924.
40. G. N. Lewis, "Valence and the Structure of Atoms and Molecules," p. 138. Chem. Catalog Co. (Tudor), New York, 1923.
41. G. N. Lewis, *J. Franklin Inst.* **226**, 293 (1938).
42. G. A. Olah, "Friedel-Crafts and Related Reactions," Vol. 1. Wiley (Interscience), New York, 1963
43. F. Fairbrother, *J. Chem. Soc., London* p. 293 (1941).
44. For review, see P. A. H. Wyatt, *Annu. Rep. Chem. Soc.* **66**, 94 (1969).
45. Cf. R. P. Bell, "The Proton in Chemistry," 2nd ed., Chapter 2. Chapman & Hall, London, 1973.
46. C. MacLean, J. H. van der Waals, and E. L. Mackor, *Mol. Phys.* **1**, 247 (1958).
47. R. Hulme and J. C. Scrutton, *J. Chem. Soc., A* p. 248 (1968).
48. R. Hulme and J. C. Scrutton, *Acta Crystallogr., Sect. A* **25**, 171 (1969).
49. R. Hulme and J. T. Szymanski, *Acta Crystallogr., Sect. A* **25**, 753 (1969).
50. D. P. N. Satchell and R. S. Satchell, *Chem. Rev.* **69**, 251 (1969).
51. J. N. Bronsted, A. Delbanco, and A. Tovborg-Jensen, *Z. Phys. Chem., Abt. A* **169**, 361 (1934).
52. For review, see R. G. Pearson, *J. Chem. Educ.* **45**, 581 and 643 (1968).
53. H.-H. Perkampus, "Wechselwirkung von π-Elektronenensystemen mit Metallhalogeniden." Springer-Verlag, Berlin and New York, 1973.
54. O. Hassel and H. Hope, *Acta Chem. Scand.* **15**, 407 (1961).
55. W. I. Aalbersberg, S. J. Hoijtink, E. L. Mackor, and W. P. Weijland, *J. Chem. Soc., London* pp. 3049 and 3055 (1959).
56. D. S. Payne, *in* "Non-aqueous Solvent Systems" (T. C. Waddington, ed.), Chapter 8. Academic Press, New York, 1965.
57. H. Schlundt, *J. Phys. Chem.* **5**, 503 (1901).
58. L. Patritzka and R. Bertram, *J. Electroanal. Chem.* **28**, 119 (1970).
59. H. Oppermann, *Z. Anorg. Allg. Chem.* **356**, 1 (1967).
60. K. Maeda, *Sci. Pap. Inst. Phys. Chem. Res., Tokyo* **67**, 143 (1973).
61. D. Zhuravlev, *J. Phys. Chem., USSR* **13**, 684 (1939).
62. K. Takeyama and T. Atoda, *Bull. Chem. Soc. Jap.* **45**, 3078 (1972).
63. L. Kolditz, *Advan. Inorg. Chem. Radiochem.* **7**, 1 (1965).
64. L. Kolditz, *in* "Halogen Chemistry" (V. Gutmann, ed.), Vol. 2, Chapter 8. Academic Press, New York, 1967.
65. J. E. Martin, unpublished work at the National Physical Laboratory, Teddington, England.
66. E. Chemouni, M.-H. Maglione, and A. Potier, *Bull. Soc. Chim. Fr.* p. 489 (1970).
67. R. M. Fuoss and L. Onsager, *J. Phys. Chem.* **62**, 1339 (1958).
68. R. M. Fuoss and C. Accascina, "Electrolytic Conductance," Chapter XV. Wiley (Interscience), New York, 1959.
69. E. M. Hanna, A. D. Pethybridge, and J. E. Prue, *Electrochim. Acta* **16**, 677 (1970).

70. K. W. Fung, G. M. Begun, and G. Mamantov, *Inorg. Chem.* **12**, 53 (1973).
71. J. C. Coutrurier, *Rev. Chim. Miner.* **7**, 565 (1970).
72. E. C. Baughan, M. G. Evans, and M. Polanyi, *Trans. Faraday Soc.* **37**, 377 (1941).
73. N. A. Bonner and W. Goishi, *J. Amer. Chem. Soc.* **79**, 3020 (1961).
74. G. A. Olah, *Advan. Phys. Org. Chem.* **4**, 305 (1964).
75. G. A. Olah, numerous later papers in *J. Amer. Chem. Soc.*
76. J. M. Brierley and G. B. Porter, unpublished work in these laboratories.
77. M. R. Symons, *Adv. Phys. Org. Chem.* **1**, 284 (1963).
78. F. Gerson and J. Heinzer, *Helv. Chim. Acta* **49**, 7 (1966); **50**, 1852 (1967).
79. C. Eischenbroich and F. Gerson, *Helv. Chim. Acta* **53**, 838 (1970).
80. H. Kainer and K. H. Hausser, *Chem. Ber.* **86**, 1563 (1953).
81. E. Clar, "Polycyclic Hydrocarbons," Vol. 1, p. 243. Academic Press, New York, 1964.
82. A. Streitwieser, Jr. "Molecular Orbital Theory for Organic Chemists," p. 178. Wiley, New York, 1961.
83. M. J. S. Dewar, *J. Amer. Chem. Soc.* **74**, 3341–3357 (1952).
84. E. L. Mackor, A. Hofstra, and J. H. van der Waals, *Trans. Faraday Soc.* **54**, 66 (1958).
85. R. J. Bruehlman and F. H. Verhoek, *J. Amer. Chem. Soc.* **70**, 1401 (1948).
86. A. F. Trotman-Dickenson, *J. Chem. Soc., London* p. 1293 (1949).
87. L. J. Andrews, *Chem. Rev.* **54**, 713 (1954).
88. M. Eigen, *Angew. Chem., Int. Ed. Engl.* **3**, 1 (1964).
89. E. S. Pysh and N. C. Yang, *J. Amer. Chem. Soc.* **85**, 2124 (1963).
90. J Bruggemann, W. Krauss, and J. Tiews, *Chem. Ber.* **85**, 315 (1952).
91. J. R. Lacher, V. D. Croy, A. Kianpour, and J. D. Park, *J. Amer. Chem. Soc.* **58**, 206 (1954).
92. J. R. Lacher, J. L. Bitner, D. J. Emery, M. E. Seffl, and J. D. Park, *J. Amer. Chem. Soc.* **59**, 615 (1955).
93. J. R. Lacher, J. L. Bitner, and J. D. Park, *J. Amer. Chem. Soc.* **59**, 610 (1955).
94. H. A. Szymanski, W. Collins, and A. Bluemle, *J. Polym. Sci., Part B* **3**, 81 (1965).
95. H. A. Szymanski, K. Broda, J. May, W. Collins, and D. Bakalik, *Anal. Chem.* **37**, 617 (1965).
96. E. Denchik, S. C. Nyburg, G. A. Ozin, and J. T. Szymanski, *J. Chem. Soc., A* p. 3157 (1971).
97. C. R. McCully, T. M. Rymarz, and S. B. Nicholson, *Advan. Chem. Ser.* **64**, 198 (1967).
98. V. Gutmann, *Z. Anorg. Allg. Chem.* **266**, 332 (1951).
99. V. Gutmann, *Monatsh. Chem.* **85**, 491 (1954).
100. P. Walden, *Z. Phys. Chem.* **43**, 385 (1903).
101. V. Gutmann, *Sv. Kem. Tidsk.* **68**, 1 (1956).
102. L. H. Andersson and I. Lindqvist, *Acta Chem. Scand.* **9**, 79 (1955).
103. H. A. Szymanski, R. Ripley, R. Fiel, W. Kinlin, L. Zwolinski, H. Drew, D. Bakalik, J. Muller, A. Bluemle, and W. Collins "Progress in Infra-Red Spectroscopy," Vol. 2. Plenum, New York, 1964.
104. L. A. Nisel'son and V. V. Mogucheva, *Russ. J. Inorg. Chem.* **11**, 77 (1966).
105. G. Jander and J. Weis, *Z. Elektrochem.* **61**, 1275 (1957).
106. G. Jander and J. Weis, *Z. Elektrochem.* **62**, 850 (1958).
107. G. Jander and J. Weis, *Z. Elektrochem.* **63**, 1037 (1959).
108. V. Lenher *et al. J. Amer. Chem. Soc.* **42**, 2498 (1920); **43**, 29 (1921); **44**, 1664 (1922); **45**, 2090 (1923); **47**, 1842 (1925).
109. G. B. L. Smith, *Chem. Rev.* **38**, 165 (1938).
110. J. Devynck and B. Tremillon, *J. Electroanal. Chem.* **23**, 241 (1969).
111. J. Devynck and B. Tremillon, *J. Electroanal. Chem.* **30**, 443 (1971).
112. V. Gutmann, *Angew. Chem., Int. Ed. Engl.* **9**, 843 (1970).

6

Cyclic Carbonates

W. H. Lee

Department of Chemistry, University of Surrey
Guildford, Surrey, England

I. Introduction

The cyclic carbonates considered in this review are listed in Table I; the systematic nomenclature is that used in the *Chemical Abstracts Index Guide* (1972).[1]

TABLE I

NOMENCLATURE OF CYCLIC CARBONATES

Formula	Nomenclature	
	Common	Systematic
$C_3H_2O_3$	Vinylene carbonate	1,3-Dioxol-2-one
$C_3H_4O_3$	Ethylene carbonate	1,3-Dioxolan-2-one
$C_4H_6O_3$	Propylene carbonate	4-Methyl-1,3-dioxolan-2-one
$C_3H_3ClO_3$	Chloroethylene carbonate	4-Chlorodioxolan-2-one
$C_4H_5ClO_3$	Chloromethyl ethylene carbonate	4-Chloromethyl-1,3-dioxolan-2-one
$C_5H_8O_3$	Butylene carbonate	4,4-Dimethyl-1,3-dioxolan-2-one
$C_7H_4O_3$	*o*-Phenylene carbonate, catechol carbonate	1,3-Benzodioxol-2-one

In this chapter the cyclic carbonates are considered mainly as chemical solvents and electrochemical media. The extensive literature concerning their polymerization is not discussed here.

The physical properties of these compounds are summarized in Table II, and their electrical properties and viscosities are given in Table III.

TABLE II

PHYSICAL PROPERTIES OF CYCLIC CARBONATES

Formula	Molecular weight	mp (°C)	bp (°C) (mm Hg)	Density (g ml^{-1}) (°C)	Refractive index (n_D) (°C)
$C_3H_2O_3$	86.05	22[a, b]	63–65 (18)[c]	1.3541 (25)[a, f]	1.4212 (20)[g]
			73–74 (32)[a, d, e]		1.4190 (25)[g]
			162 (735)[a]		1.4183 (26)[e]
$C_3H_4O_3$	88.07	36.2[h]	120 (17)[j]	1.3208 (40)[h]	1.4199 (40)[h, k]
		36.4[i]	152.5 (50)[i]	1.3155 (45)[l]	
			238 (760)*[, h]	1.3093 (50)[l]	
$C_4H_6O_3$	102.1	−49.2[m]	58–63 (0.1)[n]	1.203 (20)[m]	1.4209 (20)[m]
			92 (4.5)[n]	1.198 (25)[i]	1.4193 (25)[o]
			126 (17)[j]	1.2065 (25)[h]	
			241.7 (760)[m]	1.183 (40)[i]	
$C_3H_3ClO_3$	122.5	—	86–88 (4)[p]	1.505 (20)[s]	1.455 (20)[s]
			106–107 (11)[f, q]	1.508 (25)[f]	

TABLE II (*continued*)

Formula	Molecular weight	mp (°C)	bp (°C) (mm Hg)	Density (g ml^{-1}) (°C)	Refractive index (n_D) (°C)
			212 (760)[f,r]		
$C_4H_5ClO_3$	136.5	—	252 (760)[h]	1.4403 (25)[h]	1.4680 (25)[h]
$C_5H_8O_3$	116.1	21.2–21.5[t]	83 (0.6)t	1.1408 (15)[x]	1.4241 (15)[x]
			104.8 (10)[u]	1.136 (22.5)[v]	1.4226 (26)[t]
			240 (740)[v]	1.129 (26)[t]	
			243 (760)[w]		
$C_7H_4O_3$	136.1	119–120[y,z]	—	—	—

* Asterisk indicates with partial decomposition.

[a] M. S. Newman and R. W. Addor, *J. Amer. Chem. Soc.* **77**, 3789 (1955).

[b] W. H. Lee and A. H. Saadi, *J. Chem. Soc.*, *B* p. 5 (1966).

[c] W. K. Johnson and T. L. Patton, *J. Org. Chem.* **25**, 1042 (1960).

[d] H. C. Haas and N. W. Schuler, *J. Polym. Sci.* **31**, 237 (1958).

[e] J. M. Judge and C. C. Price, *J. Polym. Sci.* **41**, 435 (1959).

[f] M. S. Newman and R. W. Addor, *J. Amer. Chem. Soc.* **75**, 1263 (1953).

[g] H. Kwart and W. G. Vosburgh, *J. Amer. Chem. Soc.* **76**, 5400 (1954).

[h] R. F. Kempa and W. H. Lee, *J. Chem. Soc., London* p. 1936 (1958).

[i] W. S. Harris, Thesis, Radiation Lab., Univeristy of California, 1958.

[j] M. Watanabe and R. M. Fuoss, *J. Amer. Chem. Soc.* **78**, 527 (1956).

[k] R. F. Kempa and W. H. Lee, *J. Chem. Soc., London* p. 100 (1961).

[l] H. Mujamoto and Y. Watanabe, *J. Chem. Soc. Jap.* **88**, 36 (1967).

[m] Technical Bulletin, Jefferson Chemical Co., Houston, Texas, 1971.

[n] P. L. Kronick and R. M. Fuoss, *J. Amer. Chem. Soc.* **77**, 6114 (1955).

[o] L. Simeral and R. L. Amey, *J. Phys. Chem.* **74**, 1443 (1970).

[p] J. Paasivirta and S. Kleemola, *Suom. Kemistilehti B* **43**, 285 (1970).

[q] C. R. Konwarski and S. Sarel, *J. Org. Chem.* **38**, 117 (1973).

[r] R. F. Kempa and W. H. Lee, *J. Chem. Soc., London* p. 1937 (1958).

[s] "Catalog of Organic Chemicals 9." Fluka AG, CH 9470, Buchs, Switzerland, 1973.

[t] J. J. Kolfenbach, E. I. Fulmer, and L. A. Underkofler, *J. Amer. Chem. Soc.* **67**, 502 (1945).

[u] P. Chabrier, H. Najer, and R. Guidicelli, *C. R. Acad. Sci.* **238**, 108 (1954).

[v] H. K. Garner and H. J. Lucas, *J. Amer. Chem. Soc.* **72**, 5497 (1950).

[w] E. E. Walker, *J. Appl. Chem.* **2**, 470 (1952).

[x] S. Sarel, L. A. Pohoryles, and R. Ben-Shoshan, *J. Org. Chem.* **24**, 1875 (1959).

[y] D. C. De Jongh and D. A. Brent, *J. Org. Chem.* **35**, 4204 (1970).

[z] A. T. Balaban, *Rev. Roum. Chim.* **14**, 1323 (1969).

The following reviews are relevant to aspects of this study: Harris[2] (electrochemical studies in EC and PC*), Jasinski[3,4] (bibliography of reports

* EC and PC will be used as abbreviations for ethylene carbonate and propylene carbonate, respectively.

on the uses of PC in high-energy density batteries; electrochemistry and applications of PC), McComsey[5] (polarography in PC), Gutmann[6, 6a] (solubilities in EC, PC, and chloro derivatives), and Lee[7] (solubility considerations, application to PC). Technical literature is also available from some suppliers.[8]

TABLE III

ELECTRICAL PROPERTIES AND VISCOSITIES OF CYCLIC CARBONATES

	ε_s	μ (Debye)	g^d	κ (ohm^{-1} cm^{-1})	η (poise) (°C)
$C_3H_2O_3$	$126 \pm 1.0^{a,b}$	$4.45 \pm 0.01^{b,c}$ 4.51 ± 0.05^e 4.57 ± 0.05^f	$1.81^{b,}$	1.7×10^{-7b}	—
$C_3H_4O_3$	$89.6^{a,g}$ 89.1^j 89.78^h	$4.78^{c,g}$ 4.93^h	1.27^g	$0.4\text{–}0.9 \times 10^{-7}$ (40°C)^g	0.0185 (40)^i 0.01955 (40)^j 0.01478 (40)^k
$C_4H_6O_3$	$61.7^{a,g}$	$4.94^{c,g}$	1.08^g 1.04^i	2×10^{-8l} $2 \times 10^{-7j,m}$	0.02530 (25)^j 0.01916 (40)^j
$C_3H_3ClO_3$	$62.0^{a,g}$	$3.99^{c,g}$	1.45^g	—	—
$C_4H_5ClO_3$	97.5^g	4.68^g	—	—	—
$C_5H_8O_3$	—	5.27^n	—	—	—
$C_7H_4O_3$	—	4.14^o	—	—	—

[a] At 1 MHz.
[b] W. H. Lee and A. H. Saadi, *J. Chem. Soc., B* p. 5 (1966).
[c] Extrapolation of data in benzene solution by the method of Guggenheim. *Trans. Faraday Soc.* **45**, 714 (1949).
[d] Kirkwood g factor. J. G. Kirkwood, *J. Chem. Phys.* **7**, 911 (1939); F. E. Harris and B. J. Alder, *ibid.* **21**, 1031 (1953).
[e] From vapor phase microwave spectra. G. R. Slayton, J. W. Simmons, and J. H. Goldstein, *J. Chem. Phys.* **22**, 1678 (1954).
[f] K. L. Dorris, C. O. Britt, and J. E. Boggs, *J. Chem. Phys.* **44**, 1352 (1966).
[g] R. F. Kempa and W. H. Lee, *J. Chem. Soc., London* p. 1936 (1958).
[h] Benzene solution data. R. J. W. Le Févre, A. Sundaram, and R. K. Pierens, *J. Chem. Soc., London* p. 479 (1963).
[i] R. F. Kempa and W. H. Lee, *J. Chem. Soc., London* p. 100 (1961).
[j] W. S. Harris, Radiation Lab., University of California, 1958.
[k] Static viscosity in Stokes.
[l] L. M. Mukherjee and D. P. Boden, *J. Phys. Chem.* **73**, 3965 (1969).
[m] G. Pistoia, M. De Rossi, and B. Scrosati, *J. Electrochem. Soc.* **117**, 500 (1970).
[n] G. F. Longster and E. E. Walker, *Trans. Faraday Soc.* **49**, 228 (1953).
[o] A. H. Saadi, Ph.D. Thesis, University of London (1964).

II. PREPARATION: GENERAL METHODS

Ethylene and propylene carbonates are readily available commercially.[8–12]

General methods for the preparation of cyclic carbonates have been reviewed by Idris Jones,[13] and the factors affecting the formation of homologs have been investigated.[14,15] The synthesis of these compounds from epoxides and CO_2, with Ni(0) complexes as catalysts, has been reported recently.[16] The pyrolysis of 2-bromoethyl carbonate at 195°–205°C yields EC and ethyl bromide. Starting from 1,3-dibromo-2-propyl ethyl carbonate or 2,3-dibromopropyl ethyl carbonate, equilibrium is established, and 3-bromopropylene carbonate is finally obtained in 88% yield.[17]

The preparation and purification of vinylene carbonate (VC) have been investigated extensively.[18–22] The method involves monochlorination of EC and subsequent removal of a molecule of HCl, usually by treatment with a tertiary amine,[21] but sometimes by pyrolysis.[23] It is essential that all traces of HCl be removed if polymerization, sometimes violent, is to be avoided.[24] Vinylene carbonate and its precursor, monochloroethylene carbonate, are listed by several manufacturers.[25–27]

Butylene carbonate has been prepared by ester exchange,[28] and from the reaction of the glycol with phosgene[29] or *N*-dichloromethylenebenzamide.[30]

o-Phenylene (catechol) carbonate is prepared by the reaction of catechol with a solution of phosgene in toluene.[31]

III. Vinylene Carbonate (VC)

The vapor pressure of VC has been measured at several temperatures by manometry and by the Knudsen effusion technique.[32] The results and derived enthalpies are recorded in Table IV.

TABLE IV

Vapor Pressures and Enthalpies of Vinylene Carbonate[a]

Temperature (°K)	Pressure (mm Hg)	Temperature (°K)	Pressure (mm Hg)
308.2	5.12	338.2	20.84
318.7	8.42	348.2	33.06
328.0	14.07	399.8	73.31
ΔH_c^0 [b]	ΔH_f^0 (l)[b]	ΔH_f^0 (g)[b]	ΔH_{vap}[b]
−240.56	−109.91	−100.55	9.86

[a] J. K. Choi and M. J. Joncich, *J. Chem. Eng. Data* **16**, 87 (1971).
[b] ΔH values in kcal $mole^{-1}$.

The bulk magnetic susceptibility of VC from the microwave Zeeman effect is -42.8 ± 3.0.[33] The microwave spectrum has been reported.[34,35] The molecular dimensions and moments of inertia are given below.

$$r_{32} = 1.385 \pm 0.012 \text{ Å} \qquad C_3\hat{O}_2C_1 = 106°56' \pm 24'$$
$$r_{33} = 1.331 \pm 0.003 \text{ Å} \qquad O_2\hat{C}_1O_2 = 108°48' \pm 40'$$
$$r_{11} = 1.1908 \pm 0.0006 \text{ Å} \qquad C_3\hat{C}_2O_2 = 108°40' \pm 9'$$

$$I_A = 54.08 \times 10^{-40} \text{ g cm}^2$$
$$I_B = 120.67 \times 10^{-40} \text{ g cm}^2$$
$$I_C = 174.77 \times 10^{-40} \text{ g cm}^2$$

The molecule is planar with C_{2v} symmetry.[34] A ring-puckering vibration (A_2) at 250 cm^{-1} agrees with the corresponding infrared absorption at 258 cm^{-1}.[34] Infrared and Raman spectra have been reported.[36–38] For the C=O stretch,[36] $\nu_{C=O} = 1833$ cm^{-1} (for solution in CCl_4), 1830 cm^{-1} (liquid) and 1822 cm^{-1} (solid film). Dorris *et al.*[37] report a high stretching frequency in the vapor at 1867 cm^{-1}. The C=O bond length of 1.19 Å (cf. 1.15 Å in EC[39]) is shorter than in aldehydes and ketones (1.21 Å) or carboxylic acids (1.23 Å).

Ionization potentials of the carbonyl lone-pair orbitals have been measured for VC and EC and compared with CNDO/2 and INDO calculations[40] as tabulated below.

	Ionization potential (eV)			Charge on oxygen	
Compound	Obsd.	CNDO/2	INDO	CNDO/2	INDO
VC	11.91	10.24	9.35	0.37	0.43
EC	11.47	10.53	9.69	0.32	0.38

Thermodynamic data, calculated for the most part from IR measurements, are listed in Table V. In water, VC slowly deposits the polymer, forming an opalescent solution; the pH changes from 5.65 to 3.10 in about 24 hr.[41] The derived activation parameters[42] are $\Delta H^{\ddagger} = 6.89 \pm 0.02$ kcal $mole^{-1}$ and $\Delta S^{\ddagger} = -46 \pm 1.0$ cal $mole^{-1}\,°K^{-1}$.

Thermal dissociation of the molecule yields glyoxal (**II**) and orbital symmetry control of the reaction has been considered.[43]

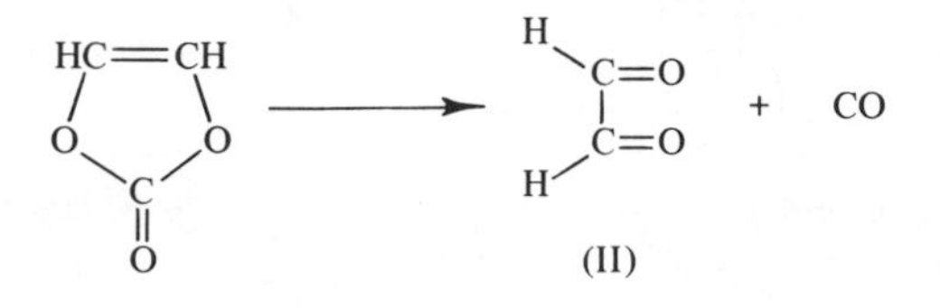

TABLE V

THERMODYNAMIC PROPERTIES OF VINYLENE CARBONATE[a,b]

Temperature (°K)	S (cal deg mole^{-1})	$-(G-H_0)/T$ (kcal deg mole^{-1})	$(H-H_0)/T$ (kcal deg mole^{-1})	C_v (cal deg mole^{-1})
200	60.60	51.26	9.34	10.58
273.15	65.09	54.37	10.73	14.51
298.15	66.60	55.33	11.27	15.85
300	66.71	55.40	11.31	15.95
400	72.55	58.96	13.59	20.74
500	78.04	62.23	15.81	24.48
600	83.13	65.30	17.83	27.33
700	87.82	68.19	19.63	29.53
800	92.15	70.91	21.23	31.27
900	96.15	73.50	22.65	32.68
1000	99.87	75.95	23.91	33.83
1500	115.15	86.60	28.55	37.36
2000	126.72	95.24	31.48	39.02
2500	135.98	102.50	33.48	39.90
3000	143.66	108.73	34.93	40.43

[a] J. R. Durig, J. W. Clark, and J. M. Casper, *J. Mol. Struct.* **5**, 67 (1970).
[b] H_0 is the zero-point enthalpy.

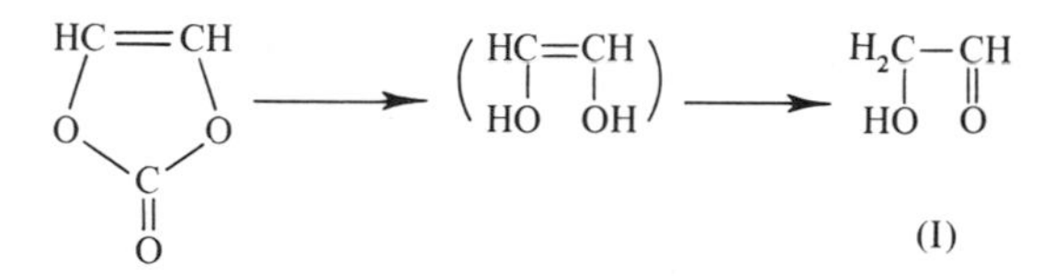

The alkaline hydrolysis of VC in 10% (by volume) ethanol–water produces glycolaldehyde (**I**). The reaction is first-order in ester concentration with velocity constants[42] as shown below.

Temperature (°K)	288	293	298
k (sec^{-1})	4.5×10^{-2}	5.5×10^{-2}	6.7×10^{-2}

The photosensitized addition of VC to olefins[44] and uracil[45] is reported. Hydrolysis of the cyclic monoolefin derivatives readily yields cyclobutane-*cis*-1,2-diols.[46] Vinylene carbonate also undergoes photosensitized addition to cyclohexene.[47] Scharf[48] has reviewed the use of mono- and dihalogen VC in synthesis.

IV. Ethylene Carbonate (EC)

The purification of EC is described by Harris[2] and Pistoia.[49]

The dielectric constant ε and the Kirkwood g factor vary with temperature, as shown in Table VI. Here ε was measured in the frequency range 1--9000 MHz.

TABLE VI

Variation of Dielectric Constant and g Factor of Ethylene Carbonate with Temperature[a, b]

	Temperature (°C)				
	25	40	50	60	70
ε_0	5.4	89.78	85.81	82.01	78.51
g	—	1.20	1.20	1.19	1.18
ε_∞	—	2.62	—	—	—

[a] R. Payne and I. E. Theodorou, *J. Phys. Chem.* **76**, 2892 (1972).

[b] The experimental results and the formula $\varepsilon_t = 85.1 - 0.408(t - 50°\text{C})$, measured at 10–20 kHz, of R. P. Seward and E. C. Vierira, *J. Phys. Chem.* **62**, 127 (1958) give ε values about 1 unit lower than those in Table VI.

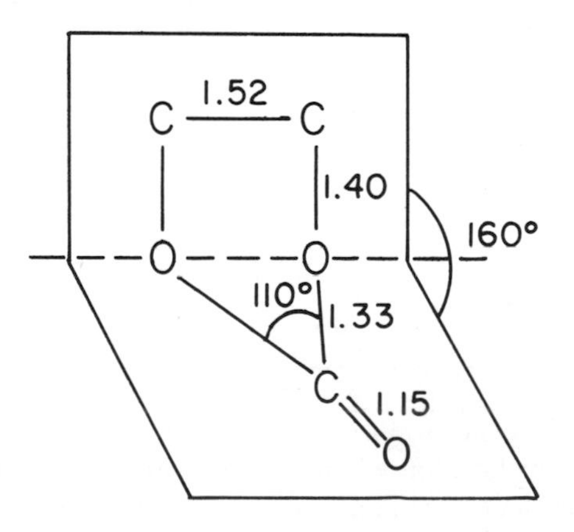

Fig. 1. Crystal structure of EC giving molecular dimensions.

According to Payne and Theodorou, the equilibrium ε values are consistent with the absence of specific intermolecular forces.[50]

The molar Kerr constant has the value[51] $10^{12}{}_mK_2 = 350.5$, and at $t = 45°C$, $\lambda = 5460$ Å, B(EC)/B(nitrobenzene) $= 0.55 \pm 0.03$.[52]

Crystal structure determination shows that the molecule is nonplanar in the solid phase (Fig. 1).[39]

From the absence of IR bands allowed by C_2 symmetry, but not by C_{2v}, it was inferred that the molecule becomes planar in the melt or when dissolved or vaporized.[53, 54] From the microwave spectrum, however, Wang *et al.*[55] conclude that the molecule is nonplanar in the vapor phase, so that the "forbidden" bands are below the observable limit. The moments of inertia are: $I_a = 62.8$, $I_b = 131.8$, and $I_c = 185.8$ amu Å^2.

From the Raman spectrum of liquid EC, Durig[56] assigned the 717 cm^{-1} band to an A1* skeletal bending vibration, similar to the 735 cm^{-1} band in VC.[57] Six such modes were assigned:[56]

A1 717	B1 529	B2 217 cm^{-1}
B1 700	B2 529	A2 (217) cm^{-1} (tentative)

The IR spectrum in the range 1800–500 cm^{-1} and the ultrasonic absorption spectrum for 50% EC in acetone (w/v) have been studied.[58, 58a] A comprehensive study of IR and Raman spectra for solid, liquid, and vapor phases makes assignment of the 24 fundamental frequencies.[59]

The fundamental C=O stretching frequencies of liquid EC and PC are higher in the IR than in the Raman spectrum. For example, for EC the

	ν_2^0 (cm^{-1})	
Temperature (°K)	Raman	IR
313	1784	1797
393	1790	1799

frequencies are shown in the tabulation. In both cases, this is attributed to some degree of alignment of molecular dipoles.[60]

The NMR spectrum shows a single sharp peak at $\tau 5.80$ due to the equivalent methylene groups.[58, 58a] The absolute signs for dipolar coupling are $+ve$ for geminal and cis-vicinal couplings and $-ve$ for trans-vicinal couplings.[59]

* The symmetries assume C_{2v} for the molecule.

TABLE VII

VAPOR PRESSURES AND ENTHALPIES OF ETHYLENE CARBONATE[a]

Temperature (°K)	Pressure (mm Hg)
273.9	2.15×10^{-3}
286.1	7.39×10^{-3}
296.6	21.72×10^{-3}

ΔH_c^{0} [b]	ΔH_f^{0} (s)[b]	ΔH_f^{0} (g)[b]	ΔH_{sub}[b]
−255.58	−163.20	−144.43	18.77

[a] J. K. Choi and M. J. Joncich, *J. Chem. Eng. Data* **16**, 87 (1971).
[b] ΔH values given in kcal $mole^{-1}$.

TABLE VIII

THERMODYNAMIC PROPERTIES OF ETHYLENE CARBONATE[a]

Temperature (°K)	C_p	S_T	$(H_T - H_0)/T$	$-(G_T - H_0)/T$
Solid				
50	31.94	18.39	12.50	5.89
100	54.18	48.40	28.54	19.86
150	66.80	72.82	39.24	33.58
200	80.57	93.85	47.79	46.06
250	97.92	113.63	56.03	57.60
298.15	117.14	132.54	64.36	68.18
309.49	122.06	137.01	66.39	70.62
Liquid				
309.49	138.07	179.91	109.29	70.62
350	144.49	197.34	112.84	84.50

[a] Data in kJ $mole^{-1}$ $°K^{-1}$. I. A. Vasil'ev and A. D. Korkhov, *Tr. Khim. Khim. Tekhnol.* Pt. I, p. 103 (1974).

The mass spectrum (at 70 eV) shows the following major peaks,[61] as tabulated below.

Group	Peak	Relative abundance (%)
M^+	88	14.3
$M-CH_2O$	58	2.0
$M-CO_2$	44	7.2
$M-HCO_2$	43	19.5
CHO (rearrangement)	29	39.8

Vapor pressures and enthalpies of EC are given in Table VII and free energy data in Table VIII.

The solubilities of inorganic compounds in EC are shown in Table IX. In both EC and PC, traces of water have a very marked effect on these solubilities.

TABLE IX

SOLUBILITIES (g/100 g SOLVENT) OF INORGANIC COMPOUNDS IN ETHYLENE CARBONATE

	Temperature (°C)				
Compound	25	30	35	40	Ref.
LiCl	—	—	—	0.83	*a*
$LiClO_4$	26.1	26.3	26.6	26.9	*a*
NaBr	—	—	—	0.31	*b*
NaI	—	—	—	37.6	*b*
KI	—	—	—	11.16	*b*
KBF_4	—	—	—	0.04	*a*
KCNS	27.1	27.4	27.7	28.0	*a*
KPF_6	17.0	17.6	18.3	18.8	*a*
$(CH_3)_4NBr$	—	—	—	0.09	*a*
$CaCl_2$	—	—	—	0.20	*b*
$CoCl_2$	3.22	—	—	—	*c*
$ZnCl_2$	—	—	—	33	*b*
$HgCl_2$	22.37	—	—	—	*c*
$HgCl_2$	—	—	—	49	*b*
$FeCl_3$	40.1	—	—	—	*c*

[a] G. Pistoia, M. De Rossi, and B. Scrosati, *J. Electrochem. Soc.* **117**, 500 (1970).
[b] W. S. Harris, Radiation Lab., University of California (1958).
[c] R. F. Kempa and W. H. Lee, *Z. Anorg. Allg. Chem.* **311**, 140 (1961).

TABLE X

OSMOTIC (φ) AND ACTIVITY (γ) COEFFICIENTS OF NONELECTROLYTES IN ETHYLENE CARBONATE[a]

Molality	Carbon tetrachloride		Dibenzyl		*o*-Dichlorobenzene		Ethylene glycol		Urea		1,3-Dimethylurea	
(m)	ϕ	γ	ϕ	γ	ϕ	γ	ϕ	γ	ϕ	γ	ϕ	γ
0.005	0.989	0.977	0.988	0.972	0.983	0.960	0.990	0.979	0.981	0.960	0.986	0.967
0.010	0.982	0.961	0.981	0.956	0.974	0.938	0.982	0.962	0.969	0.932	0.979	0.949
0.015	0.978	0.949	0.977	0.944	0.969	0.924	0.975	0.947	0.959	0.910	0.972	0.933
0.020	0.975	0.940	0.973	0.934	0.965	0.910	0.969	0.934	0.950	0.890	0.967	0.920
0.040	0.965	0.911	0.963	0.903	0.955	0.880	0.952	0.894	0.919	0.825	0.949	0.878
0.060	0.958	0.890	0.953	0.879	0.948	0.857	0.941	0.865	0.894	0.775	0.936	0.846
0.080	0.952	0.874	0.945	0.859	0.941	0.838	0.933	0.842	0.871	0.732	0.925	0.821
0.100	0.946	0.859	0.938	0.842	—	—	0.926	0.823	0.850	0.695	0.914	0.798
0.150	0.935	0.829	—	—	—	—	0.910	0.784	0.803	0.618	0.889	0.748
0.200	0.926	0.805	—	—	—	—	0.898	0.753	0.763	0.558	0.867	0.706

[a] O. D. Bonner, Si Joong Kim, and A. L. Torres, *J. Phys. Chem.* **73**, 1968 (1969).

TABLE XI

OSMOTIC (φ) AND ACTIVITY (γ) COEFFICIENTS OF ELECTROLYTES IN ETHYLENE CARBONATE[a]

Molal value	NaI		KI		CsI		NH_4I		Pr_4NI		Ethylenebis-(Me_4NI)		ZnI_2	$ZnCl_2$
(m)	ϕ	γ	ϕ	γ	ϕ	γ	ϕ	γ	ϕ	γ	ϕ	γ	ϕ	ϕ
0.0025	0.982	0.949	0.982	0.947	0.983	0.949	0.965	0.898	0.983	0.949	0.935	0.827	—	—
0.005	0.974	0.927	0.973	0.925	0.975	0.929	0.946	0.855	0.975	0.929	0.906	0.761	—	—
0.010	0.963	0.897	0.960	0.892	0.963	0.898	0.922	0.798	0.965	0.901	0.864	0.675	0.19	0.07
0.020	0.948	0.857	0.944	0.849	0.946	0.856	0.897	0.731	0.954	0.867	0.831	0.588	0.22	0.10
0.030	0.941	0.832	0.935	0.821	0.936	0.828	0.884	0.690	0.949	0.845	0.807	0.533	0.24	0.13
0.040	0.937	0.813	0.932	0.803	0.929	0.806	0.872	0.658	0.944	0.828	0.787	0.493	0.26	0.15
0.050	0.935	0.801	0.930	0.789	0.923	0.788	0.861	0.632	0.939	0.814	0.769	0.461	0.27	0.17
0.075	0.933	0.777	0.926	0.763	0.911	0.753	0.833	0.578	0.928	0.784	0.728	0.404	0.28	0.19
0.100	0.931	0.761	0.922	0.745	0.903	0.727	0.806	0.534	0.919	0.760	—	—	—	—

[a] O. D. Bonner, Si Joong Kim, and A. L. Torres, *J. Phys. Chem.* **73**, 1968 (1969).

Ethylene carbonate is a satisfactory solvent for the determination of molecular weight by freezing point depression. The cryoscopic constant was first measured by Gross and Schuerch,[62] using as solutes azobenzene, benzoic acid, and benzophenone. They obtained the mean value 7.03 ± 0.14 deg mole^{-1}. Kempa and Lee[63] used as solutes naphthalene and *p*-nitroaniline and obtained the value 7.01 ± 0.07 deg mole^{-1}. From the cooling curve using the Plato method, the latent heat of fusion was determined as 27.3 ± 0.6 cal g^{-1}, corresponding to the cryoscopic constant 6.96 ± 0.15 deg mole^{-1}.[63]

Bonner and co-workers[64] claim that these values are too large, because of the considerable supercooling which occurs in these solutions. From the latent heat of fusion, measured as 34.3 cal g^{-1}, they derived the cryoscopic constant 5.55 deg mole^{-1}. Fawzy,[65] repeating the cryoscopic determination with *p*-nitroaniline as solute, obtained the cryoscopic constant 7.03 ± 0.08 deg mole^{-1}, so that the degree of supercooling in previous work[62,63] appears to be reproducible.

From the freezing point measurements, Bonner *et al.* derived osmotic and activity coefficients of solutes in EC. The results are listed for nonelectrolytes in Table X and for electrolytes in Table XI. Viscosity data are given in Table XII.

TABLE XII

VISCOSITIES OF MOLAL SOLUTIONS IN ETHYLENE CARBONATE AND PROPYLENE CARBONATE[a]

	EC[b]		PC[b]	
Electrolyte	η_c	η_d	η_c	η_d
$LiClO_4$	5.67	7.91	6.58	8.32
KCNS	3.72	5.09	3.99	4.94
KPF_6	3.36	4.82	3.78	4.87

[a] G. Pistoia, *J. Electrochem. Soc.* **118**, 153 (1971).

[b] Temperature, 30°C. η_c, Static viscosity in centistokes. η_d, Dynamic viscosity in centipoises.

The specific conductances of some electrolyte solutions in EC are given in Table XIII; data for *saturated* solutions of five salts are shown in Table XIV. The data for alkali metal perchlorates and tetraalkylammonium salts are collected in Table XV. From the data in Table XV, assuming the Walden

TABLE XIII

SPECIFIC CONDUCTANCES OF ELECTROLYTES IN ETHYLENE CARBONATE[a]

Electrolyte	Molality (m)	κ (ohm^{-1} cm^{-1}) $\times 10^3$	Ref.
$LiClO_4$	0.330	6.07	*b*
	0.489	7.27	*b*
	0.723	7.87	*c*
	0.920	7.80	*b*
	1.361	6.30	*b*
	1.860	4.32	*b*
	2.34	2.73	*b*
	2.82	1.55	*b*
KCNS	0.925	11.40	*b*
	1.346	11.84	*c*
	1.655	11.62	*b*
	2.185	10.68	*b*
	2.600	10.07	*b*
	2.980	9.27	*b*
	3.290	8.50	*b*
	3.630	7.67	*b*
KPF_6	1.03	11.32	*c*

[a] Temperature, 25°C.
[b] G. Pistoia, *J. Electrochem. Soc.* **118**, 153 (1971).
[c] G. Pistoia, M. De Rossi, and B. Scrosati, *J. Electrochem. Soc.* **117**, 500 (1970).

TABLE XIV

CONDUCTANCES OF SATURATED SOLUTIONS IN ETHYLENE CARBONATE[a]

Compound	Molality (m)	κ (ohm^{-1} cm^{-1})	Λ (ohm^{-1} cm^2 equiv.$^{-1}$)
NaBr	0.030	9.46×10^{-4}	36
NaI	2.51	1.10×10^{-2}	4.38
KI	0.67	1.5×10^{-2}	26.8
$CaCl_2$	0.0181	8.66×10^{-5}	—
$HgCl_2$	1.81	—	—

[a] Temperature, 40°C. W. S. Harris, Radiation Lab., University of California (1958).

TABLE XV

EQUIVALENT CONDUCTANCES IN ETHYLENE CARBONATE[a]

Compound	Λ_0	Slope	c_{max} ($\times 10^4$)	$\Lambda_0 \eta$
$LiClO_4$	32.85	27.5	18.1	0.607
$NaClO_4$	38.84	29.5	16.0	0.719
$KClO_4$	41.99	33.2	13.75	0.776
$RbClO_4$	42.59	34.5	15.9	0.788
$CsClO_4$	43.59	35.9	14.0	0.807
Me_4NI	44.81	37.0	12.3	0.828
Et_4NBr	42.48	34.8	13.6	0.785
Et_4NI	42.83	35.0	13.0	0.792
Et_4NClO_4	42.13	34.4	13.7	0.779
n-Bu_4NBr	36.97	36.3	10.25	0.683
n-Bu_4NI	37.41	35.2	9.48	0.692
n-Bu_4NClO_4	36.52	35.9	10.9	0.975

[a] The plots of Λ vs. $c^{1/2}$ are linear with the given slope up to concentration c_{max}. Temperature, 40°C. R. F. Kempa and W. H. Lee, *J. Chem. Soc., London* p. 100 (1961).

product $\Lambda_0 \eta$ for the Et_4N^+ cation in EC to be 0.294, as in a number of organic solvents,[66] the following ion conductances were derived[67]:

	Li^+	Na^+	K^+	Rb^+	Cs^+	NMe_4^+	NEt_4^+	NBu_4^+	ClO_4^-	Br^-	I^-
λ_0:	7	13	15	16	17	18	(15.8)	10	26	26.5	27

Equivalent conductances and viscosities of tetrabutylammonium picrate in EC at 91°C are shown in Table XVI. Λ/c shows a minimum, as for solutions of this solute in anisole and nitrobenzene; the $\Lambda\eta$ product for the fused salt ($\eta = 0.570$, $\Lambda = 0.775$) is approximately equal to $\Lambda_0 \eta_{solvent}$.

TABLE XVI

CONDUCTANCES AND VISCOSITIES OF Bu_4N PICRATE IN ETHYLENE CARBONATE[a]

c (mole liter^{-1})	Λ (ohm^{-1} cm^2 equiv.$^{-1}$)	η ($\times 10^{-3}$) (poise)	$\Lambda\eta$
0.1	40.0	10.2	0.408
0.5	25.7	13.8	0.355
1.50	8.08	45.3	0.366
1.00	15.60	22.3	0.348
2.00	2.70	145	0.392
2.20	1.47	276	0.406

[a] Temperature, 91°C. R. P. Seward, *J. Phys. Chem.* **62**, 758 (1958).

Figure 2[67] shows the plots of Λ vs. $c^{1/2}$ for electrolytes in EC.

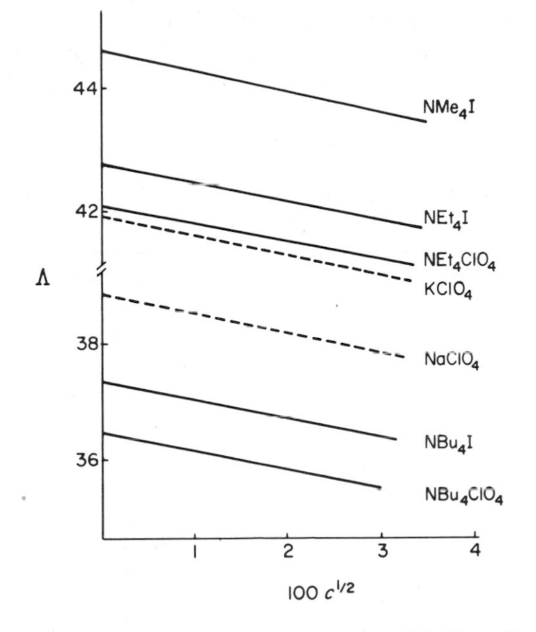

FIG. 2. Λ vs. $c^{1/2}$ for electrolytes in EC. Temperature, 40°C. From Kempa and Lee.

From conductivity studies, zinc iodide appears to be a weak electrolyte in EC. In calculating the equivalent conductances in Table XVII, the solute is assumed to behave as a bi-univalent electrolyte.[64] The two lines at 138

TABLE XVII

CONDUCTANCES OF ZINC IODIDE IN ETHYLENE CARBONATE[a]

Molality ($\times 10^3$) (m)	$c^{1/2}$ ($\times 10^2$) (equiv. liter^{-1})$^{1/2}$	κ ($\times 10^6$) (ohm^{-1} cm^{-1})	Λ (ohm^{-1} cm^2 equiv.$^{-1}$)
0	0	0.13	—
0.44	3.44	4.87	4.12
1.10	5.40	11.0	3.78
7.32	13.9	57.7	2.98
19.6	22.8	129.0	2.49
79.4	45.8	336	1.60

[a] Temperature, 40°C. O. D. Bonner, Si Joong Kim, and A. L. Torres, *J. Phys. Chem.* **73**, 1968 (1969).

and 163 cm^{-1} in the Raman spectrum of the molar solution indicate formation of a polyiodide ion.[64] The results of polarographic reduction of alkali metal cations in EC are included in Table XLIX.

Dielectric constants and densities of water–EC mixtures at 25°C and 40°C are shown in Table XVIII, and of other mixtures in Table XIX.

TABLE XVIII

DENSITIES AND DIELECTRIC CONSTANTS OF WATER–ETHYLENE CARBONATE MIXTURES[a]

W_1[b]	Density (g cm^{-3})	Dielectric constant	X_1[b]
	Temperature = 25°C		
0	1.336[c]	91.60	0
4.99	1.3160	89.90	0.2048
10.02	1.2860	88.60	0.3521
19.28	1.2595	86.52	0.5389
26.33	1.2328	85.06	0.6362
31.60	1.2139	84.23	0.6923
39.98	1.1842	83.24	0.7653
50.70	1.1484	82.20	0.8341
60.01	1.1178	81.53	0.8801
72.73	1.0785	80.58	0.9288
74.86	1.0720	80.48	0.9358
86.74	1.0363	79.58	0.9697
87.63	1.0337	79.51	0.9719
89.91	1.0269	79.31	0.9775
92.14	1.0204	79.14	0.9829
100	0.9971	78.35	1
	Temperature = 40°C		
0	1.3214[c]	90.36	0
5.01	1.3001	87.10	0.2048
9.13	1.2832	85.15	0.3296
14.98	1.2600	83.08	0.4633
22.40	1.2522	80.99	0.5854
23.25	1.2289	80.69	0.5984
34.22	1.1905	79.10	0.7179
50.51	1.1369	77.20	0.8331
64.99	1.0925	75.80	0.9008
75.37	1.0622	75.06	0.9374
90.01	1.0201	73.93	0.9778
100	0.9922	73.28	1

[a] A. D'Aprano, *Gazz. Chim. Ital.* **104**, 91 (1974).
[b] W_1, Weight % water; X_1, mole fraction water.
[c] From extrapolation of $d-W_1$ and $DC-W_1$ curves.

The interaction of water and EC has been investigated by a study of the physical chemistry of their mixtures.[68]

TABLE XIX

DIELECTRIC CONSTANTS OF ETHYLENE CARBONATE SOLUTIONS[a]

	Temperature	Wt. % EC					
Solution	(°C)	0	20	40	60	80	100
EC–benzene							
D	25	2.27	9.47	21.2	38.6	62.8	
D	40	2.24	9.03	20.0	36.1	58.5	89.1
EC–methanol							
D	25	32.6	39.1	47.4	58.6	74.0	
D	40	29.8	35.9	43.7	54.0	68.5	89.1
EC–water							
D	25	78.5	80.5	81.6	83.3	86.4	
EC–PC							
D	25	64.6	69.1	74.6	80.5	87.2	
Dilute benzene solution							
N_2[b]	—	0.50	1.00	1.50	2.00	2.50	3.00
d[b]	25	0.8754	0.8770	0.8790	0.8803	0.8829	0.8848
D	25	2.41	2.60	2.80	2.93	3.09	3.25

[a] R. P. Seward and E. C. Vieira, *J. Phys. Chem.* **62**, 127 (1958).
[b] N_2, Mole % ethylene carbonate; d, density (g cm^{-3}).

Sarel and co-workers have made an extensive study of the kinetics of hydrolysis of the cyclic carbonates.[69] In both acid and base hydrolyses, CO–O bond fission first occurs; in H_2O^{18} oxygen exchange occurs between the exocyclic oxygen and the solvent.[70] The alkaline hydrolysis at 0°C is a second-order reaction. For the series shown below the following rate constants (k_2 in liters mole^{-1} min^{-1}) were recorded.[69, 71] k_0 is the rate constant for the hydrolysis of EC.

R_1HC—CHR_1 (ring: O, O, C=O)

R_1	R_2	k_2 (K_2CO_3)	k_2 (NaOH)	k_2/k_0
H	H	10.2	—	22.1
CH_3	H	5.0	—	10.8
CH_3	CH_3	4.0	5.0	8.7

The second-order rate constants at higher temperatures (k_2 in liter mole^{-1} sec^{-1})[72] were given as:

Temperature (°C): 15 25 35

k_2: 0.99 ± 0.04 2.04 ± 0.06 4.5 ± 0.10

For the formation of the transition state[72] $\Delta H^{\ddagger} = 13.3$ kcal mole^{-1} and $\Delta S^{\ddagger} = -15.2$ cal mole^{-1} deg^{-1}.

The acid hydrolysis is slower.[73] Rate constants and activation parameters are shown in Table XX.

TABLE XX

ACID HYDROLYSIS OF CYCLIC CARBONATES[a]

R_1[b]	R_2[b]	Temperature (°C)	k_2 ($\times 10^3$)	k/k_0 25°C	k/k_0 52°C	$\Delta H^{\ddagger}$	$\Delta S^{\ddagger}$
			Water				
H	H	52	0.99	1.00	1.00	18.4	−26.40
		72	5.50	—	—	—	—
CH_3	H	52	0.61	0.48	0.61	19.7	−23.40
		60.1	1.30	—	—	—	—
		72	3.66	—	—	—	—
CH_3	CH_3	60.1	0.89	0.36	0.41	19.29	−25.36
		69	1.93	—	—	—	—
		72	2.50	—	—	—	—
			20% Dioxane–water				
CH_3	CH_3	52	1.04	—	—	14.35	−50.5
		72.9	4.22	—	—	—	—
		94.0	14.80	—	—	—	—

[a] J. Katzhendler, L. A. Poles, H. Dagan, and S. Sarel, *J. Chem. Soc., London* Part XI, p. 1037 (1971).

[b] R_1 and R_2 in this and subsequent tables refer to substituents in the structure on p. 185. k_0 is the rate constant for ethylene carbonate.

Comparison with the rates of hydrolysis of trimethylene carbonate and γ-butyrolactone shows that 5-membered cyclic carbonates are hydrolyzed more slowly than the 6-membered homologs, comparable in rate to the corresponding lactones.[69] The 5-membered cyclic carbonates were shown to be essentially insensitive to polar effects; alkyl substitution decreases the rate of reaction and increases the activation energy, mainly owing to steric effects.[74]

In the hydrolysis of PC by HCl (concentration above 1 M), the rate does not follow the Hammett acidity function h_0. Large negative entropies of activation indicate the same A2 mechanism of hydrolysis for all cyclic carbonates.[73]

The decrease in rate with a decrease in the dielectric constant of the reaction medium, illustrated in Table XXI, is taken to mean that, in alkaline hydrolysis, the rate-determining step is between a negatively charged species and the dipolar cyclic carbonate.[75]

TABLE XXI

RATES OF ALKALINE HYDROLYSES OF SOME CYCLIC CARBONATES[a]

R_1	R_2[b]	Temperature (°C)	k_2 (liter mole^{-1} min^{-1})
H	H	0	28.2
		6.4	46.2
		10.7	64.2
		16.4	94.6
		27.6	216.0
		35	405
CH_3	H	0	17.5
		6.4	24.7
		12.8	44.9
		16.8	51.0
CH_3	CH_3	6.4	17.00
		13.8	28.6
		15.6	31.30
		18.6	38.2
		27.2	65.3

			k_2[c]		k_2/k_0	
R_1	R_2	Base–substrate ratio	Water	Dioxane–water (33%)	Water	Dioxane–water
H	H	2.66	27.5	23.0	1.00	1.00
CH_3	H	4.98	16.4	15.7	0.59	0.68
CH_3	CH_3	6.52	10.1	6.0	0.36	0.26

[a] J. Katzhendler, L. A. Poles, and S. Sarel, *J. Chem. Soc.* (*B*), 1847 (1971).
[b] See footnote to Table XX. Data k_2 are mean values.
[c] Temperature, 0°C.

A recent paper[75] has summarized the previous results. Parameters relating to compounds concerned in this review are given in Table XXII.

TABLE XXII

KINETIC PARAMETERS FROM THE HYDROLYSES OF CYCLIC CARBONATES[a]

		Alkaline				Acid			
R_1	R_2	k/k_0	$\Delta H^{\ddagger}$	$\Delta S^{\ddagger}$	$\Delta G^{\ddagger}$	k/k_0	$\Delta H^{\ddagger}$	$\Delta S^{\ddagger}$	$\Delta G^{\ddagger}$
H	H	100	11.65	19.75	18.01	100	18.40	26.40	26.98
CH_3	H	46	10.73	23.96	18.52	61	19.70	23.40	27.30
CH_3	CH_3	29	10.82	24.59	18.81	43	19.29	25.36	27.54

R_1	R_2	$-\Delta(\Delta G^{\ddagger})_{acid}$	$-\Delta(\Delta G^{\ddagger})_{base}$	$E_s{}^b$	δ^b
H	H	0	0	0	0
CH_3	H	0.32	0.50	−0.214	−0.10
CH_3	CH_3	0.56	0.80	−0.374	−0.20

[a] Temperature, 52°C. J. Katzhendler, L. A. Poles, and S. Sarel, *Isr. J. Chem.* **10**, 111 (1972).
[b] $\log(k/k_0)_{acid} = E_s$; $\log(k/k_0)_{base} = E_s + 2.48\delta$.

Reactions of EC other than hydrolysis are with potassium halides,[76] amine hydrohalides,[77] and aryl biguanidines.[78]

V. Propylene Carbonate (PC)

The purification of PC has received much attention,[2,79] especially as a solvent for polarography and voltammetry.[80] Trace amounts of water in PC have been determined by differential infrared spectroscopy[81] and gas chromatography[82] and also by the use of Na–K alloy.[83] The electrochemical behavior of the intensively dried liquid has been investigated.[84] From an NMR ^{19}F study it was concluded that zeolites in this solvent are attacked by F^- ions, removing about 1% F in 2 hr.[85]

The variation with temperature of the dielectric constant and the Kirkwood g factor for PC are listed in Table XXIII.

Temperature coefficients of density, viscosity, and dielectric constant have been determined.

$$d \ln \rho/dT = -1 \times 10^{-3}, \text{ Ref. 86}$$
$$d \ln \eta/dT = -0.018, \text{ Ref. 86}$$
$$d \ln \varepsilon/dT = -0.0037, \text{ Ref. 87}$$

Simeral and Amey[88] conclude that there is little or no specific association in liquid PC, and certainly not of the n-mer type.

TABLE XXIII

VARIATION OF DIELECTRIC CONSTANT AND g FACTOR OF PROPYLENE CARBONATE WITH TEMPERATURE

Temperature (°K)	ε_0	g	Ref.	Temperature (°K)	ε_0	g	Ref.
195.15	89	0.73	*a*	298.15	64.92	—	*a*
213.15	89.3	1.04	*b*	298.15	64.6	1.08	*c*
223.15	86.1	—	*b*	303.15	63.7	—	*a*
233.15	83.0	—	*b*	308.15	62.6	1.01	*a*
243.15	80.0	—	*b*	313.15	61.42	1.02	*a*
253.15	76.9	–	*b*	313.15	61.7	—	*d*
263.15	73.9	—	*b*	318.15	60.27	—	*a*
273.15	71.0	0.99	*a, b*	323.15	59.17	—	*a*
283.15	68.4	—	*b*	328.15	58.05	—	*a*
293.15	66.1	1.17	*b*	333.15	56.89	1.02	*a*

[a] Frequency range 1–9000 MHz. R. Payne and I. E. Theodorou, *J. Phys. Chem.* **76**, 2892 (1972).

[b] At 10 kHz. g values from Fig. 2, in L. Simeral and R. L. Amey, *J. Phys. Chem.* **74**, 1443 (1970).

[c] R. P. Seward and E. C. Vieira, *J. Phys. Chem.* **62**, 127 (1958).

[d] R. F. Kempa and W. H. Lee, *J. Chem. Soc., London* p. 1936 (1958).

Cavell has studied the dielectric relaxation and loss in PC at several frequencies in the GHz range (Table XXIV).

The variation of the density of PC with temperature in the range 220.15°–293.15°C is linear:[88]

$$\rho_{\mathrm{T}} = 1.541 - 1.148 \times 10^{-3}\, T, \qquad \rho \pm 0.001 \text{ g cm}^{-3}$$

Several reports of NMR studies at 60 and 100 MHz have appeared, and chemical shifts and coupling constants have been tabulated.[58,58a,88,89] Propylene carbonate has been recommended as a good solvent for EPR studies, especially suitable for *in situ* generation of organic radicals. It shows a marked increase in resolution over acetonitrile.[90]

Within the frequency range 15–65 MHz, no relaxation is observed in the ultrasonic absorption of liquid PC.[58,58a]

An electron density map of PC is shown in Fig. 3[91]; in this CNDO/2 calculation, the ring is assumed planar, with the =O in the plane.

Infrared frequencies and assignments have appeared.[58,58a,88] In a recent comprehensive study, the frequencies of IR and Raman bands for PC, Cl–EC, Cl_2–EC, and Cl–PC have been listed and the fundamental vibra-

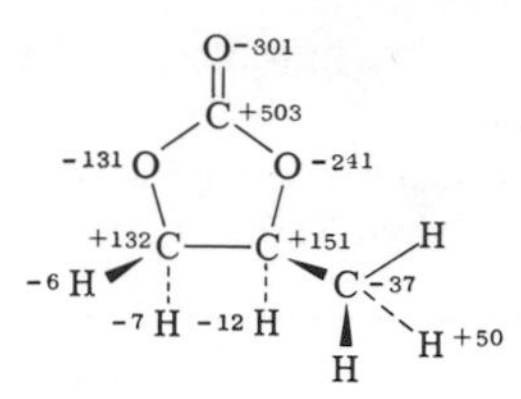

FIG. 3. Electron densities in propylene carbonate. Reprinted by permission of the *Journal of Physical Chemistry*.

tional frequencies assigned.[92] The authors conclude that these compounds possess slightly distorted planar, reasonably rigid ring structures.

TABLE XXIV

VARIATION OF RELATIVE PERMITTIVITY (ε') AND LOSS (ε'') OF PROPYLENE CARBONATE WITH FREQUENCY, AT VARIOUS TEMPERATURES, INCLUDING DERIVED LIMITING PERMITTIVITIES (ε_s AND ε_∞), DIELECTRIC RELAXATION TIMES (τ_0) AND DISTRIBUTION PARAMETERS (ξ)[a, b]

f (GHz)	ε' 2°C	ε'' 2°C	ε' 20°C	ε'' 20°C	f (GHz)	ε' 40°C	ε'' 40°C	ε' 50°C	ε'' 50°C
0.50	67.7	12.5	64.2	8.04	—	—	—	—	—
1.00	60.9	22.5	61.6	15.3	1.00	58.7	10.06	57.0	8.44
1.50	52.5	28.3	56.4	21.4	1.50	—	—	55.2	12.25
3.00	—	—	41.2	29.7	3.00	—	—	48.1	20.3
9.997	11.42	16.0	13.8	21.5	9.997	19.8	23.9	21.1	25.2
33.88	—	—	5.74	7.51	33.88	8.40	10.40	9.47	11.67

	2°C	20°C	40°C	50°C
ε_s	70.75	66.4	62.7	59.7
ε_∞	4.45	3.9	4.55	3.4
$10^{12}\tau_0$ (sec.)	74.7	46.2	33.4	31.8
ξ	0.85	0.91	0.86	0.78

[a] E. A. S. Cavell, *J. Chem. Soc.* (Faraday II), **70**, 78 (1974).

[b] These data fit the empirical equation proposed by Cole and Davidson.

$$\varepsilon'(\omega) - i\varepsilon''(\omega) = \varepsilon_\infty + (\varepsilon_s - \varepsilon_\infty)/(1 + i\omega\tau_0)\xi$$

R. H. Cole and D. W. Davidson, *J. Chem. Phys.* **18**, 1417 (1951).

The mass spectrum (at 10 eV) of PC[61] has been tabulated below.

Group	Peak	Relative abundance (%)
M^+	102	0.9
M–$C_nH_{2n}O$	72 (M-30)	
	58 (M-44)	7.5
M CO_2	58	7.2
M–HCO_2	57	22.8
Rearrangements		
C_2H_3O	43	19.4
C_2H_5	29	9.5
CHO	29	7.8

The vapor pressure of PC at several temperatures and the molar enthalpies are given in Table XXVa, and the free energy data in Table XXVb.

TABLE XXVa

VAPOR PRESSURES AND ENTHALPIES OF PROPYLENE CARBONATE[a]

Temperature (°K)	Pressure (mm Hg)
328.2	1.20
338.6	1.78
358.6	3.14
369.6	5.07

ΔH_c^0	ΔH_f^0 (l)	ΔH_f^0 (g)	ΔH_{vap}
−430.14	−151.01	−143.68	7.33

[a] J. K. Choi and M. J. Joncich, *J. Chem. Eng. Data* **16**, 87 (1971).
[b] ΔH values in kcal $mole^{-1}$.

The solubilities of electrolytes in PC and the specific conductances (and densities where available) of their saturated solutions are shown in Table XXVI.

The solubilities of NaI and KI in EC, PC, and two related solvents are compared in Table XXVII.

TABLE XXVb

FREE ENERGY DATA FOR PROPYLENE CARBONATE[a]

Temperature (°K)	C_p	S_T	$(H_T - H_0)/T$	$-(G_T - H)/T$
Solid				
50	35.61	25.84	17.19	8.65
100	65.43	62.26	35.57	26.69
150	83.35	92.34	48.63	43.71
200	99.63	118.53	59.32	59.21
224.85	107.48	130.56	64.21	66.35
Liquid				
224.85	158.14	173.39	107.04	66.35
250	159.94	190.37	112.25	78.12
298.10	167.60	219.17	120.54	98.63
310	169.96	225.75	122.38	103.37

[a] Data in kJ mole^{-1} °K^{-1}. I. A. Vasil'ev and A. D. Korkhov, *Tr. Khim. Khim. Tekhnol.* Pt. I, p. 103 (1974).

TABLE XXVI

SOLUBILITIES OF ELECTROLYTES AND SPECIFIC CONDUCTANCES OF SATURATED SOLUTIONS IN PROPYLENE CARBONATE[a]

Salt[b]	Solubility (m)	Density (g cm^{-3})	κ (ohm^{-1} cm^{-1})	Ref.
LiCl	8×10^{-4}	—	1.91×10^{-5}	*c*
LiBr	2.43	1.372	5×10^{-4}	*c*
LiI	1.365	—	—	*c*
$LiPF_6$	0.55	—	4.49×10^{-3}	*d*
$LiBF_4$	0.42	—	2.5×10^{-3}	*d*
$LiClO_4$	1.0	1.259	5.6×10^{-3}	*e*
NaCl	3×10^{-6}	—	1.56×10^{-6}	*c*
NaBr	8×10^{-2}	—	1.79×10^{-4}	*c*
NaI	1.1	1.336	5.8×10^{-3}	*c, f*
$NaPF_6$	0.86	—	6.8×10^{-3}	*d*
$NaClO_4$	2 *N*	—	—	*g*
KCl	2.68×10^{-4}	—	1.06×10^{-5}	*c, h*
KBr	6×10^{-3}	—	1.48×10^{-3}	*c*
KI	0.22	1.227	4.41×10^{-3}	*c*

TABLE XXVI (*continued*)

Salt[b]	Solubility (m)	Density ($g\ cm^{-3}$)	κ ($ohm^{-1}\ cm^{-1}$)	Ref.
KPF_6	1.2 N	—	3.12×10^{-3}	*f*
KBF_4	1.44×10^{-2}	—	1.9×10^{-4}	*c*
RbF	0.06	—	7.4×10^{-5}	*f*
$CuCl_2$	8.9×10^{-3}	—	7.07×10^{-5}	*c*
$CuBr_2$	6.25×10^{-3}	—	5.03×10^{-5}	*c*
$MgBr_2$	Very soluble[i]	—	—	*c*
$CaCl_2$	7.6×10^{-3}	—	1.74×10^{-4}	*c*
$CaBr_2$	0.75[i]	—	—	*c*
$Ca(BF_4)_2$	0.064	—	2.02×10^{-3}	*c*
$BaBr_2$	7.7×10^{-4}	—	3.98×10^{-5}	*c*
$AlCl_3$	Very soluble[i]	—	—	
$ThCl_4$	0.033	—	2.1×10^{-4}	*c*
$PbBr_2$	1.6×10^{-5}	—	2×10^{-6}	*c*
$BiCl_3$	0.77	—	—	*c*
$MnCl_2$	2.3×10^{-3}	—	6.5×10^{-5}	*c*
$MnBr_2$	1.50	—	—	*c*
$ZnCl_2$	0.118	—	—	*j*
$ZnSO_4$	0.0466	—	—	*j*
ZnO	0.0284	—	—	*j*
$CdBr_2$	4.4×10^{-3}	—	3.9×10^{-5}	*c*
CdI_2	4.5×10^{-3}	—	4×10^{-5}	*c*
CdI_2	4×10^{-4}	—	—	*j*
UCl_4	0.102	—	8.8×10^{-4}	*c*
$SnCl_4$	1.02	—	—	*c*
$CoCl_2$	6.5×10^{-3}	—	1.74×10^{-4}	*c*
$NiBr_2$	4.3×10^{-3}	—	1.39×10^{-4}	*c*

[a] Temperature, 25°C.

[b] The following salts are listed as "insoluble": NaF, NaOMe, $NaBH_4$, Na_2SiF_6, $CdCl_2$, $TiCl_2$, $SnCl_2$, $NiCl_2$, $PbCl_2$ (very slightly soluble). W. S. Harris, ref. 2.

[c] W. S. Harris, ref. 2.

[d] H. Baumann, Tech. Report APL-TD-R64-59, May 1964.

[e] W. Elliott and R. Amlie, Final Report NASA CR-72364, Sept. 1967.

[f] A. E. Lyall, H. N. Seiger, and J. Orshich, Tech. Report AFAPL-TR-68-71, July 1968.

[g] Y-C. Wu and H. L. Friedman, *J. Phys. Chem.*, **70**, 501, 2020 (1966).

[h] Solubility, 5×10^{-4} M at 25°C. I. Fried and H. Barak, *J. Electroanalyt. Chem.* **27**, 167 (1970).

[i] With slow decomposition of the solution.

[j] The values quoted are much lower than those of ref. *a*, and apply to rigorously dried PC. M. H. Fawzy, private communication.

TABLE XXVII

SOLUBILITY OF NaI AND KI IN FOUR SOLVENTS[a]

Solvent	Temperature (°C)	Dielectric constant	*g* factor	Mole fractions	
				NaI	KI
Ethylene carbonate	40	89.1	1.2	0.180	0.056
Propylene carbonate	25	64.4	1.04	0.103	0.022
Cyclopropane	25	13.5	0.65	0.128	0.002
γ-Butyrolactone	25	58.2	0.71	0.250	0.061

[a] W. S. Harris, Radiation Lab., University of California (1958).

The effects of 0.5% by volume of water, of 1 M $LiClO_4$ and $MgCl_2$ on solubilities in PC are reported in Table XXVIII.

TABLE XXVIII

EFFECTS OF WATER AND OF $LiAlCl_4$ AND $MgCl_2$ ON THE SOLUBILITIES OF ELECTROLYTES IN PROPYLENE CARBONATE[a]

PC[b]	Solubility (mole liter^{-1})						
	CuCl	$CuCl_2$	CuF_2	AgCl	HgCl	$HgCl_2$	$PbCl_2$
PC(1)	0.004	0.008	0.0002	0.0001	0.0001	s[c]	0.0001
PC(2)	0.0071	0.023	0.0006	0.0001	0.0001	s	0.0002
PC(3)	—	0.006	—	—	—	—	—
PC(4)	—	0.008	—	0.0001	—	—	0.0001
PC(5)	—	0.018	—	0.0003	—	—	0.0004
PC(6)	—	0.01	—	0.0003	—	—	0.005

[a] Temperature, 25°C. M. L. B. Rao and R. W. Holmes, *Electrochem. Technol.* **6**, 105 (1968).
[b] PC(1), anhydrous; (2) containing 0.5% water by volume; (3), containing 1 *M* $LiAlCl_4$; (4), anhydrous (second sample); (5), containing 0.01 *M* $MgCl_2$; (6), containing 0.135 *M* $MgCl_2$.
[c] s: solubility, 0.1 mole l^{-1}.

The tetraalkylammonium chlorides are soluble in PC, except for $(CH_3)_4NCl$.[93] From a potentiometric study of silver and Tl(I) halides in PC, the following formation constants have been evaluated.[94]

$K_{S_0}(\mathrm{AgCl}) = -19.87 \pm 0.02$
$K_{S_0}(\mathrm{TlCl}) = -12.4$
$\log \beta_1 = 15.15 \pm 0.15$
$\log \beta_2 = 20.87 \pm 0.15$
$\log \beta_3 = 23.39 \pm 0.06$

The main complex present in AgCl solutions in PC is $AgCl_2^-$; $K(AgCl_2^-) = 10.0$.[95] For the bromide complexes $K_{S_0}(AgBr) = -19.9$ and $K(AgBr_2^-) = 4.0$. The equilibrium constant for the triiodide equilibrium

$$I_2 + I^- \rightleftharpoons I_3^-$$

in PC at 25°C is $10^{7.6}$ (see also p. 230).[95]

The coordination of transition metal ions in PC is discussed in Gutmann's monograph.[6a] It appears that hexacoordinated species are formed with competitive anion inclusion in the coordination sphere; anion coordination is 4-fold for Co^{2+}, Ni^{2+}, Mn^{2+}, and VO^{2+} with halide, CN^-, and CNS^- (except for $MnCl_2$, which is 6-fold), and 6-fold for Ti^{3+}, V^{3+}, Cr^{3+}, and Fe^{3+}.

The surface tension of PC is 40.50 dynes cm^{-1} at 20°C; the Hildebrand solubility parameter[96] is 13.30 $cal^{1/2}$ $cm^{-3/2}$.[97] The donicities[98] of the cyclic carbonates are derived from the enthalpies of solution of $SbCl_3$ in these solvents and decrease with halogen substitution.[98a]

EC	16.4	ClEC	12.7	Cl_2EC	3.2	Cl_4EC	0.2
PC	15.1	H_2O	18.0	Pyridine	33		

Solubilities of acetylene, propadiene, diacetylene, and methyl acetylene in

TABLE XXIX

HENRY'S LAW CONSTANTS FOR GASES IN PROPYLENE CARBONATE[a]

	Temperature (°C)					
Gas	−20	0	20	40	60	80
Acetylene	6300	12670	22200	34600	50100	64200
Propadiene	5460	9340	—	26900	—	72400
Methyl acetylene	1580	3310	—	10000	—	27530
Diacetylene	120	302	—	1700	—	6450

[a] V. G. Bova, D. A. Kuznetsov, T. F. Furmer, and V. T. Sasov, *Zh. Prikl. Khim.* **47**, 687 (1974).

PC have been measured in the temperature range 250°–370°K. Henry's law constants are given in Table XXIX. Henry's law is obeyed by CO_2 in PC in the temperature range 0° to 120°C and for pressures up to 760 mm; $\Delta H_{soln} = 3500$ cal mole^{-1}.[99]

The solubilities of the thallium(I) halides in PC were measured by potentiometric titration, and their solubility products were calculated at 25°C

TABLE XXX

Solubility Products (K_s) of Thallium(I) Halides in Propylene Carbonate[a]

Halide	Solubility (mole liter^{-1} × 10^5)	K_s (× 10^{12})	pK_s PC	pK_s H_2O[b]
TlCl	2.24	5.01	11.45	3.76
TlBr	3.31	11.0	11.11	3.62
TlI	12.0	144.0	9.99	7.19

[a] Temperature, 25°C. N. Matsuura and K. Umemoto, *Bull. Chem. Soc. Jap.* **47**, 1334 (1974).

[b] J. F. Coetzee and J. J. Champion, *J. Amer. Chem. Soc.* **89**, 2513 and 2517 (1967).

(Table XXX). Single-ion medium effects were calculated from solubility and emf data, by the equations

$$\log m\gamma_{M^+} = (E^0_{solvent} - E^0_{water})/0.05916$$

and

$$\log m\gamma_X = \log m\gamma_{TlX} - \log m\gamma_{Tl^+}$$

The single-ion medium effects for PC, dimethyl sulfoxide (DMSO), and dimethylformamide (DMF) are recorded in Table XXXI.

The heats of solution of the alkali metal trifluoroacetates, tetraphenylborates, and perchlorates in PC have been measured. Standard heats of transfer from water to PC and heats of transfer of ions relative to Na^+ are summarized in Tables XXXII and XXXIII. Krishnan and Friedman[99a] have made an extensive study of enthalpies of solvation and transfer for the solvents PC and DMSO. Group solvation enthalpy parameters for nonelectrolytes are given in Table XXXIV and transfer enthalpies in Table

TABLE XXXI

SINGLE-ION MEDIUM EFFECTS WITH REFERENCE TO WATER (MOLALITY SCALE)[a]

	Solvent		
Ion	DMSO	DMF	PC
Li^+	−1.49	−0.24	+5.53
Na^+	−1.10	+0.05	+3.82
K^+	−1.20	−0.37	+2.15
Rb^+	−0.58	+0.08	+2.52
Cs^+	−0.61	−0.08	+2.42
Tl^+	−3.16	−1.74	+2.53
Ag^+	−5.11	−2.30	+3.77
Cl^-	+4.98	+6.51	+5.36
Br^-	+3.01	+4.35	+3.31
I^-	+0.75	+1.56	+0.47
ClO_4^-	−1.18	−1.24	—

[a] N. Matsuura and K. Umemoto, *Bull. Chem. Soc. Jap.* **47**, 1334 (1974).

XXXV. The extensive literature on enthalpies of transfer of electrolytes and nonelectrolytes in systems involving PC can be summarized in the following tables: ionic enthalpies of transfer from water to organic solvents, including PC (Table XXXVI); enthalpies of transfer of electrolytes and ions from DMSO and water to PC (Table XXXVII); phenyl group contributions to the enthalpy of transfer of hydrocarbons and $(Ph)_nX$ ions from DMSO to PC (Table XXXVIII); enthalpies of transfer of hydrogen-bonded donors from DMSO to PC (Table XXXIX); and energetics (ΔG_t, ΔH_t, and ΔS_t) of transfer of LiCl and NaI from water to organic solvents, including PC (Table XL).

Activity coefficients for electrolytes in PC solution have been calculated from emf measurements. The data are recorded in Table XLI. Some results for alkali halides are illustrated in Fig. 4.[99b]

TABLE XXXII

STANDARD HEATS OF TRANSFER FROM WATER TO PROPYLENE CARBONATE[a]

Cation	ΔH^0(MX)	ΔH^0(MX) − ΔH^0(NaX)
	Trifluoroacetates	
Li	8.69	3.34
Na	5.35	0
K	2.52	−2.83
Rb	1.89	−3.46
Cs	1.52	−3.83
	Tetraphenylborates	
Na	−5.93	0
K	−8.67	−2.74
Rb	−9.32	−3.39
Cs	−10.02	−4.09
	Perchlorates	
Li	−3.16	3.21
Na	−6.37	0
	Iodides	
Li	−0.10	3.12
Na	−3.22	0

[a] Data given in kcal mole^{-1}. Y.-C. Wu and H. L. Friedman, *J. Phys. Chem.* **70**, 501 (1966).

TABLE XXXIII

HEATS OF TRANSFER RELATIVE TO Na^+[a]

	ΔH^0 (kcal mole^{-1})		
Ion	aq. → PC	aq. → gas	PC → gas
Li^+	3.17	26.28	23.11
Na^+	0	0	0
K^+	−2.80	−20.18	−17.38
Rb^+	−3.43	−26.09	−22.66
Cs^+	−3.96	−30.61	−26.65

[a] Y.-C. Wu and H. L. Friedman, *J. Phys. Chem.* **70**, 508, and footnotes to Table IX therein (1966).

TABLE XXXIV

GROUP SOLVATION ENTHALPY PARAMETERS FOR NONELECTROLYTES IN PROPYLENE CARBONATE[a]

Nonelectrolyte	Solvent		
	Water	PC	DMSO
Primary alcohols			
(HO···H)	−7.68	−6.20	−8.13
(CH_2)	0.32	−0.96	−0.87
Tertiary alcohols			
(HO···H)		−5.52	−6.89
(CH_2)		−0.79	−0.78
Alkanes			
(CH_3)		−0.61	−0.46
Aromatic hydrocarbons			
(C_6H_5)		−7.5	−7.7

[a] Data given in kcal mole^{-1} at 25°C. C. V. Krishnan and H. L. Friedman, *J. Phys. Chem.* **73**, 1572 (1969).

TABLE XXXV

ENTHALPIES OF TRANSFER FROM PROPYLENE CARBONATE, DIMETHYL SULFOXIDE, AND WATER[a]

	Saturated hydrocarbons (RH)			
	Pentane	Hexane	Heptane	Cyclohexane
$N =$	3	4	5	6
$(RH)^{gas}_{PC}$	4.13	5.02	5.90	5.77
$(RH)^{gas}_{PC} + N(CH_2)^{b}_{PC}$	1.25	1.18	1.10	0.01
$(RH)^{DMSO}_{PC}$	0.56	0.66	0.78	0.60
$(RH)^{DMSO}_{PC} - N(CH_2)^{DMSO}_{PC}$	0.29	0.30	0.33	0.06
	Methane	Ethane	Propane	Butane
$(RH)^{gas}_{water}$	3.15	4.10	5.3	5.9
$(RH)^{gas}_{water} - (ROH)^{gas}_{water}$	−7.65	−8.50	−8.5	−8.9

(*continued*)

TABLE XXXV (*continued*)

	Aromatic hydrocarbons (PhX)				
	PhH	PhMe	PhEt	*m*-Xylene	*i*-PrPh
$N =$	0	1	2	2	3
$(PhX)_{PC}^{gas}$	7.72	8.52	9.34	9.39	9.90
$(PhX)_{PC}^{gas} - 0.96N^{c}$	7.72	7.56	7.42	7.47	7.02
$(PhX)_{DMSO}^{PC}$	−0.26	−0.34	−0.36	−0.35	−0.47
$(PhX)_{DMSO}^{PC} + 0.09N$	−0.26	−0.25	−0.18	−0.17	−0.20
$(PhX)_{water}^{PC}$	0.26	0.36	1.09[d]	—	—
$(PhX)_{water}^{PC} - (PhH)_{water}^{PC}$	0	0.10	0.83[d]	—	—
$(X)_{water}^{PC}$	—	−0.13[e]	3.23[f]	—	—

	Group R		
	CH_2	CH_3	$C_6H_5 \cdots H$
$(R)_{DMSO}^{PC}$	−0.09	−0.15	−0.26

[a] Data in kcal mole^{-1} at 25°C. C. V. Krishnan and H. L. Friedman, *J. Phys. Chem.* **73**, 1572 (1969).

[b] $(RH)_{PC} + N(CH_2)_{PC} = 2(CH_3)_{PC}$, a test of the $(RH)_{PC}$ data and of $(CH_2)_{PC} = -0.96$ (Table XXXIV).

[c] $(PhX)_{PC} + N(CH_2)_{PC} = (PhH)_{PC}$, with $(CH_2)_{PC} = -0.96$.

[d] $X = CH_2OH$.

[e] Estimated as $(\text{1-pentanol})_{water}^{PC} - (\text{1-butanol})_{water}^{PC}$.

[f] Estimated as $(CH_3OH)_{water}^{PC}$.

TABLE XXXVI

IONIC ENTHALPIES OF TRANSFER[a]

Ion	$H_2O \rightarrow DMSO$	$H_2O \rightarrow PC$	$H_2O \rightarrow D_2O$
$CH_3NH_3^+$	−6.82	−3.44	0.03
$C_2H_5NH_3^+$	−5.79	−2.50	0.00
$C_3H_7NH_3^+$	−5.30	−2.09	−0.06
$C_4H_9NH_3^+$	−5.05	−1.94	−0.11
$C_5H_{11}NH_3^+$	−5.01	−1.79	−0.15
$C_6H_{13}NH_3^+$	−5.00	−2.17	−0.19
$C_7H_{15}NH_3^+$	−4.95	−2.24	−0.22
$C_8H_{17}NH_3^+$	−4.92	−1.99	−0.24
$(CH_3)_2NH_2^+$	−4.61	−1.73	
$(C_2H_5)_2NH_2^+$	−2.14	0.60	
$(C_3H_7)_2NH_2^+$	−1.15	1.55	

TABLE XXXVI (*continued*)

Ion	$H_2O \rightarrow DMSO$	$H_2O \rightarrow PC$	$H_2O \rightarrow D_2O$
$(C_4H_9)_2NH_2^+$	−0.57	1.78	
$(C_5H_{11})_2NH_2^+$	−0.71	1.64	
$(CH_3)_3NH^+$	−3.28	−2.08	
$(C_2H_5)_3NH^+$	−0.30	0.79	
$(C_3H_7)_3NH^+$	1.62	2.37	
$(C_4H_9)_3NH^+$	2.76	3.15	
NH_4^+	−9.81	−5.11	
$(C_2H_5)_3CH_3N^+$	0.07	−0.70	
$(C_2H_5)_3C_3H_7N^+$	1.85	0.81	
$(C_4H_9)_3CH_3N^+$	3.78	2.39	
$(C_4H_9)_3C_3H_7N^+$	5.65	4.14	

[a] Data given in kcal mole^{-1} at 25°C. C. V. Krishnan and H. L. Friedman, *J. Phys. Chem.* **74**, 3900 (1970). For each solvent pair the ionic enthalpies of transfer are based on the assumption that Ph_4As^+ and Ph_4B^- have the same enthalpy of transfer. The footnote to Table II [C. V. Krishnan and H. L. Friedman, *J. Phys. Chem.* **74**, 3902 (1970)] records the agreement between salts of these cations with different anions.

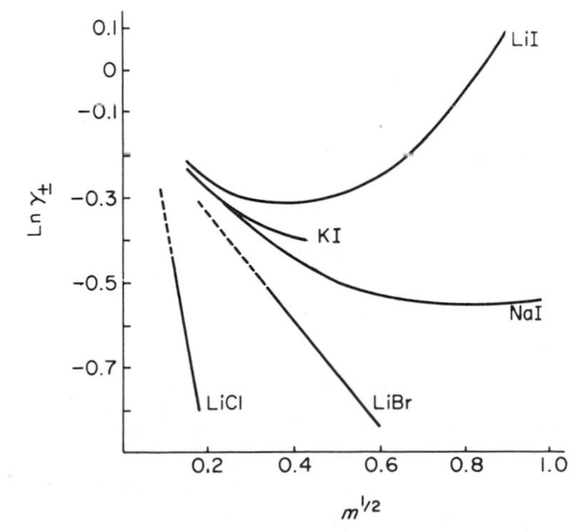

FIG. 4. Natural log of mean molal activity coefficient vs. (molality)$^{1/2}$. Temperature, 25°C; solvent, PC. From Salomon.[99b]

TABLE XXXVII

ENTHALPIES OF TRANSFER TO PROPYLENE CARBONATE[a,c]

Ion	Single-ion values[b]	Cl^- 1.82	Br^- 2.41	I^- 2.27	ClO_4^- 0.67	BPh_6^- −0.65
			From dimethyl sulfoxide			
Li^+	7.04	9.03[d]	9.45[d]	8.97[d]	—	—
		(8.86)	(9.45)	(9.31)		
Na^+	4.18	—	—	6.45[e,,]	—	3.53[e,f]
				(6.45)		(3.53)
K^+	3.10	—	5.49[d,e]	5.40[d,e]	—	2.45[e]
			(5.51)	(5.37)		(2.45)
Rb^+	2.14	—	—	—	—	1.49[e]
Cs^+	1.31	—	—	—	—	0.66[e]
Me_4N^+	−0.23	—	2.22	2.04	0.45	—
			(2.18)	(2.04)	(0.44)	
Et_4N^+	−0.84	0.93	1.63	1.38	0.17	—
		(0.98)	(1.57)	(1.43)	(0.17)	
Pr_4N^+	−1.23	0.65	1.19	0.98	−0.53	—
		(0.59)	(1.18)	(1.04)	(−0.56)	
Bu_4N^+	−1.66	0.16	0.75	0.58	−0.92	—
		(0.16)	(0.75)	(0.61)	(−0.99)	
Am_4N^+	−2.03	−0.21	0.35	0.18	−1.31	—
		(−0.21)	(0.38)	(0.24)	(−1.36)	
Hex_4N^+	−2.38	—	0.11	—	−1.79	—
			(0.03)		(−1.71)	
Bu_4P^+	−1.53	—	0.88	—	—	—
Me_3PhN^+	0.48	—	—	2.75	—	—
Ph_4P^+	−0.99	0.83	—	—	—	—
Ph_4As^+	−0.65	1.17	1.78	—	−0.04	—
		(1.17)	(1.76)		(0.02)	
Ph_4Sb^+	1.13	—	3.54	—	—	—

Ion	Single-ion values[g]	Cl^- 6.31	Br^- 3.24	I^- −0.78	ClO_4^- −3.93	BPh_4^- −3.49	$CF_3CO_2^-$ 7.79
				From water			
Li^+	0.73	—	—	—	—	—	—
Na^+	−2.44	—	—	—	—	—	—
K^+	−5.24	—	—	—	—	—	—
Rb^+	−5.87	—	—	—	—	—	—
Cs^+	−6.40	—	—	—	—	—	—
Me_4N^+	−3.89	—	—	—	—	—	—
Et_4N^+	0.17	—	—	1.98	−3.77	—	—
				(2.01)	(−3.76)		
Pr_4N^+	2.79	9.08	6.05	—	—	—	—
		(9.10)	(6.03)				

TABLE XXXVII (*continued*)

Ion	Single-ion values[g]	Cl^- 6.31	Br^- 3.24	I^- −0.78	ClO_4^- −3.93	BPh_4^- −3.49	$CF_3CO_2^-$ 7.79
Bu_4N^+	4.39	10.52 (10.70)	7.72 (7.53)	—	—	—	—
Am_4N^+	4.79	11.05 (11.10)	8.18 (8.13)	—	—	—	—
Hex_4N^+	4.24	—	—	—	0.31	—	—
Ph_4P^+	−3.22	3.09 (3.09)	—	—	−6.87 (−7.15)	—	—
Ph_4As^+	−3.49	2.82 (2.82)	−0.05 (−0.25)	—	−6.64 (−7.42)	—	—
Ph_4Sb^+	−2.29	—	0.95	—	—	—	—
Bu_4P^+	4.41	—	7.65	—	—	—	—
Me_3PhN^+	5.55	—	—	4.77	—	—	—

[a] Data in kcal/mole. Temperature, 25°C. C. V. Krishnan and H. L. Friedman, *J. Phys. Chem.* **73**, 3934 (1969).

[b] Single-ion values are based on the convention that the enthalpies of transfer of Ph_4As^+ and Ph_4B^- are equal.

[c] Values in parentheses are derived by adding the ionic enthalpies; the remainder are experimental values.

[d] These experimental values are obtained by combining ionic differences from the water → PC table with data on $(MX)_{water}^{DMSO}$ from R. F. Rodewald, K. Mahendran, J. L. Bear, and R. Fuchs, *J. Amer. Chem. Soc.* **90**, 6698 (1968).

[e] These data make use of heats of solution in PC. Y.-C. Wu and H. L. Friedman, *J. Phys. Chem.* **70**, 501, 2020 (1966).

[f] These data make use of heats of solution in DMSO. E. M. Arnett and D. R. McKelvey, *J. Amer. Chem. Soc.* **88**, 2598 (1966).

[g] The anion values and the cation values from Li^+ to Et_4N^+ are taken from Wu and Friedman (ref. *e*) adjusted to accord with the convention $(Ph_4As^+)_{water}^{PC} = (Ph_4B^-)_{water}^{PC}$.

TABLE XXXVIII

PHENYL GROUP CONTRIBUTIONS TO ENTHALPY OF TRANSFER TO PROPYLENE CARBONATE FROM DIMETHYL SULFOXIDE[a,b]

Hydrocarbons[c]	Me_3PhN^+	Ph_4As^+	Ph_4B^-	Ph_4P^+	Ph_4Sb^+
−0.17 to −0.26	0.75	−0.16	−0.16	−0.25	0.28

[a] Temperature, 25°C. Data given in kcal $mole^{-1}$.

[b] C. V. Krishnan and H. L. Friedman, *J. Phys. Chem.* **73**, 3934 (1969).

[c] C. V. Krishnan and H. L. Friedman, *J. Phys. Chem.* **73**, 1572 (1969).

TABLE XXXIX

ENTHALPIES OF TRANSFER OF HYDROGEN-BOND DONORS TO PROPYLENE CARBONATE FROM DIMETHYL SULFOXIDE[a],[b]

$R_3N{-}H^+$	$R_2NH_2^{+}$[b]	RNH_3^+	NH_4^+	ROH[c]	H_2O[c]
1.64 ± 0.05	1.59 ± 0.15	1.16 ± 0.05	1.18 ± 0.15	1.93 ± 0.05	1.64 ± 0.05

[a] Extrapolated to zero alkyl substituent. Temperature, 25°C. Data given in kcal/mole^{-1} per hydrogen bond.
[b] C. V. Krishnan and H. L. Friedman, *J. Phys. Chem.* **74**, 3900 (1970).
[c] C. V. Krishnan and H. L. Friedman, *J. Phys. Chem.* **73**, 1572 (1969).

TABLE XL

ENERGETICS OF TRANSFER FROM WATER TO ORGANIC SOLVENTS

Solvent	ΔG_t^0 (kcal mole^{-1})	ΔH_t^0 (kcal mole^{-1})	ΔS_t^0 (cal mole^{-1} deg^{-1})	Temperature (°C)
		LiCl (molarity scale)		
DMSO[a]	4.75	−3.00	−26.01	25
PC[b]	14.74	6.98	−26.01	25
NMF[c]	3.20	−4.22	−24.89	25
CH_3OH[d]	4.80	−2.60	−24.15	25
H_2O–20% dioxane[e]	0.055	−0.585	−2.15	25
		NaI (molality scale)		
PC[f]	7.58	−2.09	−33.59	15
PC[f]	7.92	−2.04	−33.39	25
PC[f]	8.25	−1.98	−33.19	35
PC[f]	8.58	−1.91	−32.99	45

[a] W. H. Smyrl and C. W. Tobias, *J. Electrochem. Soc.* **115**, 33 (1968).
[b] M. Salomon, *J. Phys. Chem.* **73**, 3299 (1969).
[c] O. Poponych, *Anal. Chem.* **38**, 558 (1966); J. N. Butler, *J. Phys. Chem.* **72**, 1017 (1968).
[d] D. Feakins, B. C. Smith, and L. Thakur, *J. Chem. Soc., A* p. 714 (1966).
[e] H. S. Harned and B. B. Owen, "The Physical Chemistry of Electrolytic Solutions." Van Nostrand-Reinhold, Princeton, New Jersey, 1958.
[f] M. Salomon, *J. Electroanal. Chem.* **25**, 1 (1970).

TABLE XLI

MEAN MOLAL ACTIVITY COEFFICIENTS IN PROPYLENE CARBONATE[a]

Molality	15°C	25°C	35°C	45°C
		For LiCl		
0.01631	0.6093	0.6046	0.6057	0.6038
0.01772	0.6135	0.6137	0.6099	0.6095
0.01932	0.5732	0.5776	0.5729	0.5699
0.02003	0.5554	0.5647	0.5658	0.5675
0.02439	0.5173	0.5202	0.5175	0.5157
0.02701	0.4863	0.4879	0.4842	0.4813
0.02718	0.4786	0.4801	0.7783	0.4743
0.02771	0.4726	0.4754	0.4747	0.4744
0.02956	0.4528	0.4540	0.4510	0.4493
0.03156	0.4430	0.4446	0.4418	0.4393
0.03197	0.4391	0.4420	0.4416	0.4422
		For LiBr		
0.64829	0.3593	0.3476	0.3380	0.3275
0.35385	0.4453	0.4340	0.4241	0.4140
0.31715	0.4596	0.4515	0.4452	0.4380
0.28684	0.4790	0.4710	0.4647	0.4553
0.23449	0.5103	0.5010	0.4957	0.4881
0.20425	0.5306	0.5219	0.5159	0.5090
0.19384	0.5409	0.5325	0.5272	0.5198
0.14837	0.5781	0.5747	0.5687	0.5636
0.12840	0.6105	0.6029	0.5962	0.5884
		For LiI[b]		
0.8688	1.1608	1.0862	1.0683	1.0422
0.6131	0.8953	0.8737	0.8564	0.8401
0.5818	0.8532	0.8369	0.8410	0.8295
0.4890	0.8292	0.8109	0.7982	0.7872
0.4724	0.8334	0.8145	0.8050	0.7934
0.4056	0.8031	0.7893	0.7790	0.7696
0.3748	0.7887	0.7767	0.7706	0.7635
0.2803	0.7596	0.7494	0.7445	0.7360
0.1889	0.7475	0.7377	0.7359	0.7318
0.1342	0.7609	0.7539	0.7531	0.7517
0.1178	0.7341	0.7268	0.7258	0.7224
0.09644	0.7502	0.7452	0.7430	0.7406
0.09279	0.7590	0.7517	0.7423	0.7360
0.06801	0.7471	0.7421	0.7438	0.7455
0.05201	0.7621	0.7573	0.7542	0.7495
0.04780	0.7694	0.7638	0.7644	0.7638
0.02748	0.7821	0.7850	0.7861	0.7888

(*continued*)

TABLE XLI (*continued*)

Molality	15°C	25°C	35°C	45°C
		For NaI[c]		
1.0282	0.6116	0.5763	0.5405	0.5068
0.9286	0.5986	0.5708	0.5401	0.5117
0.9205	0.5945	0.5728	0.5447	0.5175
0.8323	0.5933	0.5733	0.5474	0.5221
0.8010	0.5869	0.5704	0.5456	0.5222
0.5882	0.5806	0.5681	0.5509	0.5336
0.4312	0.5993	0.5830	0.5650	0.5485
0.3633	0.5964	0.5848	0.5728	0.5609
0.2332	—	0.6125	—	—
0.2290	—	0.6313	—	—
0.1523	—	0.6462	—	—
		For KI[b]		
0.1953	—	0.6622	—	—
0.1556	—	0.6776	—	—
0.1002	—	0.6964	—	—
0.05336	—	0.7266	—	—
0.03768	—	0.7821	—	—
0.02825	—	0.7944	—	—

[a] M. Salomon, *J. Phys. Chem.* **73**, 3299 (1969). m/mole kg^{-1}.
[b] M. Salomon, *J. Electroanalyt. Chem.* **26**, 319 (1970).
[c] Idem, ibid. **25**, 1 (1970).

Viscosities of some electrolyte solutions in PC are recorded in Table XLII. Further viscosity measurements are reported by Mukherjee and Boden[100] for dilute solutions of Li^+ and $NR_4{}^+$ halides and perchlorates, by Mukherjee, Boden, and Lindauer[101] for dilute solutions of i-Am_4NI, i-Am_4B^- i-Am_4N^+, KI, $KClO_4$, and by Jansen and Yeager[102] for alkali metal halides, perchlorates, and $NR_4{}^+$ salts.

Miscellaneous specific conductances in PC are collected in Table XLIII. Several of these data refer to the composition of maximum conductance at 25°C. The limiting equivalent conductance Λ_0 and related parameters are shown in Table XLIV; K_A is the ion-pair association constant and J_2 the coefficient of the concentration term $(c\gamma)^{3/2}$ in the theoretical equations used to analyze the conductance data.[103,104] All data were taken at 25°C. Limiting conductances λ_0 for single ions at 25°C are listed in Table XLV.

TABLE XLII

VISCOSITIES OF SOME ELECTROLYTE SOLUTIONS IN PROPYLENE CARBONATE

Electrolyte	Concentration (*m*)	Temperature (°C)	Viscosity Dynamic (CP)[g]	Viscosity Static (CS)[g]	Ref.
LiBr	2.4	25	86.1	—	*a*
$LiClO_4$	0.25	25	3.6	—	*b*
$LiClO_4$	1.00 *M*	25	7.2	—	*c*
$LiClO_4$	1.00	30	8.32	6.58	*d*
$LiAlCl_4$	0.75	25	4.42	—	*e*
$LiAlCl_4$	1.25	25	7.57	—	*e*
NaI	1.1	25	9.6	—	*a*
$NaBF_4$	0.25	25	2.72	—	*b*
KI	0.2	25	3.12	—	*a*
KCNS	1.00	30	4.94	3.99	*d*
KPF_6	0.75	25	4.6	—	*f*
KPF_6	1.00	25	5.7	—	*f*
KPF_6	1.00	30	4.87	—	*d*

[a] W. S. Harris, Thesis, Radiation Lab., University of California (1958).
[b] G. McDonald, K. Murphy, and J. Gower, Report TID 4500-SG-CR-67-2855. 1967.
[c] R. Keller *et al.*, 4th Quarterly Report, Contract NAS 3-8521. 1967.
[d] G. Pistoia, *J. Electrochem. Soc.* **118**, 153 (1971).
[e] M. Eisenberg, 4th Quarterly Report, Contract N00017-68-C-1401. 1968.
[f] W. Elliott and R. Amlie, Final Report, NASA CR-72364. Nat. Aeron. Space Admin., Washington, D.C., 1967.
[g] (CP): centipoise; (CS): centistokes.

Cations appear to be strongly solvated and anions weakly solvated in PC. Tetraalkylammonium halides appear to be unassociated in dilute solution. Of the alkali metal halides, where solubility is sufficient for conductance measurements, only LiCl and LiBr appear associated (Fig. 5).

Very little work has been done on the solubilities and conductances of electrolytes other than uni-univalent in PC; indeed, this is true of organic solvents in general. From a current study[65] it seems that bi-univalent electrolytes are certainly weak in PC; probably several species are in equilibrium.[64, 65]

$$MX_2 \rightleftharpoons MX^+ + X^- \rightleftharpoons M^{2+} + X^-$$
$$2MX_2 \rightleftharpoons MX^+ + MX_3^- \rightleftharpoons M^{2+} + MX_4^{2-}$$

TABLE XLIII

SPECIFIC CONDUCTANCES IN PROPYLENE CARBONATE AT 25°C

Salt	Concentration	κ ($ohm^{-1}\ cm^{-1}$)	Ref.
$AlCl_3$	1.2 *m*	10^{-2} (max)	*a*
$AlCl_3$	1.01 *N*	7×10^{-3}	*b*
$LiAlCl_4$	0.75 *m*	5.1×10^{-3}	*a*
$LiClO_4$	1.0 *m*	5.6×10^{-3} (max)	*c*
$CaBr_2$	0.75 *m*	3.3×10^{-4}	*d*
KPF_6	0.5 *m*	6.8×10^{-3} (max)	*c*
KCNS	1.5 *m*	8.9×10^{-3} (max)	*c*
$Mg(ClO_4)_2$	0.4 *m*	4.7×10^{-3}	*e*
$ZnCl_2$	0.000525 *M*	3.627×10^{-6}	*f*
$ZnCl_2$	0.001071 *M*	4.952×10^{-6}	*f*
$ZnCl_2$	0.001975 *M*	7.372×10^{-6}	*f*
$ZnBr_2$	0.000204 *M*	1.764×10^{-6}	*f*
$ZnBr_2$	0.000565 *M*	3.343×10^{-6}	*f*
$ZnBr_2$	0.001062 *M*	5.584×10^{-6}	*f*
ZnI_2	0.000352 *M*	4.034×10^{-6}	*f*
ZnI_2	0.000724 *M*	5.5030×10^{-6}	*f*
ZnI_2	0.001397 *M*	8.026×10^{-6}	*f*
ZnI_2	0.002042 *M*	10.922×10^{-6}	*f*

[a] D. P. Boden, *Proc. Annu. Power Sources Conf.* **20**, 63 (1966).
[b] R. Keller, J. N. Foster, and J. M. Sullivan, *NASA Contract. Rep.* CR-72106 (1966).
[c] G. Pistoia, M. De Rossi, and B. Scrosati, *J. Electrochem. Soc.* **117**, 500 (1970).
[d] W. S. Harris, Thesis, Radiation Lab., University of California (1958).
[e] J. Chilton and G. Cook, Tech. Report ASD-TDR-62-837 (1962).
[f] M. H. Fawzy and W. H. Lee, to be published.

Transport numbers for $LiClO_4$ and KPF_6 are given in Table XLVI.

The concentration dependence of ^{23}Na chemical shifts in PC has been measured for the anions BPh_4^-, ClO_4^-, SCN^-, and I^-. These results, together with the estimated frequencies (cm^{-1}) of solvation bands in PC, are given in Table XLVII.

The deposition of metals from solutions of their salts in PC has been investigated. Dousek *et al.*[84] obtained 100% faradaic efficiency in depositing potassium from KPF_6 in intensively dried PC, at platinum, gold, and nickel cathodes. The efficiency for redissolution was 25–30%.

TABLE XLIV

EQUIVALENT CONDUCTANCES IN PROPYLENE CARBONATE

Salt	Λ_0	K_a	a (Å)	J_2	Ref.
		Alkali halides and perchlorates			
LiCl	27.5	557	2.53	55.1	*a*
	25.0	—	—	—	*b*
	26.1	437	—	—	*c*
LiBr	27.35	19	2.66	57.0	*a*
	25.3	5.6	—	—	*c*
$LiClO_4$	26.08	—	2.75	57.5	*a*
	26.1	—	—	—	*b*
	27.33	—	—	130, 78	*d*
NaI	28	—	—	—	*e*
$NaClO_4$	28.4	—	—	—	*b*
	27.89	—	—	99, 104	*d*
KI	29.49	—	—	—	*d*
	31	—	—	—	*e*
	30.75	—	3.70	85	*f*
$KClO_4$	30.75	—	3.175	72.5	*f*
	29.6	—	—	97, 98	*d*
$RbClO_4$	30.34	—	—	170, 89	*d*
	31.0	—	—	—	*b*
$CsClO_4$	31.6	2.4	—	79	*b*
	31.1	—	—	—	*d*
$AgClO_4$	30.9	—	—	—	*b*
$TlClO_4$	32.0	—	—	—	*b*
		Trifluoroacetates and MPF_6			
$NaCF_3CO_2$	27.2	189 ± 58	—	—	*g*
KCF_3CO_2	29.04	42.3 ± 0.2	—	—	*g*
$RbCF_3CO_2$	29.57	26.7 ± 0.1	—	—	*g*
$CsCF_3CO_2$	30.21	18.1 ± 0.04	—	—	
$NaPF_6$	26.55	—	—	—	*i*
KPF_6	29.5	—	—	—	*i*
	29.8	—	—	—	*d*
KCNS	33.28	—	—	—	*d*
$LiAsF_6$	22.2	84.7	—	—	*j*
		Tetraalkylammonium salts			
Me_4NClO_4	32.61	2.6	—	271, 237	*d*
Et_4NCl	31.64	—	4.73	108	*a*
Et_4NClO_4	32.6	—	3.60	83	*a, b*
	32.06	—	—	—	*f*
	31.63	1.1	—	—	*d*
Pr_4NCl	28.7	2 ± 0.1	—	—	*g*
Pr_4NClO_4	28.9	1.8	—	272, 320	*d*
Bu_4NCl	29.42	1.9 ± 0.1	—	—	*g*
n-Bu_4NBr	28.65	—	3.53	76	*a, b, f*
	27.89	2.4	—	210, 241	*d*

(*continued*)

TABLE XLIV (*continued*)

Salt	Λ_0	K_a	a (Å)	J_2	Ref.
n-Bu_4NI	25.9	—	—	—	*c*
	27.33	2.5	—	264, 300	*d*
n-Bu_4NNO_3	29.42	1.9 ± 0.1	—	—	*g*
n-Bu_4NClO_4	28.2	—	5.0	—	*a, b, f*
	27.43	2.25	—	229, 223	*d*
Bu_4NPic	21.6	—	—	—	*b*
Bu_4NSCN	30.3	—	—	—	*c*
n-Bu_4NPh_4B	17.14	—	—	—	*h*
i-Am_4NI	26.95	—	3.30	70	*f*
	26.52	2.8	—	—	*d*
i-Am_4N, i-Am_4B	16.37	5.0	5.22	78	*f*
	16.35	5.0	—	354, 292	*d*

[a] L. M. Mukherjee and D. P. Boden, *J. Phys. Chem.* **73**, 3965 (1969).
[b] J. Courtot-Coupez and M. L'Her, *C. R. Acad. Sci., C* **271**, 357 (1970).
[c] U. Mayer, V. Gutmann, and A. Lodzinska, *Monatsh. Chem.* **104**, 1045 (1973).
[d] M. L. Jansen and H. L. Yeager, *J. Phys. Chem.* **77**, 3089 (1973).
[e] W. S. Harris, Thesis, Radiation Lab., University of California (1958).
[f] L. M. Mukherjee, D. P. Boden, and R. Lindauer, *J. Phys. Chem.* **74**, 1942 (1970).
[g] M. L. Jansen and H. L. Yeager, *J. Phys. Chem.* **78**, 1380 (1974).
[h] R. M. Fuoss, J. B. Berkowitz, E. Hirsch, and S. Petrucci, *Proc. Nat. Acad. Sci. U. S.* **44**, 27 (1958); R. M. Fuoss and E. Hirsch, *J. Amer. Chem. Soc.* **82**, 1013 (1960).
[i] A. E. Lyall, H. N. Seiger, and J. Orshich, Tech. Report, AFAPL-TR-68-71 (1968).
[j] H. V. Venkatasetty, *J. Electrochem. Soc.* **122**, 245 (1975).

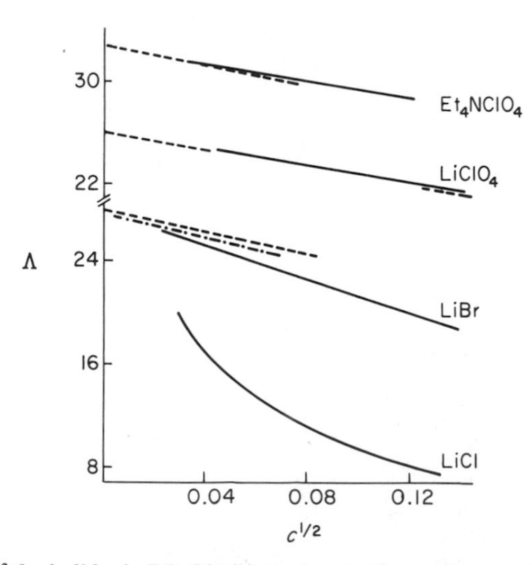

FIG. 5. Λ vs. $c^{1/2}$ for halides in PC. Limiting tangents shown as ––––; that for LiCl –·–·–. Temperature, 25°C. From Mukherjee and Boden.[100]

Harris[2] investigated the deposition of metals from PC solutions. Li was deposited from LiBr, and Na and K from their halides; no deposition of Mg, Ca, or Al was obtained.

Butler[105] discussed the setting up of reference electrodes in aprotic organic solvents including PC. Earlier work with glass electrodes in this solvent shows

TABLE XLV

SINGLE-ION CONDUCTANCES IN PROPYLENE CARBONATE

Ion	λ_0	Ref.	Ion	λ_0	Ref.
Li^+	7.30	*a*	Me_4N^+	14.6	*c*
	6.5	*b*	Et_4N^+	13.28	*a*
	8.89	*c*		13.0	*b*
Na^+	8.8	*b*		13.18	*c*
	9.45	*c*	Pr_4N^+	10.46	*c*
K^+	11.97	*a*	n-Bu_4N^+	9.39	*a*
	10.5	*b*		8.6	*b*
	11.17	*c*		8.98	*c*
Rb^+	11.4	*b*		8.57	*d*
	11.90	*c*		9.3	*c*
Cs^+	12.0	*b*	i-Am_4N^+	8.185	*a*
	12.66	*c*		8.17	*c*
Ag^+	11.3	*b*	PF_6^-	18.6	*b*
Tl^+	12.4	*b*	CNS^-	22.12	*c*
Cl^-	20.20	*a*	Pic^-	13.0	*b*
	18.5	*b*	AsF_6^-	13.3	*g*
Br^-	19.26	*a*	i-Am_4B^-	8.185	*a*
	20.0	*b*		8.17	*c*
	18.91	*c*	Ph_4B^-	8.57	*d*
I^-	18.78	*a*	$CF_3CO_2^-$	17.87	*f*
	18.35	*c*			
ClO_4^-	18.78	*a*			
	19.6	*b*			
	18.44	*c*			

[a] L. M. Mukherjee, D. P. Boden, and R. Lindauer, *J. Phys. Chem.* **74**, 1942 (1970); based on $\lambda^0(i\text{-}Am_4N^+) = \lambda^0(i\text{-}Am_4B^-) = 8.185$ and t°_{Li} in $LiClO_4 = 0.28$.

[b] J. Courtot-Coupez and M. L'Her, *C. R. Acad. Sci., C* **271**, 357 (1970); based on $\lambda^0(Bu_4N^+) = \lambda^0(BPh_4^-) = 8.57$, ref. (d).

[c] M. L. Jansen and H. L. Yeager, *J. Phys. Chem.* **77**, 3089 (1973); based on $\lambda^0(i\text{-}Am_4N^+) = \lambda^0(i\text{-}Am_4B^-)$.

[d] R. M. Fuoss and E. Hirsch, *J. Amer. Chem. Soc.* **82**, 1013 (1960).

[e] U. Mayer, V. Gutmann, and A. Lodzinska, *Monatsh. Chem.* **104**, 1045 (1973).

[f] M. L. Jansen and H. L. Yeager, *J. Phys. Chem.* **78**, 1380 (1974).

[g] H. V. Venkatasetty, *J. Electrochem. Soc.* **122**, 245 (1975); from $\Lambda_0(LiAsF_6) = 22.2$, with $\lambda_0(Li^+) = 8.9$, ref. (c).

TABLE XLVI

TRANSPORT NUMBERS IN PROPYLENE CARBONATE[a]

Salt	Concentration	t^+	Ref.
$LiClO_4$	0	0.39	*b*
		0.28	*c*
		0.32	*d*
	0.02 *N*	0.24	*c*
	0.12 *N*	0.18	*c*
	1.0 *N*	0.19 ± 0.08	*e*
		0.20	*f*
KPF_6	0.45 *m*	0.11	*g*, *h*

[a] Temperature, 25°C.

[b] J. Gabano, Y. Jumel, and J. Laurent, *in* "Power Sources" (D. Collins, ed.), Vol. 2, p. 255. Pergamon, Oxford, 1968.

[c] L. M. Mukherjee, D. P. Boden, and R. Lindauer, *J. Phys. Chem.* **74**, 1942 (1970).

[d] J. P. Hoare and C. R. Wiese, *J. Electrochem. Soc.* **121**, 83 (1974).

[e] R. Keller, 4th Quarterly Report, Contract NAS 3-8521 (1967).

[f] A. N. Dey and M. L. B. Rao, Final Report, Contract DA-44-009-AMC 1537(T) (1967). As quoted by Jasinski, *Adr. Electrochem. E'Chem. Eng.*, No. 8 (1971).

[g] H. N. Seiger, A. Lyall, and R. C. Shair, *Proc. Int. Conf. Power Sources, 6th*, 1968, p. 267 (1968).

[h] A. Lyall, H. Seiger, and J. Orshich, Tech. Report AFAPL-TR-68-71 1968

the emf to be linear to $\log(M^+)$ between 10^{-2} and 10^{-5} M.[94,106,107] The following slopes (mV/log M^+) were recorded at 27°C: Li^+, 53; Na^+, 52; K^+53 (theoretical $2.303RT/F = 59.5$).[107] Calomel and Zn^{2+}/Zn electrodes are said to be unsatisfactory in PC; Tl(I), Pb, and Cd are possible halide-ion reversible electrodes.[108] The Nernst behavior of the Li^+/Li couple in PC has been established,[109] and a Li ribbon electrode in 1 *M* $LiClO_4$ solution proved satisfactory for general electrochemical applications.[110]

The I^-/I_3^- reference electrode has been investigated in PC solution. Two separate equilibria exist[111]: if $[I^-] > [I_2]$, $I_3^- \rightleftharpoons I^-$; and if $[I^-] < [I_2]$, $I_2 \rightleftharpoons I_3^-$.

Matsuura *et al.* have measured the standard electrode potentials of metal–metal ion couples in PC (Table XLVIII).

Electrochemical charge and discharge characteristics at copper electrodes in 1 *M* $LiClO_4$ solution in PC show that oxidation produces sparingly soluble CuCl and reduction of this product gives soluble chlorocuprate complexes.[112]

Anodic stripping of Li, deposited on Mg from the $LiClO_4$ solution, shows both elementary and alloyed Li to be present; the alloying energy is about 7 kcal $mole^{-1}$.[113]

A nonaqueous carbon paste electrode has been used satisfactorily in voltammetric studies.[114] Electrochemical decomposition of PC at a graphite cathode in 1 *M* $LiClO_4$ electrolyte proceeds with almost 100% efficiency.[115]

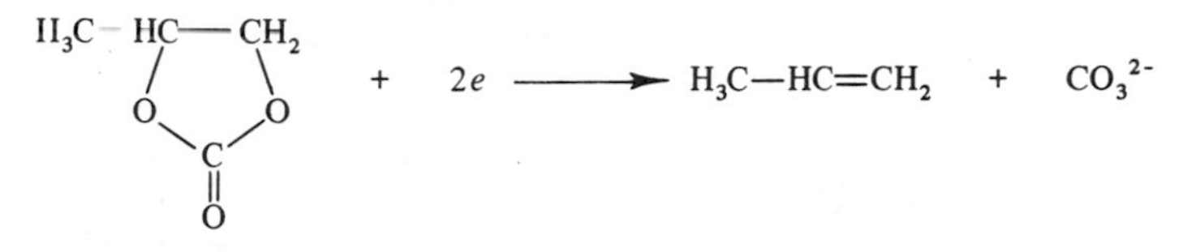

In high-temperature electrode studies, $LiClO_4$ and acetylene black in PC exploded at 215°C.[116]

TABLE XLVII

^{23}Na CHEMICAL SHIFTS AND FREQUENCIES OF SOLVATION BANDS IN PC[a]

Molality (*m*)	^{23}Na chemical shifts (ppm)			
	$NaBPh_4$	$NaClO_4$	NaSCN	NaI
0.5	9.34	9.50	6.61	6.60
0.3	9.22	9.58	6.32	7.13
0.10	9.34	9.49	7.09	8.50
0.01	9.45[b]	9.04	8.85[c]	9.13[c]

	Frequencies (cm^{-1}) of solvation bands in PC					
	Li^+	NH_4^+	Na^+	K^+	Rb^+	Cs^+
ClO_4^-	397	184	186	—	—	—
Ph_4B^-	397	—	184	—	—	—
NO_3^-	401[d]	180[e]	188	144[d]	115[d]	112[d]
Br^-	385	—	—	—	—	—
I^-	—	184	181	141[d]	—	—

[a] M. S. Greenberg, D. M. Weid, and A. I. Popov, *Spectrochim. Acta*, **29***A*, 1927 (1973).
[b] Free Na^+.
[c] Indicative of contact ion pairs. *K* (ion-pair association) for NaCNS = $36.6 \pm 10.6\ m^{-1}$.
[d] $\pm 6\ cm^{-1}$; remaining data $\pm 4\ cm^{-1}$.
[e] Shoulder.

TABLE XLVIII

STANDARD POTENTIALS OF METAL ION COUPLES IN PROPYLENE CARBONATE[a,b]

Ion	E^0 vs. SCE	E^0	
		Rb scale	Water[c]
Li	−2.906	−2.943	−3.27
Na	−2.691	−2.728	−2.954
K	−3.002	−3.039	−3.166
Rb	−2.980	−3.017	−3.166
Cs	−2.986	−3.023	−3.164
Tl	−0.402	−0.439	−0.577
Ag	+0.813	+0.776	+0.553

[a] Temperature, 25°C.

[b] N. Matsuura, K. Umemoto, and Z. Takeuchi, *Bull. Chem. Soc. Jap.* **47**, 806 and 813 (1974). The estimated liquid junction potential against the aqueous phase is +0.037 V.

[c] G. Charlot, "Selected Constants. Oxidation-Reduction Potentials." Pergamon, Oxford, 1958.

Diffusion coefficients have been measured in PC solution from the change in weight of a porous disc filled with solution and suspended in the pure solvent.[117] Values obtained at 25°C were:[118] 1 *M* $LiClO_4$, 2.58×10^{-6} cm^2 sec^{-1}; 0.7 *M* LiCl + 1 *M* $AlCl_3$, 3.04×10^{-6} cm^2 sec^{-1}.

McComsey and Spritzer have published a preliminary report on polarography in PC (note that in Table I of their paper, p. 429, "Dielectric Constant" should read "Static Viscosity").[5] The choice of salt bridge, reference electrode, and supporting electrolyte is discussed,[119] and the polarographic behavior of acids in PC is reported.

A platinum mesh reference electrode was used in the polarographic investigation of polymethine dyes in PC.[120]

The oxidation of halide ions in PC at a rotating platinum anode has been investigated. Oxidation of Cl^- is a single-stage process, those of Br^- and I^- are 2-stage. The voltages are referred to the bisbiphenyl-Cr(I) scale.[121]

$$2Cl^- \rightarrow Cl_2 \qquad E = 1.85 \text{ V}$$
$$3I^- \rightarrow I_3^- \qquad E = 1.08 \text{ V}$$
$$2I_3^- \rightarrow 3I_2 \qquad E = 1.38 \text{ V}$$
$$3Br^- \rightarrow Br_3^- \qquad E = 1.50 \text{ V}$$
$$2Br_3^- \rightarrow 3Br_2 \qquad E = 1.85 \text{ V}$$

TABLE XLIX

POLAROGRAPHIC POTENTIALS OF ALKALI METAL CATIONS IN ORGANIC SOLVENTS[a,b]

	d.c. polarography			a.c. polarography		
	$-E_{1/2}$ (V)	I_d	Slope (mV)	$-E_p$ (V)	$\delta E_p/2$ (mV)	$i_p/i_d\tau^{1/2}$
			Ethylene carbonate			
Li^+	2.009	1.27	73	2.046	140	10
Na^+	1.861	1.28	70	1.862	150	29
K^+	1.986	1.45	70	1.986	130	33
Rb^+	2.009	1.60	75	2.009	135	29
Cs^+	2.011	1.59	77	2.046	170	27
			Propylene carbonate			
Li^+	2.001	0.94	77[c] 82[d]	2.039	155	11
Na^+	1.848	1.08	65 65	1.854	120	44
K^+	1.965	1.11	63 62	1.970	110	45
Rb^+	1.989	1.16	63 67	1.991	120	42
Cs^+	1.981	1.22	60 72	1.982	125	43
			Dimethylformamide			
Li^+	2.385	1.65	91	2.425	170	2
Na^+	2.035	1.83	62	2.035	105	34
K^+	2.055	1.84	63	2.060	108	35
Rb^+	2.045	1.91	62	2.050	105	42
Cs^+	2.011	2.04	58	2.013	98	39
			Dimethyl sulfoxide			
Li^+	2.578	0.93	90	2.688	260	2
Na^+	2.075	1.24	60	2.075	120	38
K^+	2.093	1.34	63	2.095	115	45
Rb^+	2.075	1.33	60	2.075	120	46
Cs^+	2.047	1.46	61	2.050	120	49
			Water			
Li^+	2.332	2.38	56	2.339	105	38
Na^+	2.113	3.02	62	2.114	105	40
K^+	2.136	3.44	61	2.140	135	35
Rb^+	2.131	3.66	62	2.140	138	35
Cs^+	2.108	3.77	65	2.128	140	39

[a] $E_{1/2}$ and E_p are half-wave and peak potentials, respectively, measured against aqueous SCE. I_d is the diffusion current constant, $i_d/cm^{2/3}t^{1/6}$. $\delta E_p/2$ is the half-width of the a.c. current peak. $i_p/i_d\tau^{1/2}$ is the reversibility factor, with drop time at a.c. peak $= \tau$ sec. Temperature, 25°C.

[b] N. Matsuura, K. Umemoto, M. Waki, Z. Takeuchi, and M. Omoto, *Bull. Chem. Soc. Jap.* **47**, 806 (1974).

[c] Values obtained from maximum current polarogram.

[d] Values obtained from mean current polarogram.

Half-wave (d.c.) and peak (a.c.) potentials have been measured against aqueous SCE for alkali metal ions in EC, PC, DMF, DMSO, and water. These data are listed in Table XLIX.

Primary batteries using PC as solvent have been described.[122] $LiClO_4$ electrolyte and Li, CuF_2, and NiS electrodes appear to give greater efficiency in EC than in PC; the lower temperature limit in EC may be reduced by adding some PC or another salt.[123]

Secondary cells have also been described.[124,124a] Two such systems are Li–$LiAlCl_4$–$CuCl_2$[124] and Li–$LiClO_4$–graphite.[124a] Neither fulfills the requirement for the maintenance of charge for extended periods of time stressed by Selim and Bro[125] in their observations on rechargeable Li electrodes in PC.

The kinetics of hydrolysis of PC under varying conditions of pH and temperature has been considered under EC (pp. 12 and 13). Fuoss *et al.* used PC as a solvent in an extended study of quaternization kinetics (e.g., pyridine + *n*-BuBr).[126]

Propylene Carbonate Mixtures

Water and PC are not miscible in all proportions; separation of phases occurs between 0.036 and 0.700 mole fraction of water.[127]

Meredith and Tobias[128] studied the conductances of KCl and R_4NI in PC–water emulsions.* They found the data to fit a modified form of the Maxwell equation

$$K_m = \left(\frac{2+2Xf}{2-Xf}\right)\left(\frac{2+(2X-1)f}{2-(X+1)f}\right)$$

where $X = (K_d - 1)/(K_d + 2)$ for spherical particles of two sizes with volume fractions f_1 and f_2; $f = f_1 + f_2$ and $K_m = k_m/k_d$, $K_d = k_d/k_c$. k_m is the effective conductance of the system, k_c that of the continuous phase, and k_d that of the disperse phase.

In the reduction of water in PC at a Pt disc electrode with Et_4NClO_4 as supporting electrolyte, OH^- is formed. It reacts with the solvent to form HCO_3^- and propane diol.[129] At lower vapor pressures of water, propylene is evolved.[129a] If ions are present, they are usually solvated preferentially by the water, which is not decomposed until present in excess above the solvation requirements.[130] The relative solvation of ions by the two constituents is considered by L'Her *et al.*[127] (p. 139).

* I.e., salts more soluble in the aqueous, or the organic, phase respectively.

TABLE L

PROPERTIES OF WATER–PROPYLENE CARBONATE SOLUTIONS: WATER-RICH PHASE[a, b, c]

X	Y	ε	ε^{ex}	n	n^{ex} ($\times 10^4$)	η (cP)	η^{ex} ($\times 10^4$)	d_4^{25} (g/cm^3)
0	0	78.5	—	1.3327	—	0.8903	—	0.99707
0.0025	0.0117	79.2	0.8	1.3338	9	0.9114	171	1.0000
0.0050	0.0231	78.9	0.7	1.3350	19	0.9267	283	1.0028
0.100	0.0454	78.6	0.75	1.3373	37	0.9590	526	1.0085
0.0200	0.0877	78.3	1.0	1.3411	66	1.0278	1052	1.0183
0.0250	0.1077	78.1	1.1	1.3433	84	1.0641	1335	1.0237
0.0307	0.1298	77.9	1.2	1.3457	103	1.0922	1524	1.0289

X	R (cm^3)	R^{ex} ($\times 10^4$)	P (cm^3)	P^{ex} ($\times 10^4$)	g	V_M (ml/mole)	V^{ex} ($\times 10^3$)	$\bar{V}_W$	$\bar{V}_{PC}$
0	3.7134	—	17.3948	—	2.6	18.065	—	—	—
0.0025	3.7579	−11	17.5516	−29	2.6	18.225	−10	18.07	81.58
0.0050	3.8022	−4	17.7020	−122	2.5	18.384	−19	18.07	81.58
0.0100	3.8908	−9	18.0013	−323	2.5	18.697	−41	18.07	81.58
0.0200	4.0662	−39	18.6197	−527	2.4	19.342	−67	18.07	81.58
0.0250	4,1552	−40	18.9153	−764	2.4	19.651	−93	18.07	81.58
0.0307	4.2595	−13	19.2661	−897	2.3	20.018	−108	18.07	81.57

[a] Temperature, 25°C.

[b] X, mole fraction of PC; Y, volume fraction of PC; ε, dielectric constant at 1.5 MHz; ε^{ex}, excess dielectric constant; n, refractive index; η, viscosity; R, molar refraction; P, polarization; g, Kirkwood tor, evaluated g facaccording to T. B. Hoover, *J. Phys. Chem.* **73**, 57 (1969). V_M, Molar volume of mixture; $\bar{V}_W$, $\bar{V}_{CP}$, partial molar volumes of constituents water and PC, respectively.

[c] J. Courtot-Coupez and M. L'Her, *C. R. Acad. Sci., C* **275**, 103 (1972).

The physical properties of water–PC solutions are recorded in Table L (water-rich phase) and LI (PC-rich phase). Molar and partial molar enthalpies of mixing of water and PC are given in Table LII.

Activity coefficients of transfer are derived from free energies of transfer:[131]

$$\Delta G_t(i) = -1.36 p^E \Gamma^S(i)$$

$p^E \Gamma^S(i)$ is the cologarithm of the activity coefficient and is positive if the ion is more strongly solvated in the medium S than in water. These activity coefficients of transfer are recorded for a number of cations in Table LIII.

TABLE LI

PROPERTIES OF WATER–PROPYLENE CARBONATE SOLUTIONS: PC-RICH PHASE[a, b]

X	Y	ε	ε^{ex}	n	n^{ex} ($\times 10^4$)	η (cP)	η^{ex} ($\times 10^4$)	d_4^{25} (g/cm^3)
0.700	0.9166	65.1	−0.7	1.4138	198	2.1360	1169	1.1835
0.800	0.9496	65.0	−0.4	1.4163	136	2.1767	−368	1.1893
0.850	0.9639	65.0	−0.2	1.4170	99	2.2153	−457	1.1919
0.900	0.9769	64.9	−0.15	1.4181	66	2.2911	−505	1.1946
0.925	0.9831	64.9	−0.1	1.4188	52	2.3279	−540	1.1960
0.950	0.9889	64.8	−0.05	1.4192	34	2.3707	−516	1.1972
0.975	0.9946	64.7	−0.0	1.4198	18	2.4518	−108	1.1985
1	1	65.7	—	1.4202	—	2.5029	—	1.1997

X	R (cm^3)	R^{ex} ($\times 10^4$)	P (cm^3)	P^{ex} ($\times 10^4$)	g	V_M (ml/mole)	V^{ex} ($\times 10^3$)	$\bar{V}_W$	$\bar{V}_{PC}$
0.700	16.2232	285	62.0473	−603	1.3	64.950	−38	17.94	85.09
0.800	18.0052	274	68.4902	−49	1.2	71.702	11	17.95	85.14
0.850	18.8794	101	71.7098	209	1.2	75.073	30	17.96	85.15
0.900	19.7671	63	74.9044	217	1.2	78.422	28	17.95	85.14
0.925	20.2166	100	76.4952	156	1.2	80.088	18	17.95	85.13
0.950	20.6567	43	78,0922	157	1.2	81.763	18	17.95	85.12
0.975	21.1041	60	79.6792	59	1.2	83.428	7	17.95	85.11
1	21.5439	—	81.2702	—	1.2	85.097	—	17.95	85.10

[a] See Table L for list of symbols used. Temperature, 25°C.

[b] J. Courtot-Coupez and M. L'Her, *C. R. Acad. Sci., Ser.* C **275**, 195 (1972).

Polarographic half-wave potentials of ions in PC–water mixtures are reported in Table LIV. Partial pressures, activity coefficients, and excess Gibbs free energies of the PC–water system are shown in Table LV. The equation

$$\Delta G_{\text{excess}} = 2.303RT(\text{B}x_1 + \text{C}x_1^2)$$

applies to these mixtures [PC(1)H_2O(2)], with the following parameters.[132]

x_1	B	C	σ (cal mole^{-1})
>0.682	1.03	−0.54	1
<0.04	1.86	−0.87	1

TABLE LII

ENTHALPIES OF MIXING WATER–PROPYLENE CARBONATE[a]

Mole fraction PC	ΔH_M (J mole^{-1})	ΔH_{PC}[d]	ΔH_W[d]
0.0025[b]	3.6	—	—
0.0049[b]	6.3	—	—
0.0095[b]	18.4	—	—
0.0180[b]	39.7	—	—
0.0219[b]	54.0	—	—
0.0261[b]	70.7	—	—
0.716[c]	1324	362	3764
0.807[c]	1118	564	3285
0.854[c]	901	343	4386
0.902[c]	697	123	5974
0.926[c]	544	50	6750
0.951[c]	385	10	7353
0.975[c]	185	−2	7632

[a] Temperature, 25°C.
[b] J. Courtot-Coupez and M. M. L'Her, *C. R. Acad. Sci.*, *C* **275**, 103 (1972).
[c] J. Courtot-Coupez and M. M. L'Her, *C. R. Acad. Sci.*, *C* **275**, 195 (1972).
[d] Partial molar enthalpies.

The dissociation of the acids $HFSO_3$, $H_2S_2O_7$, $HSbCl_6$, and H_3PF_6 has been followed by conductance measurements in PC. The acid strengths are in the order[133]: HPF_6(fully dissociated) > $HSbCl_6$ > $H_2S_2O_7$ > $HFSO_3$. The H^0 values for solutions of $H_2S_2O_7$ are shown in Table LVI. x_1 is the mole fraction of PC, and σ is the standard deviation of the calculated value from the experimental value. (See Fig 6 for a comparison of PC–H_2O and DMSO–H_2O).

The physical properties of mixtures of PC and DMSO are listed in Table LVII; the interaction between these molecules appears to be relatively weak.[131] The enthalpies of mixing are recorded in Table LVIII.

The dielectric constants of mixtures of PC and EC are given in Table XIX.

Breivogel and Eisenberg[134] measured the conductances of solutions of $AlCl_3$ and LiCl–$AlCl_3$ in PC by a d.c. method; they interpreted the minimum in equivalent conductance (Figs. 3 and 4, ibid, p. 463) as the result of several equilibria.

$$3AlCl_3 \rightleftharpoons AlCl_2^+ + Al_2Cl_7^-$$
$$4AlCl_3 \rightleftharpoons AlCl_4^- + Al_3Cl_8^+$$

TABLE LIII

ACTIVITY COEFFICIENTS OF TRANSFER OF CATIONS FROM WATER TO PROPYLENE CARBONATE[a]

Mole fraction PC	Ag^+, Tl^+, H^+			Alkali metal ions				
	${}^{P}E\Gamma^{S}_{(Ag^+)}$	${}^{P}E\Gamma^{S}_{(Tl^+)}$	${}^{P}E\Gamma^{S}_{(H^+)}$	${}^{P}E\Gamma^{S}_{(Li^+)}$	${}^{P}E\Gamma^{S}_{(Na^+)}$	${}^{P}E\Gamma^{S}_{(K^+)}$	${}^{P}E\Gamma^{S}_{(Rb^+)}$	${}^{P}E\Gamma^{S}_{(Cs^+)}$
0	—	—	—	—	—	—	—	—
0.0025	0.1	0.0	0.1	—	0.1_5	0.1	0.0_5	0.1
0.005	0.0_5	0.0_5	0.1	—	0.1	0.0_5	0.1_5	0.0
0.010	0.0_5	0.2	0.2	−0.1	0.1_5	0.0	0.1	0.0
0.020	0.3	0.2_5	0.3	0.2_5	0.4	0.2	0.2_5	0.4
0.025	0.2	0.3	—	—	—	—	—	—
0.031	0.3	0.3_5	0.4	-0.2_5	0.3_5	0.4	0.3	0.4
0.700	1.1	1.2	—	0.4	0.5	1.2	1.5	1.8
0.800	0.2	1.3_5	—	−0.7	−0.1	0.6	1.0	1.2
0.850	0.0	1.2_5	—	−0.7	−0.1	0.7	1.0	1.3
0.900	−0.4	1.2	—	−0.9	−0.5	0.6_5	1.0_5	1.3_5
0.925	−0.6	1.1	—	-1.2_5	−0.5	0.6	1.0	1.3
0.950	−0.9	1.0_5	—	−1.7	−0.6	0.6	0.9_5	1.3
0.975	−1.1	1.0	—	−2.1	−0.6	0.6	0.9_5	1.4
1	-1.6_5	0.8_5	—	−2.2	−0.8	0.7	1.1	1.5

[a] Temperature, 25°C. M. L'Her, D. Morin-Bozec, and J. Courtot-Coupez, *J. Electroanalyt. Chem.* **55**, 133 (1974).

TABLE LIV

POLAROGRAPHIC HALF-WAVE POTENTIALS IN WATER–PROPYLENE CARBONATE SOLUTIONS[a]

Mole fraction PC	$E_{1/2}(Li^+)$	$E_{1/2}(Na^+)$	$E_{1/2}(K^+)$	$E_{1/2}(Rb^+)$	$E_{1/2}(Cs^+)$
0	−2.477	−2.248	−2.279	−2.274	−2.246
0.0025	—	−2.257	−2.284	−2.277	−2.252
0.005	—	−2.255	−2.282	−2.283	−2.247
0.010	−2.473	−2.257	−2.281	−2.279	−2.247
0.020	−2.492	−2.270	−2.290	−2.288	−2.268
0.031	−2.492	−2.269	−2.301	−2.292	−2.270
0.700	−2.501	−2.279	−2.351	−2.362	−2.350
0.800	−2.437	−2.242	−2.316	−2.332	−2.317
0.850	−2.438	−2.242	−2.319	−2.332	−2.321
0.900	−2.423	−2.219	−2.317	−2.336	−2.325
0.925	−2.404	−2.217	−2.312	−2.331	−2.323
0.950	−2.377	−2.313	−2.312	−2.330	−2.323
0.975	−2.352	−2.211	−2.313	−2.330	−2.331
1	−2.349	−2.200	−2.323	−2.340	−2.334

[a] Data given in volts. Temperature, 25°C. M. L'Her *et al.*, Table LIII.

Densities and specific conductances of solutions of alkali metal chlorides in 1 *M* $AlCl_3$ solution in PC have been measured at 25°C and 35°C.[135] The results are interpreted in terms of ion solvation and complex ion ($AlCl_4^-$) formation, together with structural changes of the solvent. Assuming the solvation,[136]

$$AlCl_3 + \tfrac{3}{2}PC \rightarrow \tfrac{1}{4}Al(PC)_6^{3+} + \tfrac{3}{4}AlCl_4^-$$

the alkali metal ions react

$$xMCl + \tfrac{1}{4}Al(PC)_6^{3+} + \tfrac{3}{4}AlCl_4^- \rightarrow xM^+ + \frac{(1-x)}{4}Al(PC)_6^{3+} + \frac{(3+x)}{4}AlCl_4^- + \frac{3x}{2}PC$$

Electrode kinetics in the system $AlCl_3$–PC have been studied by the same authors.[137] Kinetic parameters are given for the deposition and dissolution of solid alkali metals (Li to Cs) in PC–$AlCl_3$–MCl at 25°C.[137]

The mixed solvent PC–methanol has been used for the titration of organic bases by $HClO_4$.[138] The 21 bases investigated range in p*K*a (of the conjugate acids in water) from 2.50 (*m*-nitraniline) to 11.12 (piperidine).

The solubility of silver halides in PC–tetrahydrothiophene (THT) mixtures has been investigated recently.[139] Equilibrium constants for the formation of Ag halide complexes are defined in the usual way.

$$\begin{array}{rl}
AgX \rightarrow Ag^+ + X^- & K_{s0} \\
Ag^+ + 2X^- \rightarrow AgX_2^- & \beta_2 \\
Ag^+ + 3X^- \rightarrow AgX_3^- & \beta_3 \\
AgX + X^- \rightarrow AgX_2^- & K_{s2},\ K_{s2} = \beta_2 K_{s0} \\
AgX + 2X^- \rightarrow AgX_3^{2-} & K_{s3},\ K_{s3} = \beta_3 K_{s0}
\end{array}$$

These equilibrium constants are listed in Table LIX. Free energies of transfer of single ions from water to PC–THT and other aprotic solvents are collected in Table LX.

TABLE LV

PARTIAL PRESSURES, ACTIVITY COEFFICIENTS, AND EXCESS GIBBS FREE ENERGIES OF THE PROPYLENE CARBONATE–WATER SYSTEM[a]

Mole fraction H_2O	P (mm Hg)	P_{PC} (mm Hg)	P_{H_2O} (mm Hg)	γ_{PC}	γ_{H_2O}	ΔG^E (cal mole^{-1})
0.00	0.053	0.0530[b]	0.00	1.000	—	0
0.01	2.23	0.0524	2.18	1.000	9.17	13
0.02	4.26	0.0520	4.21	1.000	8.85	26
0.03	6.07	0.0516	6.02	1.004	8.44	39
0.05	9.15	0.0510	9.10	1.013	7.65	64
0.08	12.76	0.0495	12.71	1.015	6.68	98
0.12	16.46	0.0484	16.41	1.038	5.75	143
0.16	19.10	0.0472	19.05	1.060	5.01	184
0.20	20.77	0.0463	20.72	1.092	4.36	216
0.24	22.05	0.0456	22.00	1.132	3.86	248
0.28	22.98	0.0449	22.94	1.177	3.45	275
0.318	23.35	0.0446	23.31	1.234	3.08	297
0.96	23.35	0.0446	23.31	21.05	1.021	84
0.97	23.45	0.0388	23.41	24.4	1.015	64
0.98	23.54	0.0344	23.51	32.5	1.009	46
0.99	23.65	0.0223	23.63	42	1.004	25
1.00	23.78	0.000	23.78	—	1.000	0

[a] Temperature, 25°C. S. Y. Lam and R. L. Benoit, *Can. J. Chem.* **52**, 718 (1974).
[b] A value of 0.0530 mm Hg was used for PC.

TABLE LVI

HAMMETT ACIDITY FUNCTIONS FOR SOLUTIONS IN PROPYLENE CARBONATE[a]

Indicator	Disulfuric acid		Hammett function, H^0
	$[H_2S_2O_7]$ (m)	$[H^+]$	
Me-4-dinitro-2,6-aniline	3.3×10^{-3}	2.8×10^{-3}	−4.3
	6.9×10^{-3}	5.2×10^{-3}	−4.6
	1.1×10^{-2}	7.4×10^{-3}	−4.9
Dinitro-2,4-aniline	3.7×10^{-3}	3.1×10^{-3}	−5.1
	4.7×10^{-3}	3.8×10^{-3}	−5.1
	7.0×10^{-3}	5.3×10^{-3}	−5.5
	1.7×10^{-2}	1.0×10^{-2}	−5.5
N,*N*-DiMe-trinitro-2,4,6-aniline	3.4×10^{-3}	2.9×10^{-3}	−4.0
	6.0×10^{-3}	4.6×10^{-3}	−4.1

[a] M. L'Her and J. Courtot-Coupez, *J. Electroanal. Chem.* **48**, 265 (1973).

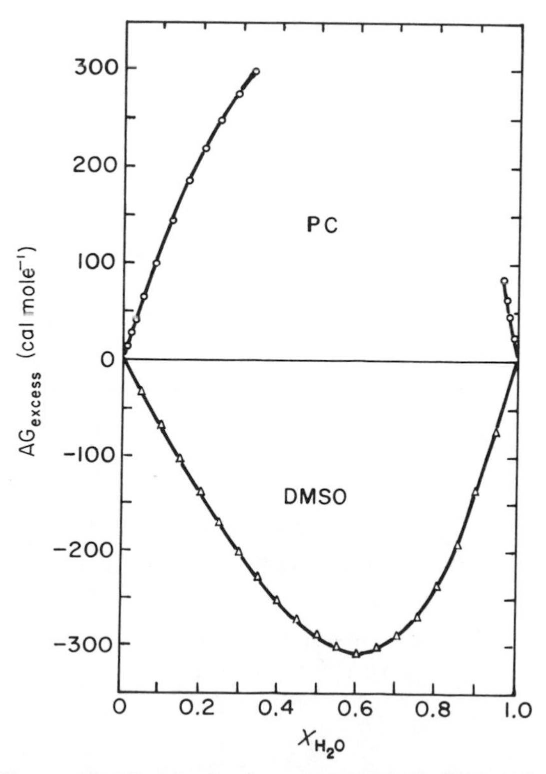

FIG. 6. Excess free energies of mixing for the systems PC–H_2O and DMSO–H_2O. $t = 25°C$. Lam and Benoit.[132] Reprinted by permission of the *Canadian Journal of Chemistry*.

TABLE LVII

PROPERTIES OF DIMETHYL SULFOXIDE–PROPYLENE CARBONATE

X_{DMSO}	ε	ε^{ex}	n	n^{ex} ($\times 10^4$)	η (CP)	η^{ex} ($\times 10^3$)	d_4^{25}	d_4^{25ex} ($\times 10^4$)
0	64.7	0	1.4202	0	2.5024	0	1.1997	0
0.05	64	0.08	1.4223	−7.5	2.4437	−33	1.1944	−0.8
0.1	63.3	0.13	1.4239	−20	2.3814	−69	1.1890	−2.6
0.2	62	0.41	1.4281	−35	2.2739	−125	1.1787	−1.2
0.3	60.5	0.53	1.4331	−42	2.1988	−148	1.1679	−4.8
0.4	59.1	0.81	1.4379	−51	2.1197	−175	1.1573	−6.4
0.5	57.5	0.94	1.4431	−57	2.0593	−183	1.1469	−6
0.6	55.7	0.92	1.4491	−54	2.0206	−170	1.1363	−7.6
0.7	53.8	0.85	1.4551	−51	1.9869	−152	1.1259	−7.2
0.8	51.8	0.77	1.4620	−39	1.9647	−122	1.1155	−6.8
0.9	49.5	0.48	1.4683	−33	1.9734	−61	1.1053	−4.4
1	{46.9; 46,6 (12)}	0	1.4773	0	1.9829	0	1.0953	0

X_{DMSO}	R (cm^3)	R^{ex} ($\times 10^3$)	P (cm^3)	P^{ex} ($\times 10^2$)	g	V_M (ml mole^{-1})	V^{ex} ($\times 10^3$)	$\bar{V}_{PC}$	$\bar{V}_{DMSO}$
0	21.508	0	81.269	0	1.22	85.097	0	85.1	—
0.05	21.444	−4	80.635	8	1.21	84.472	63	85.1	72.5
0.1	21.356	−15	79.996	16	1.21	83.848	128	85.1	72.4
0.2	21.207	−27	78.678	27	1.20	82.547	203	85.1	72.2
0.3	21.089	−7	77.359	38	1.19	81.259	292	85.2	72.1
0.4	20.946	−14	76.008	47	1.17	79.933	343	85.3	71.9
0.5	20.801	−21	74.607	50	1.16	78.569	355	85.4	71.7
0.6	20.677	−8	73.179	50	1.14	77.193	356	85.6	71.6
0.7	20.533	−16	71.704	46	1.12	75.778	318	85.8	71.5
0.8	20.406	−5	70.191	37	1.09	74,336	253	86	71.4
0.9	20.233	−4	68.610	23	1.06	72.854	147	86.3	71.3
1	20.137	0	66.954	0	1.01	71.330	0	—	71.3

[a] Temperature, 25°C. J. Courtot-Coupez and C. Madec, *C. R. Acad. Sci., Ser. C* **277**, 15 (1973).

VI. Chloro–Substituted Cyclic Carbonates

The preparation of these compounds was outlined by Idris Jones[13]; a list of references and patents is given by Paasivirta.[58a]

TABLE LVIII

ENTHALPIES OF MIXING DIMETHYL SULFOXIDE–PROPYLENE CARBONATE[a]

ΔH	X_{DMSO}							
(cal mole^{-1})	0	0.079	0.21	0.40	0.50	0.60	0.86	1
ΔH_M	0	23	59	86	80	70	53	0
ΔH_{DMSO}[b]	360	—	212	137	86	59	16	0
ΔH_{PC}[b]	0	—	18	53	75	88	275	441

[a] Temperature, 25°C. J. Courtot-Coupez and C. Madec, *C. R. Acad. Sci., Ser. C* **277**, 15 (1973).
[b] Partial molar enthalpies.

Proton chemical shifts and coupling constants (from ^{13}C satellite spectra) at 60 MHz have been measured for mono-, di-, and trichloroethylene carbonate and their conformations considered.[58a] From three different skeleton NMR signals, association constants K_n for 1:1 complexes with benzene have been determined for monochloro- and monoacetoxyethylene carbonate.[140a]

		Cl–EC	CH_3CO_2–EC
Signal 4	K_n	$= 0.37 \pm 0.05$	0.80 ± 0.02
Signal 5	K_n	$= 0.44 \pm 0.06$	0.78 ± 0.03
Signal 5′	K_n	$= 0.39 \pm 0.03$	0.79 ± 0.02

The nonplanar ("half-chair") structure deduced for solid and liquid EC is maintained in the monochloro compound.[140b]

The NMR and IR spectra and the ultrasonic absorption of chloromethyl-EC (pure, ex Fluka Chemicals[8]) have been reported.[58, 58a] Two of the three possible isomers, resulting from internal rotation of the CH_2Cl group, were identified from the IR spectrum, and the ultrasonic relaxation data (for 50% solution in acetone) were consistent. The third rotational isomer is not,however, precluded. Infrared and Raman spectra of monochloro-, dichloro-, and chloromethyl-EC are reported, and fundamental vibrational frequencies are assigned.[92]

Chloro-substitution in these compounds reduces their solvent donicities.[6, 6a]

TABLE LIX

EQUILIBRIUM CONSTANTS FOR SILVER HALIDE COMPLEXES IN PROPYLENE CARBONATE–TETRAHYDROTHIOPHENE MIXTURES[a]

System	Halide	$-\log K_{s0}$	$\log \beta_2$	$\log \beta_3$	$\log K_{s2}$	$\log K_{s3}$
PC–0.09 *M* THT[b]	AgCl	12.07 ± 0.02	13.083 ± 0.003	15.01 ± 0.03	1.01 ± 0.02	2.94 ± 0.03
	AgBr	12.50 ± 0.02	13.22 ± 0.04	15.29 ± 0.05	0.72 ± 0.05	2.79 ± 0.05
	AgI	13.73 ± 0.05	14.49 ± 0.05	—	0.77 ± 0.07	—
PC–0.56 *M* THT[c]	AgCl	9.40 ± 0.02	10.538 ± 0.002	12.13 ± 0.01	1.14 ± 0.02	2.74 ± 0.02
	AgBr	10.02 ± 0.03	11.06 ± 0.01	11.97 ± 0.11	1.04 ± 0.04	1.95 ± 0.12
	AgI	10.93 ± 0.10	12.33 ± 0.08	—	1.40 ± 0.13	—
PC–1.56 *M* THT[c]	AgCl	8.3 ± 0.35	9.30 ± 0.07	10.74 ± 0.04	1.00 ± 0.36	2.44 ± 0.35
PC[d]	AgCl	19.87	20.87	23.39	1.00	3.52
	AgBr	20.5	21.2	22	0.7	1.5
	AgI	21.8	22.8	—	1.0	—

[a] Temperature, 25°C. M. Salomon, *J. Phys. Chem.* **79**, 429 (1975).
[b] Ionic strength = 0.01 molar (*M*).
[c] Ionic strength = 0.10 *M*.
[d] Ionic strength = 0.10 *M*; J. N. Butler, *Anal. Chem.* **39**, 1799 (1967); J. Courtot-Coupez and M. L'Her, *Bull. Soc. Chim. Fr.* p. 675 (1969).

TABLE LX

FREE ENERGIES OF TRANSFER OF IONS FROM WATER TO APROTIC SOLVENTS[a]

Solvent	ΔG_t (ion) (kcal mole^{-1})						
	Ag^+	Cl^-	Br^-	I^-	$AgCl_2^-$	$AgBr_2^-$	AgI_2^-
PC	5.3	8.6	6.0	2.7	1.5	−0.8	−7.1
PC–0.09 *M* THT[b]	−5.7	8.8	6.0	2.6	1.4	−1.4	−5.0
PC–0.56 *M* THT[b]	−9.5	9.0	6.4	2.6	1.4	−1.4	−5.9
PN[c]	−2.0	9.5	7.0	3.5	1.2	−1.2	−4.7
DMSO	−7.6	8.7	5.6	1.6	0.6	−2.6	−6.6

[a] Temperature, 25°C. M. Salomon, *J. Phys. Chem.* **79**, 429 (1975). All data refer to molar scale.

[b] Data uncorrected for activity effects in solution. Remaining data corrected for these effects; see M. Salomon and B. K. Stevenson, *J. Phys. Chem.* **77**, 3002 (1973).

[c] Propionitrile.

VII. BUTYLENE CARBONATE

This compound is prepared by ester exchange.[28] Its mass spectrum (at 70 eV) is tabulated below.[61]

Group	Peak	Relative abundance (%)
M^+	116	3.2
$M-C_nH_{2n}O$	72	0.4
$M-CO_2$	72	0.4
$M-HCO_2$	71	2.4
Rearrangements		
C_3H_5O	57	2.4
C_2H_3O	43	30
C_3H_7	43	10
CHO	29	5.7
C_2H_5	29	10.7

Chemical shifts and coupling constants are listed for both the meso and *dl* forms.[28]

Rate and thermodynamic constants for the hydrolysis of butylene carbonate are given below at 52°C.[74]

	k/k_0[a]	$\Delta H^{\ddagger}$ (kcal mole^{-1})	$\Delta S^{\ddagger}$ (cal mole^{-1} deg^{-1})	$\Delta G^{\ddagger}$ (kcal mole^{-1})
Alkaline	29	10.82	24.59	18.81
Acid	43	19.29	25.36	27.54

[a] k_0 is the rate constant for EC at 52°C.

VIII. *o*-Phenylene Carbonate

o-Phenylene carbonate is prepared by the interaction of catechol and phosgene in toluene.[31,141] Mass spectrum data (at 70 eV) are tabulated below.[61]

Group	Peak	Relative abundance (%)
M^+		
cis	142	0.2
trans	142	0.5
$M-CO_2$		
cis	98	0.7
trans	98	0.5
$M-HCO_2$		
cis	97	0.7
trans	97	0.3
Rearrangements		
C_5H_9		
cis	69	5.8
trans	69	26.6
C_3H_3		
cis	55	7.3
trans	55	2.5

In pyrolysis mass spectrometry, fulven-6-one is produced with an ionization

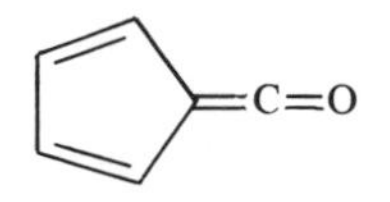

potential of 8.97 ± 1 V.[142]

Pyrolysis begins with the evolution of CO_2, followed by CO.[143] Molecular orbital calculations were designed to rationalize this result; bond lengths and angles for minimization of the energy of the molecule are tabulated below.[144]

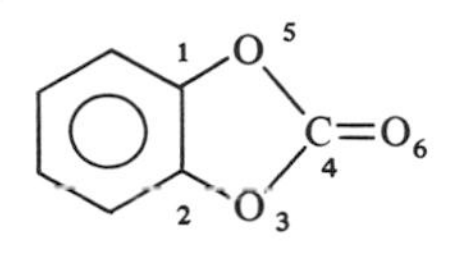

Bond:	2–3	3–4	4–5	4–6	1–5	
Length (Å):	1.360	1.312	1.312	1.233	1.360	
Angle:	$\widehat{512}$	$\widehat{123}$	$\widehat{234}$	$\widehat{345}$	$\widehat{451}$	$\widehat{546}$
	108°	108°	105°	114°	105°	123°

Wentrup has given a thermochemical explanation of the preferential loss of CO_2 on pyrolysis compared with the loss of SO by *o*-phenylene sulfite.[145] Photodecomposition of *o*-phenylene carbonate gives *o*-benzoquinone.[146] The

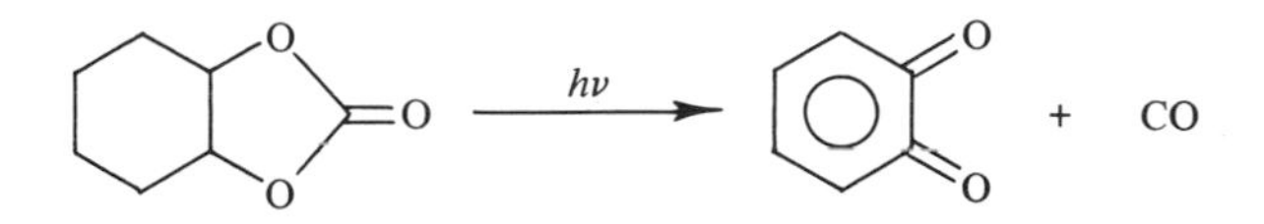

major organic products of gas-phase pyrolysis are indene and naphthalene.[147]

o-Phenylene carbonate is stable in aqueous and acid solutions, but is rapidly hydrolyzed under basic conditions.[141] For hydrolysis by NaOH, the first-order rate constant at 20°C is $k_{20} = 1.27 \pm 0.04\ \mathrm{sec}^{-1}$.

Synthetic uses of *o*-phenylene carbonate are considered by Ohme and Preuschof.[148]

IX. Ethylene Carbonate

The cryoscopic parameters of EC have been redetermined[151]; the results, together with other recent values, are listed below:

mp (°K)	k_{cryos}, K kg mole^{-1}	ΔH_{fusion}, cal g^{-1}	Ref.
309.52	5.32 ± 0.03	35.80 ± 0.20	151
	5.55	34.3	64
	5.27(5)	36.083 ± 0.009	152

TABLE LXI

DOMAIN OF ELECTROACTIVITY OF ELECTROLYTES IN ANHYDROUS EC[a, b]

Electrolyte	Electrode	Anodic limit (V)	Cathodic limit (V)	Domain (V)
$LiClO_4$	Pt	2.5	−3.3	5.8
	Ag	0.4	−3.3	3.7
$NaClO_4$	Pt	2.5	−3.4	5.9
$KClO_4$	Pt	2.5	−3.3	5.8
Et_4NClO_4	Pt	2.5	−2.8	5.3
Bu_4NClO_4	Pt	2.5	−3.7	6.2
Et_4NCl	Pt	0.5	−2.8	3.3
	Ag	−0.6	−2.8	2.2
KPF_6	Pt	3.5	−3.3	6.8

[a] At 40°C, current density 1 mA cm^{-2}.
[b] J.-Y. Cabon, M. L'Her, and M. Le Demezet, *Bull. Soc. Chim. de France*, p. 1020 (1975).

The domain of electroactivity in anhydrous EC, for a number of electrolytes, is reported in Table LXI. Potentials are recorded as limiting values for current densities 1 mA cm^{-2}. High potentials are possible in this solvent, ranging from −3.7 V (reduction of Bu_4N^+) to +3.5 V (oxidation of PF_6^-).

TABLE LXII

FORMATION CONSTANTS AND SOLUBILITY PRODUCTS OF SILVER COMPLEXES IN EC AT 40°C[a, b]

	$p\beta_{21}$	$p\beta_{43}$	$p\beta_{31}$	pK_p	pK_s
BPh_4^-					13
IO_3^-					16.1
Cl^-	−17.4			16.9	17.15
Br^-	−18			17.6	17.8
I^-	−19.7	−57.4		18.4	18.8
SCN^-	−13.8		−16.8	15.5	14.6

[a] J.-Y. Cabon, M. L'Her, and M. Le Demezet, *Comptes. Rend. Ser. C* **280**, 819 (1975).
[b] $\beta_{21} = (AgX_2^-)/(Ag^+).(X^-)^2$
$\beta_{31} = (Ag(SCN)_3^{2-})/(Ag^+)\cdot(SCN^-)^3$
$\beta_{43} = K_{43}.\beta_{21}^2$, $K_{43} = (Ag_3I_4^-)/(Ag^+)\cdot(AgI_2^-)^2$
$K_s = (X^-).(Ag^+)$, etc.
$K_p = (AgX_2^-).(Ag^+)$, etc.

TABLE LXIII

PHYSICAL PROPERTIES OF EC–WATER MIXTURES AT 40°C[a, b]

X_{EC}	ε	ε^{ex}	n	$10^4 n^{ex}$	η	$10^3 \eta^{ex}$	ρ_4^{40}	$\rho_4^{40(ex)}$
0	73.4	0	1.3295	0	0.6531	0	0.9922	0
0.1	77.0	−0.9	1.3573	190	0.7783	92	1.0922	0.0671
0.2	78.2	−2.7	1.3753	281	0.9034	143	1.1558	0.0978
0.3	79.2	−3.7	1.3858	296	1.0287	148	1.1974	0.1084
0.4	80.3	−4.2	1.3941	292	1.1539	102	1.2286	0.1047
0.6	83.1	−3.5	1.4085	259	1.4042	0	1.2714	0.0917
0.8	86.1	−1.9	1.4147	144	1.6546	40	1.2988	0.0432
1.0	89.1	0	1.4180	0	1.9050	0	1.3214	0

X_{EC}	g	V_M	V_M^{ex}	$\bar{V}_{H_2O}$	$\bar{V}_{EC}$	R	$10^4 R^{ex}$	P	P^{ex}
0	2.54	18.15	0	64.69	—	3.6977	0	17.42	0
0.1	2.02	22.90	0.10	65.05	18.14	5.0073	126	22.03	−0.09
0.2	1.76	27.70	−0.14	65.48	18.12	6.3169	286	26.66	−0.15
0.3	1.63	32.59	−0.11	66.13	18.08	7.6265	214	31.38	−0.14
0.4	1.54	37.46	−0.09	66.54	18.00	8.9360	266	36.09	−0.14
0.6	1.46	47.22	−0.02	66.68	17.96	11.5553	1063	45.55	−0.08
0.8	1.44	57.01	+0.05	66.68	17.96	14.1744	928	55.02	0.02
1.0	1.43	66.64	0	66.68	17.96	16.7936	0	64.44	0

[a] Symbols: ε, dielectric constant (1.5 MHz); n, refractive index; η, viscosity, centipoise; ρ, density, g ml^{-1}; g, Kirkwood g-factor, calculated by the method of T. B. Hoover, *J. Phys. Chem.* **73**, 57 (1969); V_M, molar volume ml $mole^{-1}$; $\bar{V}_{H_2O'}$, $\bar{V}_{EC}$, partial molar volumes; R, molar refraction, cm^3; P, molar polarization, cm^3; A^{ex}, excess property of the property A, calculated from the law of ideal mixtures, except for ε^{ex}. This is calculated from

$$\varepsilon^{ex} = \varepsilon - [\varepsilon_{H_2O}(1-Y) + \varepsilon_{EC} Y]$$

where Y is the volume fraction of PC in the mixture: D. Decrooq, *Bull. Soc. Chim., France*, p. 127 (1974).

[b] J.-Y. Cabon, M. L'Her, and M. Le Demezet, *Comptes. Rend. Ser. C* **280**, 747 (1975).

The stability constants of some silver complexes in EC have been determined by potentiometric titration, at 40°C (Table LXII). The high values for solubility products and formation constants show the high stability of these complexes, and confirm that ionic species in EC are feebly solvated.

In EC the system $(I_2 + I^-)$ forms a stable ion I_3^- and a less stable I_5^-. From potentiometric measurements in EC solutions at 40°C, stability constants have been derived[149]: $p\beta_{11}(I_3^-) = -6.8 \pm 0.1$ and $p\beta_{21}(I_5^-) = -8.6 \pm 0.1$. For comparison $p\beta$ values for some solvents concerned in this review are tabulated below.

	Water	MeOH	EtOH	DMF	DMSO	EC[b]	PC[c]
$-p\beta_{11}$[a]	2.9	4.2	4.1	7.8	5.9	6.8	7.8
$-p\beta_{21}$[a]	1	—	1.8	—	—	8.6	9.7

[a] Data refer to 25°C except for EC (40°C).
[b] This paper.[149]
[c] Courtot-Coupez and L'Her.[149a]

The carbonylation of butyl stannyl alkoxides by 5-membered ring carbonates has been investigated, e.g.

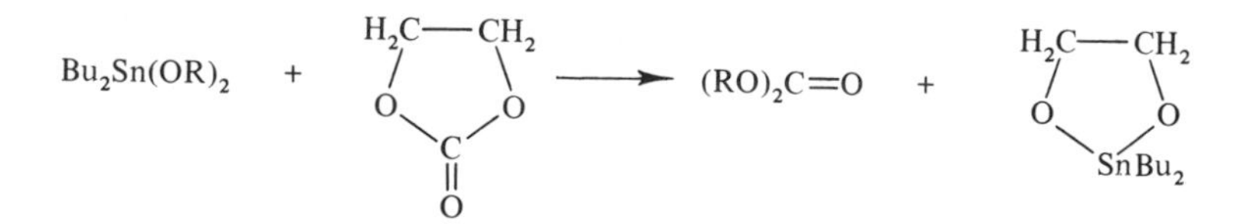

Optimum temperatures and yields for various groups R are listed.[153]

An extensive series of measurements of the physical properties of EC – water mixtures at 40°C is reported. The results, at those compositions for which data are complete, are shown in Table LXIII.

X. Propylene Carbonate

An extensive study has been made of viscosities and conductances for moderately concentrated solutions of Pr_4NBr, Bu_4NBr, and Bu_4NI in PC over the temperature range −50°C to 125°C.[150] Molar conductances, viscosities, and densities were described excellently by polynomials ($Y = \Lambda$, η, or ρ):

$$Y = C_1 + C_2 m^{1/2} + C_3 m + C_4 m^{3/2} \tag{1}$$

The coefficients for Pr_4NBr are recorded in Table LXIV. Log Λ vs. $m^{1/2}$ is linear over the concentration range 0.05 to 0.40 m except at the temperatures $\leqslant$ 0°C.

The number of moles of solvent displaced per mole of solute (from which partial molal volumes may be derived) is given by Δ:

$$\Delta = 1000(\rho_{\text{solv.}} - \rho_{\text{soln.}})/M_1 c + M_2/M_1 \tag{2}$$

where c is molarity, ρ is density, and M_1 and M_2 are molecular weights of

TABLE LXIV

COEFFICIENTS OF MOLALITY (m) [EQ. (1)] FOR SOLUTIONS OF Pr_4NBr IN PC[a]

Temperature (°C)	C_1	C_2	C_3	C_4	Mean σ (%)[b]
		Λ			
−50	1.6916	−3.3816	2.8227	−1.033	0.15
0	16.947	−25.514	22.118	−12.14	0.23
25	29.590	−38.005	47.666	−8.98	—
125	101.79	−119.67	72.180	−20.23	0.05
		η			
−50	52.255	13.977	33.809	144.96	0.37
0	4.559	0.237	2.601	3.59	0.17
25	2.499	0.0671	0.6841	0.601	0.09
125	0.6721	−0.0290	0.4355	−0.065	0.06
		ρ			
−50	1.28081	−0.00252	−0.01052	−0.02586	0.02
0	1.22569	−0.00248	−0.03080	0.00708	0.01
25	1.17310	−0.00057	−0.01759	0.00430	0.00
125	1.09310	−0.00035	−0.00311	−0.00002	0.00

[a] J. F. Casteel, J. R. Angel, H. B. McNeeley, and P. G. Sears, *J. Electrochem. Soc.* **122**, 319 (1975).
[b] $\sigma\% = (Y_{expt} - Y_{calc})/Y_{expt} \times 100$, $Y = \Lambda, \eta$, or ρ.

TABLE LXV

Δ VALUES [EQ. (2)] FOR SOLUTIONS OF Pr_4NBr IN PC[a]

Molality (m)	Δ		
	−50°C	25°C	125°C
0.0765	2.81	2.79	2.62
0.1550	2.81	2.80	2.63
0.2380	2.84	2.79	2.63
0.3200	2.87	2.79	2.63
0.4115	2.86	2.79	2.63

[a] J. F. Casteel, J. R. Angel, H. B. McNeeley, and P. G. Sears, *J. Electrochem. Soc.* **122**, 319 (1975).

solvent and solute, respectively. Δ values for the solute Pr_4NBr are reported in Table LXV (data for Bu_4NBr and Bu_4NI are also given[150]).

It is concluded that there is no significant ion–solvent interaction in these systems.

The absorption of ultrasonic waves by PC and PC–water mixtures has been studied at 25°C.[154] For PC the mean value of the absorption coefficient α/f^2, over the frequency range 10 to 60 MHz, is $45 \pm 2 \times 10^{-17}$ sec^2 cm^{-1}. In the miscible regions of the PC–water system, α/f^2 varies from 20–40 sec^2 cm^{-1} ($\times 10^{-17}$), in the region 0–0.032 mole fraction PC; and from 69–44 sec^2 cm^{-1} ($\times 10^{-17}$), in the region 0.72 to 1.00 mole fraction PC. In the latter region a minimum of $\alpha/f^2 = 42.5 \times 10^{-17}$ sec^2 cm^{-1} occurs at 0.89 mole fraction PC. The large increase in α/f^2 as the mole fraction of PC approaches 0.72 from above agrees with the micro separation of components preceding the miscibility gap, as postulated by Courtot-Coupez and L'Her.[155]

The solubility of carbon dioxide in PC has been measured,[156] and the optimum pressure for the extraction of CO_2 by PC has been established.[157]

Procedures for the purification of PC were compared by gas chromatography and by UV spectroscopy. After treatment with $KMnO_4$, passage through a column of alumina, and fractionation, the highest resistivity measured (5th fraction) was 7×10^8 ohm cm. This fraction was estimated to contain 1.1×10^{-2}% EC and was free from 1,2-propanediol.[158]

The equivalent conductance of $LiAsF_6$ in PC at 25°C is reported:[159]

c	0	10^{-4}	10^{-3}	10^{-2}	10^{-1}	1.0	equiv. liter^{-1}.
Λ_c	22.2	21.2	20.4	19.5	16.4	5.2	ohm^{-1} cm^2

These values were taken from Figure 2, p. 246 of the reference. The ion pair dissociation constant for $LiAsF_6$ in PC is 11.8×10^{-3}.

A study of the reversibility of halide ion electrodes of Cd, Tl(I), Pb and Hg (as the calomel electrode) shows that Tl(I) is the most satisfactory for Cl^-; Cd and Hg are of very limited use.[160] All the metals form stable Br^- and I^- reversible electrodes; however, interaction of the halides of Pb, Tl(I) and Cd (in that order) with PC – alkali halide solutions necessitates caution in the interpretation of their electrode potentials.

The dissociation constants (from conductance measurements), K_d, and solubility products S of some nitrates in PC have been measured[161]:

	$LiNO_3$	$NaNO_3$	KNO_3
$10^3 K_d$	3.14	8.35	19.2
10^5 S	41	1.2	2.5

Alkali metal perchlorates, and $Et_4N.NO_3$, are strong electrolytes in PC.

Stokes's radii for ions in PC (and in dimethylformamide and dimethyl sulfoxide) have been estimated; the radii of ions in PC are given in Table LXVI.

TABLE LXVI

"APPARENT" STOKES' RADII FOR IONS IN PC AT 25°C[a]

	λ_0	r_s, Å		λ_0	r_s, Å
Li^+	9.61	3.37	Bu_4N^+	9.325	3.47
Na^+	10.45	3.10	NH_4^+	14.45	2.24
K^+	11.79	2.75	Cl^-	15.94	2.03
Rb^+	14.05	2.31	Br^-	18.75	1.73
Cs^+	14.74	2.20	I^-	19.09	1.70
Me_4N^+	14.85	2.18	ClO_4^-	18.28	1.77
Et_4N^+	13.82	2.34	NO_3^-	22.12	1.47

[a] N. Matsuura, K. Umemoto, and Y. Takeda, *Bull. Chem. Soc. Jap.* **48**, 2253 (1975). The ion conductances λ_0 were calculated on the assumption $\lambda_0(Bu_4N^+) = \lambda_0(Bu_4B^-)$. Stokes' radii from the equation

$$r_s = e.E/6\pi\eta\lambda_0$$

The relationship

$$r_s - 2r_c/3 = 0.31(r_c - 3.8)^2 - 0.3$$

holds for the ions in PC, where r_c is the crystal ionic radius.

The extraction of Cu(II) from aqueous thiocyanate solutions into PC, and its subsequent spectrophotometric determination by atomic absorption, has been investigated.[162] The optimum pH range is 0–5.0. The metals Hg, Zn and Co form extractable thiocyanate species, but up to 14.5 mg of these elements in 50 ml of sample does not reduce the accuracy of the determination.

Tungsten(VI) has also been extracted by PC.[163] The optimum pH is about 2.70, 89% W being extracted. A method for the separation of W and Mo is worked out.

Propylene carbonate is one of the solvents considered in detail (the others are dimethyl formamide, acetonitrile and methyl formate) in a comprehensive review of the properties of nonaqueous electrolytes.[164]

XI. PROPYLENE CARBONATE–WATER

Liquid-vapor equilibrium in the PC—water system has been studied.[165] Typical results are at $t = 27°C$

H_2O, mass %	0.263	0.345	0.476	0.797	0.986	1.02	2.11
p, mm Hg	9.27	10.47	9.81	12.34	14.20	12.90	17.75
H_2O, mass %	3.06	3.38	7.50				
p, mm Hg	19.73	21.10	25.67				

The phase diagram for this system was verified (Fig. 7).[166]

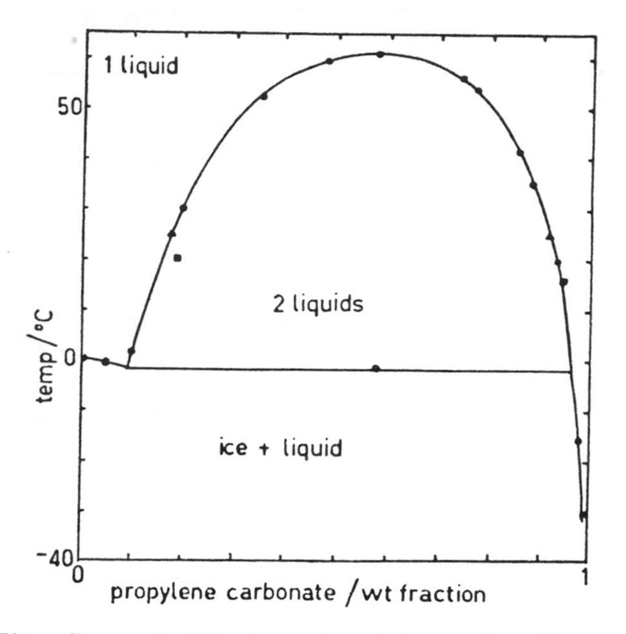

FIG. 7. Phase diagram for PC-water system at 1 atmosphere. Upper critical solution temperature = 61.1°C at 57.5 mole% PC.[166] (Reprinted by permission of the Journal of Chemical Engineering Data.)

The solvation of molecules and anions in PC-water mixtures has been investigated.[167] Solubility products and stability constants of some silver halides in these mixtures are listed in Table LXVII, and data relating to the iodine redox system in Table LXVIII.

In the water-rich mixtures, halide ions are solvated by water, and addition of PC has no effect; the introduction of water to PC-rich mixtures has a very marked effect, and increases the solvating power of the medium. Iodine (and ferrocene) are preferentially solvated by the organic solvent, and the highly polarizable anions I_3^- and BPh_4^- behave similarly.

Hydration constants for the process: $I + H_2O = I.H_2O$ (K_h) have been determined on the basis $K_h(Na^+) = K_h(NO_3^-)$:[161]

	Li^+	Na^+	K^+	NO_3^-
K_h	6.5	2.5	0.5	2.4

TABLE LXVII

SOLUBILITY PRODUCTS (pK_s), STABILITY CONSTANTS ($p\beta_{21}$) AND PRECIPITATION CONSTANTS (pK_p) OF SOME SILVER SALTS AND ANIONS IN PC–WATER MIXTURES[a,b]

Mole fraction PC:		0	0.031	0.700	0.800	0.900	0.950	0.970	1.00
AgCl(s)	pK_s	9.5	9.6	13.6	14.1	16.1	17.6	18.6	20.0
AgBr(s)	pK_s	12.1	12.0	15.5	16.0	17.6	18.7	19.2	20.5
AgI(s)	pK_s	15.8	15.8	—	—	—	—	—	21.8
$AgB(Ph)_4$	pK_s	17.2	—	10.8	11.4	12.0	12.6	12.6	12.5
$AgCl_2$	$p\beta_{21}$	−4.7	—	—	−13.5	−15.9	−17.7	−19.0	−20.9
	pK_{s2}	4.8	—	—	0.6	0.2	−0.1	−0.4	−0.9
	pK_p	14.3	—	—	14.7	16.3	17.5	18.2	19.1
$AgBr_2^-$	$p\beta_{21}$	−7.1	—	—	−15.5	−17.5	−18.8	−19.7	−21.2
	pK_{s2}	5.0	—	—	0.5	0.1	−0.1	−0.5	−0.7
	pK_p	17.1	—	—	16.5	17.7	18.6	18.7	19.8

[a] M. L'Her, D. Morin-Bozec, and J. Courtot-Coupez, *J. Electroanalyt. Interfacial Chem.* **61**, 99 (1975).
[b] $pK_{s2} = (AgX_2^-)/(X^-), = \beta_{21} K_s$.

TABLE LXVIII

STANDARD POTENTIALS OF REDOX SYSTEMS OF IODINE, AND STABILITY CONSTANTS OF I_3^- ($p\beta_{11}$), IN PC–WATER MIXTURES[a]

Mole fraction PC:	0	0.700	0.800	0.900	0.950	1.00
E^0 (I^-/I_3^-), V		−0.185	−0.190	−0.234	−0.255	−0.292
E^0 (I_3^-/I_2), V	0.398	0.379	0.404	0.393	0.393	0.388
$p\beta_{11}$	−2.9	−6.5	−6.8	−7.2	−7.4	−7.7

[a] M. L'Her, D. Morin-Bozec, and J. Courtot-Coupez, *J. Electroanalyt. Interfacial Chem.* **61**, 99 (1975).

XII. PROPYLENE CARBONATE–$AlCl_3$

The free energies of solvation of the alkali metal chlorides (Li to Cs) in 1 molar $AlCl_3$–PC solvent have been determined from emf measurements. Individual-ion free energies, determined by the method of Latimer, Pitzer and Slansky,[168] are given in Table LXIX.

TABLE LXIX

FREE ENERGIES OF SOLVATION OF IONS IN WATER, PC AND 1 MOLAR $AlCl_3$–PC

	$\Delta G^{\circ}_{solvation}$		
	$AlCl_3$–PC[a]	PC[b]	H_2O[b]
Li^+	−98.4	−95.0	−97.8
Na^+	−76.6	−71.9	−72.4
K^+	−59.6	−56.6	−54.9
Rb^+	−53.8	−54.5	−50.7
Cs^+	−51.7	−52.7	−46.7
Cl^-	−90.0	−88.2	−100.3

[a] J. Jorné and C. W. Tobias, *J. Phys. Chem.* **78**, 2576 (1974).
[b] M. Salomon, *J. Phys. Chem.* **74**, 2519 (1970).

TABLE LXX

FREE ENERGIES OF TRANSFER OF INDIVIDUAL IONS FROM WATER TO 1 MOLAR $AlCl_3$–PC, AND TO PURE PC; STANDARD ENTROPIES OF TRANSFER OF ALKALI METAL CHLORIDES FROM WATER TO 1 MOLAR $AlCl_3$–PC

	$\Delta G^{\circ}_{tr}(M)_{H_2O-AlCl_3-PC}$[a] (kcal mole^{-1})	$\Delta G^{\circ}_{tr}(M)_{H_2O-PC}$[b] (kcal mole^{-1})
Li^+	−0.6	+2.8
Na^+	−4.2	+0.5
K^+	−4.7	−1.7
Rb^+	−3.1	−3.8
Cs^+	−5.0	−6.0
Cl^-	+10.3	+12.1

	$\Delta S^{\circ}_{AlCl_3-PC}$[c]	$\Delta S^{\circ}_{H_2O}$[c]	Δ°_{tr}[c]	$\Delta^{\circ}_{tr}(vol)$[d]
LiCl	−18.4	−0.6	−17.8	−18.7
NaCl	−16.1	+4.9	−21.0	−21.9
KCl	+9.2	+12.0	−2.8	−3.7
RbCl	+55.3	+15.8	+39.5	+38.6
CsCl	+64.7	+14.7	+50.0	+49.1

[a] J. Jorné and C. W. Tobias, *J. Phys. Chem.* **78**, 2576 (1974). Aqueous values from ref. (b) below.
[b] M. Salomon, *J. Phys. Chem.* **74**, 2519 (1970).
[c] J. Jorné and C. W. Tobias, *J. Phys. Chem.* **78**, 2576 (1974). Entropies in cal mole^{-1} °K^{-1}, and on molal basis.
[d] Entropies on volume basis.

The free energies of transfer of ions, and the standard entropies of transfer of the chlorides, from water to $AlCl_3$–PC mixtures, are reported in Table LXX.

The thermodynamic properties of alkali metals in $AlCl_3$–PC mixtures have been studied. Activity coefficients of the chlorides in 1 molar $AlCl_3$–PC solution are shown in Table LXXI, and the standard oxidation potentials of the metals in various solvents are compared in Table LXXII.

TABLE LXXI

ACTIVITY COEFFICIENTS OF THE ALKALI METAL CHLORIDES IN 1 MOLAR $AlCl_3$–PC SOLVENT AT 25°C[a]

	$-\ln \gamma_{MCl}$				
Molal	LiCl	NaCl	KCl	RbCl	CsCl
1.0	4.051	—	3.808	2.768	3.738
0.5	3.892	3.181	3.623	3.990	4.455
0.25	3.719	2.652	3.626	3.970	4.398
0.10	2.863	1.713	3.423	2.885	3.273
0.05	2.604	—	—	2.143	—
0.01	1.427	0.442	1.193	0.961	1.263
0.005	0.962	—	—	—	—
0.0025	0.989	—	0.631	0.412	0.872

[a] J. Jorné and C. W. Tobias, *J. Electrochem. Soc.* **122**, 624 (1975).

TABLE LXXII

STANDARD OXIDATION POTENTIALS IN VARIOUS SOLVENTS[a]

	Aceto-nitrile	*N*-Methyl formamide	Formamide	Ethylene glycol
Li	3.23	3.124		2.996
Na	2.87	2.807		2.686
K	3.16	3.021	2.872	2.897
Rb	3.17		2.855	
Cs	3.16	2.987		
Reference electrode	H_2/H^+[b, c]	Ag/AgCl[d]	H_2/H^+[e]	Ag/AgBr[f]

(*continued*)

TABLE LXXII (*continued*)

	PC			PC–AlCl$_3$(lm)	Water
Li	1.851	1.842	1.845	2.045	3.045
Na			1.619	1.885	2.714
K			1.934	2.116	2.925
Rb				2.116	2.925
Cs				2.122	2.923
Reference electrode	Tl/TlCl[g]	Tl/TlBr[g]	Tl/TlI[h,i]	Tl/TlCl[a]	H_2/H^+[j]

[a] J. Jorné and C. W. Tobias, *J. Electrochem. Soc.* **122**, 624 (1975).
[b,c] V. A. Pleskov, *Usp. Khim.* **16**, 254 (1947); *idem, Zhur. Fiz. Khim.* **22**, 351 (1958).
[d] H. Lund, *Acta Chem. Scand.* **11**, 491 (1957).
[e] T. Pavlopoulos and H. Strehlow, *Z. Phys. Chem.* (*Frankfurt*) **2**, 89 (1954).
[f] K. K. Kundu, A. K. Rakshit, and M. N. Das, *Electrochimica Acta* **17**, 1921 (1972).
[g] M. Salomon, *J. Phys. Chem.* **73**, 3299 (1969).
[h,i] M. Salomon, *J. Electroanalyt. Chem.* **25**, 1 (1970); *ibid.* **26**, 319 (1970).
[j] W. M. Latimer, "Oxidation Potentials." Prentice Hall, New York (1952).

XIII. Propylene Carbonate–Dimethyl Sulfoxide

The polarography of the lanthanide ions Eu^{3+}, Yb^{3+} and Sm^{3+} in PC–DMSO mixtures shows that, in all cases, the passage from PC to DMSO is energetically favorable.[169] The stability and stoichiometry of DMSO–Eu^{3+}, Eu^{2+}, Y^{3+} and Y^{2+} complexes in PC are reported; some half-wave potentials are reported in Table LXXIII.

The polarographic reduction of the trifluoromethanesulfonates of Pb^{2+}, Cd^{2+} and Tl^+, in PC–DMSO solvent, has been examined.[170] The reduction waves are reversible, except for Cd and Pb in mixtures containing less than 0.2% by volume of DMSO. By comparison of the half-wave potentials with that of the cobaltocene-cobalticinium couple, transfer activity coefficients $\Gamma(M^{2+})^{Solvent}_{H_2O}$, were obtained. For the theory of the method, see Strehlow.[171]

	$E_{M^{2+}}$ (mV)			$\log \Gamma(M^{2+})^{Solvent}_{H_2O}$		
	Tl^+	Cd^{2+}	Pb^{2+}	Tl^+	Cd^{2+}	Pb^{2+}
DMSO	380 ± 4	224 ± 4	373 ± 4	−5.5	−12.2	−13.7
PC	714 ± 4	880 ± 4	944 ± 4	+0.2	+9.9	+6.9

In binary solvents the half-wave potentials shift towards more negative values, and increase in DMSO content reduces drastically the transfer coefficients.

TABLE LXXIII

POLAROGRAPHY IN DMSO–PC MIXTURES: SOME RESULTS FOR Eu AND Yb IONS[a, b]

	Vol. % DMSO	$E_1(1/2)$ (mV)	$E_2(1/2)$ (mV)
Eu	100	51	−1257
	50	71	−1200
	0.8	285	−833
	0	1020	−730
Yb	100	−622	−1353
	40	−592	−1284
	0.8	−431	−962
	0	+345	−770

[a] J. Massaux and G. Duyckaerts, *J. Chem. Soc. Belge*. **84**, 519 (1975). Other mixture compositions, and the more complex results for Sm^{3+}, are tabulated in this paper.

[b] $E_1(1/2)$ relates to $M^{3+} + e \rightarrow M^{2+}$

$E_2(1/2)$ relates to $M^{2+} + 2e \rightarrow M^0$

The solubility of acetylene in PC–acetone mixtures has been recorded from −20°C to 120°C.[172] The absorption coefficient and the mole fraction of solute are listed for two mixtures: 15 wt.% PC, and 61.2 wt.% PC.

A calorimetric study of the systems *o*-xylene–PC and *o*-xylene pyrrolidine has been made.[173] Symmetrical curves of ΔH_{mixing} against mole% composition are obtained; limiting values for PC systems are:

	0°	15°	25° C
PC solute, *o*-xylene solvent	1000	1200	1300 cal mole^{-1}
o-xylene solute, PC solvent	1050	1150	1250 cal mole^{-1}

REFERENCES

1. *Chemical Abstracts Index Guide*, **76** (1972).
2. W. S. Harris, Thesis (UCRL-8381), Radiation Lab., University of California, Berkeley 1958.
3. R. Jasinski, *J. Electroanal. Chem.* **15**, 89 (1967).
4. R. Jasinski, *Advan. Electrochem. Electrochem.* **8**, 253–335 (1971).
5. H. J. McComsey, Jr. and M. S. Spritzer, *Anal. Lett.* **3**, 427 (1970).
6. V. Gutmann, *Angew. Chem.* **9**, 843 (1970).

6a. V. Gutmann, "Coordination Chemistry in Non-Aqueous Solutions," p. 142. Springer-Verlag, Berlin and New York, 1968.

7. L.-H. Lee, *J. Paint Technol.* **42**, 365 (1970).
8. Technical Bulletin. Jefferson Chemical Co., Houston, Texas, "Catalog of Organic Chemicals No. 9." Fluka AG, CH 9470, Buchs, Switzerland, 1973.
9. Aldrich Chemical Co., 940 West St., Milwaukee, Wisconsin.
10. R. N. Emmanuel, 264 Water Road, Alperton, Middlesex, England.
11. B. D. H. Chemicals Ltd., Broom Road, Poole BH 12 4NN, England.
12. Chemische Werke Huls AG, Recklinghausen, Germany.
13. J. Idris Jones, "Chemistry Research Reports," pp. 37–38. National Physical Lab., Teddington, England, 1953; also pp. 35–36 (1954).
14. B. J. Ludwig and E. C. Piech, *J. Amer. Chem. Soc.* **73**, 5779 (1951).
15. S. Sarel, L. A. Pohoryles, and R. Ben-Shoshan, *J. Org. Chem.* **24**, 1873 (1959).
16. R. J. De Pasquale, *Chem. Commun.* p. 157 (1973).
17. R. G. Pews, *Chem. Commun.* p. 119 (1974).
18. M. S. Newman and R. W. Addor, *J. Amer. Chem. Soc.* **75**, 1263 (1953); **77**, 3789 (1955).
19. F. W. Breitbeil, J. J. McDonnell, T. A. Marolewski, and D. T. Denerlein, *Tetrahedron Lett.* p. 4627 (1965).
20. N. D. Field and J. R. Schaefgen, *J. Polym. Sci.* **58**, 533 (1962).
21. J. M. Judge and C. C. Price, *J. Polym. Sci.* **41**, 435 (1959).
22. C. R. Kowarski and S. Sarel, *J. Org. Chem.* **38**, 117 (1973).
23. W. K. Johnson and T. L. Patton, *J. Org. Chem.* **25**, 1042 (1960).
24. J. Idris Jones, private communication.
25. See refs. 8–10.
26. K and K, 121 Express St., Plainview, New York.
27. Schuchardt G.m.b.H., Munchen, Germany; Koch-Light Ltd., Colnbrook SL3 OBZ, England.
28. F. A. L. Anet, *J. Amer. Chem. Soc.* **84**, 747 (1962).
29. J. J. Kolfenbach, E. I. Fulmer, and L. A. Underkofler, *J. Amer. Chem. Soc.* **67**, 502 (1945).
30. J. Burkhardt, R. Feinauer, E. Gulbins, and K. Hamann, *Chem. Ber.* **99**, 1912 (1966).
31. R. S. Hanslick, W. F. Bruce, and A. Mascitti, *Org. Syn., Collect. Vol.* **4**, 788 (1963); D. C. De Jongh and D. A. Brent, *J. Org. Chem.* **35**, 4204 (1970).
32. G. W. Thomson, *in* "Techniques in Organic Chemistry" (A. Weissberger, ed.), Vol. I, Part I, p. 461. Wiley (Interscience), New York, 1952.
33. R. C. Benson and W. H. Flygare, *J. Chem. Phys.* **58**, 2366 (1973).
34. G. R. Slayton, J. Simmons, and J. H. Goldstein, *J. Chem. Phys.* **22**, 1678 (1954).
35. W. F. White and J. E. Boggs, *J. Chem. Phys.* **54**, 4714 (1971).
36. J. L. Hales, J. Idris Jones, and W. Kynaston, *J. Chem. Soc., London* p. 618 (1957).
37. K. L. Dorris, J. E. Boggs, A. Danti, and L. L. Altpeter, *J. Chem. Phys.* **46**, 1191 (1967).
38. J. R. Durig, J. W. Clark, and J. M. Casper, *J. Mol. Struct.* **5**, 67 (1970).
39. C. J. Brown, *Acta Crystallogr.* **7**, 92 (1954).
40. A. D. Bain and D. C. Frost, *Can. J. Chem.* **51**, 1254 (1973).
41. V. F. Katanskaya and O. M. Klimova, *Vysokomol. Soedin., Ser. A* **9**, 1889 (1967).
42. W. H. Lee and A. H. Saadi, *J. Chem. Soc.*, Section (B) p. 5 (1966).
43. J. Fleischauer and H. D. Scharf, *Tetrahedron Lett.* **12**, 1119 (1972).
44. W. Hartmann, *Chem. Ber.* **101**, 1643 (1968); J. Daub and V. Trautz, *Tetrahedron Lett.* p. 3265 (1970).
45. D. E. Bergstrom and W. C. Agosta, *Tetrahedron Lett.* p. 1087 (1974).
46. F. W. Breitbeil, J. J. McDonnell, T. A. Marolewski, and D. T. Denerlein, *Tetrahedron Lett.* p. 4627 (1965).
47. W. Hartmann, H. G. Heine, H. M. Fischler, and D. Wendisch, *Tetrahedron* **29**, 2333 (1973).

48. H. D. Scharf, *Angew. Chem.* **86**, 567 (1974).
49. G. Pistoia, M. De Rossi, and B. Scrosati, *J. Electrochem. Soc.* **117**, 500 (1970).
50. R. Payne and I. E. Theodorou, *J. Phys. Chem.* **76**, 2892 (1972).
51. R. J. W. Le Fèvre, A. Sundaram, and R. K. Pierens, *J. Chem. Soc., London* p. 479 (1963).
52. J.-P. Gosse and B. Rose, *C. R. Acad. Sci., Ser. C* **267**, 927 (1967).
53. C. L. Angell, *Trans. Faraday Soc.* **52**, 1178 (1956).
54. A. Simon and G. Heintz, *Chem. Ber.* **95**, 2333 (1962).
55. J. Wang, C. O. Britt, and J. E. Boggs, *J. Amer. Chem. Soc.* **87**, 4950 (1965).
56. J. R. Durig, G. L. Coulter, and D. W. Wertz, *J. Mol. Spectrosc.* **27**, 285 (1968).
57. K. L. Dorris, C. O. Britt, and J. E. Boggs, *J. Chem. Phys.* **44**, 1352 (1966).
58. R. A. Pethrick, E. Wyn-Jones, P. C. Hamblin, and R. F. M. White, *J. Chem. Soc., A* p. 1852 (1969); B. Arbuzov, Yu. Samitov, and R. Aminov, *Zh. Strukt. Khim.* **5**, 538 (1964); P. Swinton and G. Gatti, *J. Magn. Resonance* **8**, 293 (1972).
58a. J. Paasivirta, *Suom. Kemistilehti B* **41**, 364 (1968).
59. B. Fortunato, P. Mirone, and G. Fini, *Spectrochim. Acta, Part A* **27**, 1917 (1971).
60. G. Fini, P. Mirone, and B. Fortunato, *J. Chem. Soc., Faraday II Trans.* **69**, 1243 (1973).
61. P. Brown and C. Djerassi, *Tetrahedron* **24**, 2949 (1968).
62. S. K. Gross and C. Schuerch, *Anal. Chem.* **28**, 277 (1956).
63. R. F. Kempa and W. H. Lee, *Talanta* **9**, 325 (1962).
64. O. D. Bonner, Si Joong Kim, and A. L. Torres, *J. Phys. Chem.* **73**, 1968 (1969).
65. M. H. Fawzy, private communication.
66. H. Ulich and E. J. Birr, *Z. Angew. Chem.* **41**, 443 (1928).
67. R. F. Kempa and W. H. Lee, *J. Chem. Soc., London* p. 100 (1961).
68. H. Mujamoto and Y. Watanabe, *J. Chem. Soc. Jap.* **88**, 36 (1967).
69. L. A. Pohoryles and S. Sarel, *C. R. Acad. Sci.* **245**, 2321 (1957).
70. S. Sarel, I. Levin, and L. A. Pohoryles, *J. Chem. Soc., London* Pt. V, p. 3079 (1960).
71. L. A. Pohoryles, I. Levin, and S. Sarel, *J. Chem. Soc., London* Pt. VI, p. 3082 (1960).
72. W. H. Lee and A. H. Saadi, *J. Chem. Soc., B* p. 1 (1966).
73. I. Levin, L. A. Pohoryles, S. Sarel, and V. Usieli, *J. Chem. Soc., London* Pt. VIII, p. 3949 (1963).
74. J. Katzhendler, L. A. Poles, and S. Sarel, *Isr. J. Chem.* **10**, 111 (1972).
75. J. Katzhendler, L. A. Poles, and S. Sarel, *J. Chem. Soc., B* Pt. XII, p. 1847 (1971).
76. E. D. Bergmann and I. Shahak, *J. Chem. Soc., Org. Trans.* p. 899 (1966).
77. T. Yoshino, S. Inaba, and H. Komura, *Bull. Chem. Soc. Jap.* **47**, 405 (1974).
78. W. Furukawa, T. Yoshida, and S. Hayashi, *Bull. Chem. Soc. Jap.* **47**, 2893 (1974).
79. R. F. Nelson and R. N. Adams, *J. Electroanal. Chem.* **13**, 184 (1967); R. Jasinski and S. Kirkland, *Anal. Chem.* **39**, 1663 (1967).
80. J. B. Headridge, "Electrochemical Techniques for Inorganic Chemists." Academic Press, New York, 1969.
81. R. Jasinski and S. Carroll, *Anal. Chem.* **40**, 1908 (1968).
82. S. G. Meibuhr, B. E. Nagel, and R. Gatrell, *Energy Convers.* **10**, 29 (1970).
83. J. Jansta, F. P. Dousek, and J. Řiha, *J. Electroanal. Chem.* **38**, 445 (1972); F. P. Dousek, J. Jansta, and J. Řiha, *Chem. Listy* **67**, 427 (1973).
84. F. P. Dousek, J. Řiha, and J. Jansta, *J. Electroanal. Chem.* **39**, 217 (1972).
85. J. C. Synott, D. R. Cogley, and J. N. Butler, *Anal. Chem.* **44**, 2247 (1972).
86. Y.-C. Wu and H. L. Friedman, *J. Phys. Chem.* **70**, 501 (1966).
87. M. Watanabe and R. M. Fuoss, *J. Amer. Chem. Soc.* **78**, 527 (1956).
88. L. Simeral and R. L. Amey, *J. Phys. Chem.* **74**, 1443 (1970).
89. H. Finegold, *J. Phys. Chem.* **72**, 3244 (1968).
90. R. F. Nelson and R. N. Adams, *J. Electroanal. Chem.* **13**, 184 (1967).

91. H. L. Yeager, J. D. Fedyk, and R. J. Parker, *J. Phys. Chem.* **77**, 2407 (1973).
92. R. A. Pethrick and A. D. Wilson, *Spectrochim. Acta, Part A* **30**, 1073 (1974).
93. I. Fried and H. Barak, *J. Electroanal. Chem.* **27**, 167 (1970).
94. J. N. Butler, *Anal. Chem.* **39**, 1799 (1967).
95. R. L. Benoit, *Inorg. Nucl. Chem. Lett.* **4**, 723 (1968).
96. J. H. Hildebrand and R. L. Scott, "Solubilities of Non-Electrolytes." Van Nostrand-Reinhold, Princeton, New Jersey, 1950.
97. L.-H. Lee, *J. Paint Technol.* **42**, 369 (1970).
98. V. Gutmann and U. Mayer, *Monatsh. Chem.* **100**, 2048 (1969).
98a. V. Gutmann and K. H. Wegleitner, *Z. Phys. Chem.* (*Frankfurt am Main*) [N.S.] **77**, 77 (1972).
99. T. I. Bondareva, D. A. Kuznetsov, I. E. Furmer, and S. F. Shakhova, *Tr. Mosk. Khim.-Tekhnol. Inst.* No. 56, p. 187 (1967).
99a. C. V. Krishnan and H. L. Friedman, *J. Phys. Chem.* **73**, 1572 (1969).
99b. M. Salomon, *J. Electroanal. Chem.* **25**, 1 (1970).
100. L. M. Mukherjee and D. P. Boden, *J. Phys. Chem.* **73**, 3965 (1969).
101. L. M. Mukherjee, D. P. Boden, and R. Lindauer, *J. Phys. Chem.* **74**, 1942 (1970).
102. M. L. Jansen and H. L. Yeager, *J. Phys. Chem.* **77**, 3089 (1973).
103. R. M. Fuoss and K.-L. Hsia, *Proc. Nat. Acad. Sci. U.S.* **59**, 1550 (1967).
104. R. Fernandez-Prini, *Trans. Faraday Soc.* **65**, 3311 (1969).
105. J. N. Butler, *Advan. Electrochem. Electrochem. Eng.* **7**, 77 (1970).
106. J. N. Butler, *Advan. Electrochem. Electrochem. Eng.* **7**, 97 (Fig. 3).
107. J. E. McClure and T. B. Reddy, *Anal. Chem.* **40**, 2064 (1968).
108. R. C. Murray, *Diss. Abstr. Int. B* **32**, 5721 (1972).
109. D. Boden, H. Buhner, and V. Spera, Report Contract DA-28-043-AMC-01394 (E). 1965.
110. B. Burrows and R. Jasinski, *J. Electrochem. Soc.* **115**, 365 (1968).
111. E. Sutzkover. Y. Nemirovsky, and M. Ariel, *J. Electroanal. Chem.* **38**, 107 (1972).
112. M. L. B. Rao, *J. Electrochem. Soc.* **114**, 13 (1967).
113. M. M. Nicholson, *J. Electrochem. Soc.* **121**, 734 (1974).
114. L. Marcoux, K. B. Prater, and R. N. Adams, *Anal. Chem.* **37**, 1446 (1965).
115. A. N. Dey and B. P. Sullivan, *J. Electrochem. Soc.* **117**, 222 (1970).
116. R. Jasinski and S. Carroll, *J. Electrochem. Soc.* **117**, 218 (1970).
117. F. T. Wall, P. F. Grieger, and C. W. Childer, *J. Amer. Chem. Soc.* **74**, 3562 (1952).
118. J. M. Sullivan, D. C. Hanson, and R. Keller, *J. Electrochem. Soc.* **117**, 779 (1970).
119. H. J. McComsey, Jr. and M. S. Spritzer, *Anal. Lett.* **3**, 433 et seq. (1970).
120. E. P. Jacobs and F. Dörr, *Ber. Bunsen ges. Phys. Chem.* **76**, 1271 (1972).
121. V. Gutmann and O. Duschek, *Monatsh. Chem.* **104**, 654 (1973).
122. H. N. Seiger, A. E. Lyall, and R. C. Shair, *in* "Power Sources" (D. H. Collins, ed.), Vol. 2, p. 267. Pergamon, Oxford, 1970; J. Broadhead, *ibid.*, Vol. 4, p. 469 (1973); D. Linden, N. Wilburn, and E. Brooks, *ibid.*, p. 483.
123. G. Pistoia, *J. Electrochem. Soc.* **118**, 153 (1971).
124. A. N. Dey and M. L. B. Rao, *in* "Power Sources" (D. H. Collins, ed.), Vol. 2, p. 231. Pergamon, Oxford, 1968.
124a. J. S. Dunning, W. H. Tiedermann, L. Hsueh, and D. N. Bennion, *J. Electrochem. Soc.* **118**, 1886 (1971).
125. R. Selim and P. Bro, *J. Electrochem. Soc.* **121**, 1457 (1974).
126. P. L. Kronick and R. M. Fuoss, *J. Amer. Chem. Soc.* **77**, 6114 (1955).
127. M. L'Her, D. Morin-Bozec, and J. Courtot-Coupez, *J. Electroanal. Chem.* **55**, 133 (1974).
128. R. E. Meredith and C. W. Tobias, *J. Electrochem. Soc.* **108**, 286 (1961).
129. B. Gosse and A. Denat, *J. Electroanal. Chem.* **56**, 129 (1974).
129a. B. Gosse, *C. R. Acad. Sci., Ser. C* **279**, 249 (1974).

130. A. N. Dey, *J. Electrochem. Soc.* **114**, 823 (1967).
131. J. Courtot-Coupez and C. Madec, *C. R. Acad. Sci., Ser. C* **277**, 15 (1973).
132. S. Y. Lam and R. L. Benoit, *Can. J. Chem.* **52**, 718 (1974).
133. M. L'Her and J. Courtot-Coupez, *J. Electroanal. Chem.* **48**, 265 (1973).
134. F. Breivogel and M. Eisenberg, *Electrochim. Acta* **14**, 459 (1969).
135. J. Jorné and C. W. Tobias, *J. Phys. Chem.* **78**, 2521 (1974).
136. R. Keller, Final Report, Contract NAS 3-8521. 1969.
137. J. Jornè and C. W. Tobias, *J. Electrochem. Soc.* **121**, 994 (1974).
138. N. A., Vlasov and N. A. Baranov, *Neftepererab. Neftekhim.* (*Moscow*) p. 36 (1970).
139. M. Salomon, *J. Phys. Chem.* **79**, 429 (1975).
140a. J. Paasivirta, *Suom. Kemistilehti*, **41**, 364 (1968).
140b. J. Paasivirta and S. Kleemola, *Suom. Kemistilehti* **43**, 285 (1970).
141. A. H. Saadi, Ph.D. Thesis, University of London (1964).
142. H. F. Grutzmacher and J. Hubner, *Justus Liebigs Ann. Chem.* **748**, 154 (1971).
143. D. C. De Jongh and D. A. Brent, *J. Org. Chem.* **35**, 4204 (1970).
144. D. C. De Jongh and M. L. Thomson, *J. Org. Chem.* **37**, 1135 (1972).
145. C. Wentrup, *Tetrahedron Lett.* p. 2919 (1973).
146. O. L. Chapman and C. L. McIntosh, *J. Chem. Soc., D* p. 383 (1971).
147. D. C. De Jongh, R. Y. Fossen, and A. Dekovich, *Tetrahedron Lett.* p. 5045 (1970).
148. R. Ohme and H. Preuschof, *J. Prakt. Chem.* [4] **313**, 626 (1971).
149. J. Y. Cabon, M. L'Her, and M. Le Demezet, *C. R. Acad. Sci., Ser. C* **280**, 357 (1975).
149a. J. Courtot-Coupez and M. L'Her, *C. R. Acad. Sci., Ser. C* **266**, 1286 (1968).
150. J. F. Casteel, J. R. Angel, H. B. McNeeley, and P. G. Sears, *J. Electrochem. Soc.* **122,** 319 (1975).
151. P. T. Thompson, R. E. Taylor, and R. H. Wood, *J. Chem. Thermodyn.* **7**, 547 (1975).
152. I. A. Vasil'ev and A. D. Korkhov, *Zhur. Fiz. Chim.* **47**, 1527 (1973).
153. S. Sakai, S. Furusawa, H. Matsumaga and T. Fujinami, *J. Chem. Comms.* **265** (1975).
154. L. Werblan, J. Lesinski, and L. Skubiszak, *Roczniki Chim.* **49**, 221 (1975).
155. J. Courtot-Coupez and M. L'Her, *Comptes Rend., Ser. C* **275**, 103, 145 (1972).
156. L. K. Kaplan and S. F. Shakhova, *Khim. Prom.*, No. 4, p. 281 (1975).
157. L. K. Kaplan, S. F. Shakhova, R. S. Veranyan, and N. M. Pantjukhina, *Khim. Prom.* No. 7, p. 519 (1975).
158. A. Denat, B. Gosse, and J.-P. Gosse, *J. Chim. Phys.* **72**, 343 (1975).
159. H. V. Venkatasetty, *J. Electrochem. Soc.* **122**, 245 (1975).
160. R. C. Murray and D. A. Aikens, *Electrochimica Acta*, **20**, 259 (1975).
161. R. L. Benoit, D. Lahaie, and G. Boire, *Electrochmiica Acta.* **20**, 377 (1975).
162. B. G. Stephens and H. L. Felkel, *Analyt. Chem.* **47**, 1676 (1975).
163. C. Luca and A. Sapatino, *Rev. Chim.* (*Bucharest*) **26**, 251 (1975).
164. R. Keller, J. N. Foster, D. C. Hanson, J. F. Hon, and J. S. Muirhead, NASA CR-1425, North American Rockwell Corpn., Canoga Park, Calif. August 1969.
165. U. P. Zubchenko, L. K. Kaplan and S. F. Shakhova, *Zhur. Fiz. Chim.* **49**, 1342 (1975).
166. N. F. Catherall and A. G. Williamson, *J. Chem. Eng. Data* **16**, 335 (1971).
167. M. L'Her, D. Morin-Bozec, and J. Courtot-Coupez, *J. Elecroanalyt. Chem.* **61**, 99 (1975).
168. W. M. Latimer, K. S. Pitzer, and C. M. Slansky, *J. Chem. Phys.* **7**, 108 (1939).
169. J. Massaux and G. Dyckaerts, *J. Chem. Soc. Belge* **84**, 519 (1975).
170. J. Massaux and G. Duyckaerts, *J. Electroanalyt. Chem.* **59**, 311 (1975).
171. H. Strehlow, "Electrode Potentials in Non-Aqueous Solvents," *in* "The Chemistry of Non-Aqueous Solvents" (J. J. Lagowski, ed.), Vol. I., Ch. 4. Academic Press, New York 1966.
172. M. A. Nikitin, D. D. Zykov, and V. M. Shleinikov, *Izvest. Vyssh. Ucheb. Zaved.* **18**, 571 (1975).
173. E. I. Shcherbina, A. E. Tenenbaum, and L. L. Gurarii, *Zhur. Fiz. Chim.* **49**, 516 (1975).

7

Sulfolane

JUKKA MARTINMAA

Department of Wood and Polymer Chemistry
University of Helsinki, Helsinki, Finland

I. Introduction

Although sulfolane (trivial name of tetrahydrothiophene 1,1-dioxide or tetramethylene sulfone) (**I**) is no longer a newcomer among nonaqueous solvents, its appearance in routine work or scientific reports has not been as frequent as that of some more common dipolar aprotic solvents such as dimethylformamide or acetonitrile. This may partly be due to the fact that it did not become commercially available until 1959[1].

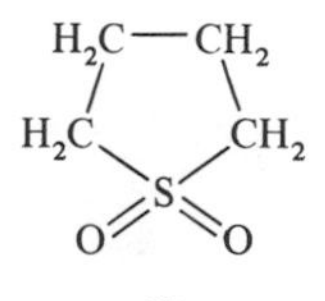

(I)

In the past few years the interest in sulfolane has encompassed both pure and applied chemistry. Applications of sulfolane in pure chemistry are primarily derived from its properties such as chemical and thermal stability, high polarity, and very low autoprotolysis constant. Thus, it has received wide interest as a solvent for electrochemical investigations of various kinds and as a medium in which to carry out many reactions which are affected by the solvent in more reactive media. Its low vapor pressure, together with its relatively low toxicity, makes it convenient to use.

The relatively high viscosity, hygroscopicity, and melting point are disadvantages of sulfolane. The melting point is compensated for by a very high cryoscopic constant. The price which is considerably higher for sulfolane than for the more common dipolar aprotic solvents is a further drawback.

In applied chemistry, sulfolane is known as an important technical solvent in many extraction processes.[1] In the petrochemical industry it is useful because of its ability to extract selectively aromatic hydrocarbons from aliphatic ones and to absorb waste gases. Numerous applications have also appeared in the patent literature for sulfolane in photographic emulsions, fabrics manufacturing, wood chips impregnated before cooking, and polymer plasticizing.

The aim of this review is to discuss the solvent characteristics of sulfolane. Therefore, particular attention is paid to the purification of sulfolane, as well as to those methods useful in control of its purity.

II. Preparation

The first report on the preparation of sulfolane appeared in 1910 by von Braun and Trümpler,[2] who carried out a permanganate oxidation of tetrahydrothiophene to yield a sulfone. They did not give any data on its properties.

Later, its preparation was reported in 1916 by Grischkevitsch-Trochimovsky.[3] He also carried out a permanganate oxidation of tetrahydrothiophene, which was previously prepared by the action of sodium sulfide on 1,4-dibromo- or 1,4-diiodobutane.

At present the commercial manufacture of sulfolane is based on the Diels–Alder addition of butadiene and sulfur dioxide. The course of the reaction, as suggested by Staudinger,[4] can be directed to the formation of a monomeric sulfone, 3-sulfolene (2,5-dihydrothiophene 1,1-dioxide or butadiene sulfone), in the presence of a polyhydric phenol or to the formation of a polysulfone in the presence of a peroxide. This reaction is dependent on the concentration of the reactants and the temperature.[5] It can also be utilized in the preparation of sulfones from other substituted dienes.

For example,[6,7] butadiene and SO_2, in a mole ratio of 1:2, are allowed to react in the presence of 1% hydroquinone (with respect to butadiene) at 100°C for 12 hr in a steel bomb or at room temperature for two to three weeks in pressure bottles. Yields of 3-sulfolene (mp 64.5°–65.5°C) in this procedure vary from 80 to 85%.

3-Sulfolene is then hydrogenated to give sulfolane. This can be carried out quantitatively in a 25% aqueous–alcoholic solution using colloidal palladium as catalyst.[6,7] The principal reactions associated with sulfolane synthesis are given in Eq. 1. 3-Sulfolene can also be hydrogenated to sulfolane by Pt in ethanol,[8,9] colloidal Pd in water,[10] Pd–C in ethanol,[11] or Raney-Ni in

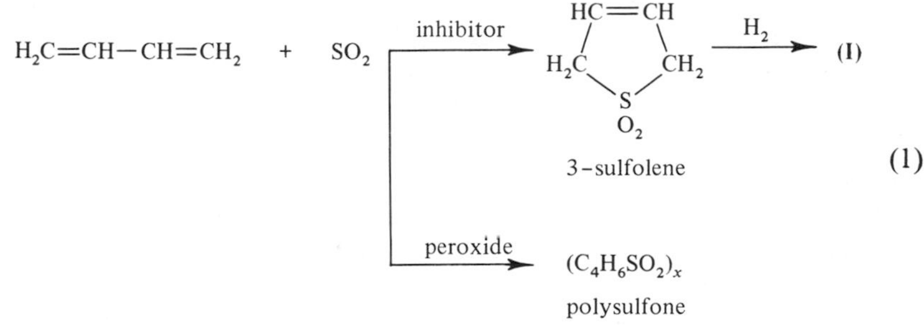

(1)

ethanol.[12] Numerous alternative methods for 3-sulfolene hydrogenation are also described in the patent literature.[13–20]

Further methods for sulfolane preparation are also known. In a study by Birch and McAllan,[21] thiophene was hydrogenated by sodium in the presence of methanol in liquid ammonia to give 2,5- and 2,3-dihydrothiophene as primary reaction products. The former, 2,5-dihydrothiophene, was then oxidized in acetic acid with an excess of hydroperoxide to give 3-sulfolene. The double bond of 3-sulfolene may, however, oxidize further in this procedure and traces of the corresponding diol are formed.

In addition, several alternative methods, based on oxidation of tetrahydrothiophene, have been reported.[22–28] In these methods various oxidants in the presence or absence of catalysts have been employed in different reaction media.

III. Purification

Pure sulfolane is a colorless and odorless solid melting above 28°C. Several purification methods, including numerous patents,[29, 30] have been developed for this solvent.

As a rule, purification procedures have been tailored to the intended use of the solvent. The technical solvent prepared according to Eq. 1 may contain, besides appreciable amounts of water (up to 3%), various impurities originating from the different stages of Eq. 1. Consequently, the crude sulfolane, usually exhibiting a dark reddish brown color and a freezing point below or near room temperature, contains 3-sulfolene as the main impurity. 3-Sulfolene undergoes addition reactions at the double bond (e.g., hydrogenation[10, 31]) and begins to decompose at temperatures above 75°C[5, 7] (quantitatively above 120°C[4]) into its components, butadiene and SO_2. The presence of small amounts of polymeric impurities also causes the color in technical grade sulfolane.[32, 33] The cryoscopic constant of sulfolane is about 65 deg kg $mole^{-1}$, thus making possible the estimation of the approximate content of impurities from the depression of freezing point.[34–37] It is known, for example, that 1.3% water by weight reduces the freezing point to 17.2°C.[33]

A simple distillation is not sufficient to remove 3-sulfolene or its cleavage products from technical sulfolane.[37] Some kind of appropriate pretreatment is, therefore, recommended before the distillation procedure, examples of which are given below.

The first method involves heating with solid sodium hydroxide[38, 39] or potassium hydroxide[34] at 100°–180°C for a day. This method is made more effective by simultaneous bubbling with dry nitrogen to remove the volatile decomposition products. The solvent is then decolorized with activated charcoal (Norit A) at 80°–90°C for 6 hr, filtered, and passed through a 4-ft column packed with a mixed bed of ion exchange resins (Amberlyst-15 and Amberlyst A-21) after which the solvent is ready for vacuum distillation.[39]

In a second method, air oxidation of the hot (160°–200°C) solvent for 12 hr is employed,[40, 41] followed by agitation under vacuum.[42] The solvent is then treated with activated charcoal as noted above.

The following method is suitable if a solvent stable toward oxidation is desired.[43] One liter of commercial sulfolane is slowly added to a mixture of 500 g crushed ice, 50 g $KMnO_4$, and 50 g concentrated sulfuric acid in a

5-liter flask. The mixture is swirled occasionally until all the impurities have been completely oxidized. After filtration, the aqueous solution is decolorized by adding sodium pyrosulfite. It is extracted three times with 500 cm^3 of dichloromethane and the extract dried with $MgSO_4(H_2O)$. After this, the dichloromethane is distilled off and the remaining sulfolane treated with 45 g P_2O_5 for a day.

Technical sulfolane is also conveniently prepurified by fractional crystallization.[44–47] One half of the batch can be purified this way to a colorless product solid at room temperature. The residual half must be submitted to a more effective treatment as described above. Additional pretreatment methods have also been reported.[48–51]

In the distillation of the prepurified solvent one should take into account that it solidifies at temperatures below 29°C, it is hygroscopic, and it requires a pressure of less than 1 torr for boiling below 100°C, the recommended distillation temperature. To achieve a fair result, a system with good fractionating properties, e..g, a 0.5-m spinning band column,[52] a fractionating column packed with small glass rings,[36] or a 1.5-m long Podbielniack column packed with glass helices,[53] should be employed. More than one distillation is usually needed depending on the effectiveness of the prepurification. In a very popular method of sulfolane purification the distillation is carried out using sodium hydroxide pellets.[54–67]

To reduce the moisture content, vacuum distillations using P_2O_5 or calcium hydride, are widely employed. Sulfolane can be further dried with molecular sieves,[68,69] calcium chloride,[70] or calcium sulfate.[39] Rigorously dried benzene can also be used (10% by volume) to azeotrope the residual moisture.[50]

Special purification schemes have been developed for electrochemical studies,[39,45,49,54] acid–base work,[71,72] and nitronium salt nitrations.[37]

The purification of sulfolane can be studied and its purity controlled by various means such as UV and visible spectroscopy,[35] NMR spectroscopy,[73] gas chromatography,[68] polarography,[45,48] and potentiometric titration.[39] Pure sulfolane should have a melting point of 28.4°C or more. It should have a maximum conductivity of about 1×10^{-8} Ω^{-1} cm^{-1} and be clear in ultraviolet light above 250 nm.[63] It should give no coloration with concentrated (100%) sulfuric acid.[37,62] The maximum water content should not exceed 10 ppm as determined by Karl Fischer titration,[39,40] although this method has been reported[37] to be unsuitable because of the formation of a sulfolane–iodine complex.

Small amounts of water do not greatly affect the aprotic properties of dipolar aprotic solvents.[69] However, water is an undesirable source of protons in many experiments using sulfolane. Some examples include the protonation of reduction intermediates,[49] titrations with perchloric acid,[74] nitrations with nitronium salts,[37] conductivity measurements,[36] and solvolytic studies. The uptake of water during experiments is very difficult to avoid owing to the high hygroscopicity of sulfolane. For the same reason the purified product should be stored under dry nitrogen. The recovery of the used solvent merits consideration.[62,75]

IV. Physical Properties

The physical properties of sulfolane are summarized in Table I.

The low values for the heat of fusion and for the entropy of fusion, together with the large heat of vaporization, suggest that solid sulfolane may be classified as plastic crystal. This means that the structure of the solid phase is very similar to that of the liquid, i.e., the molecules still have rotational freedom. On the other hand, the so-called globular substances usually exhibit a heat of fusion of less than 5 cal/g. Globular molecules have either a center of symmetry (CCl_4, cyclohexane) or an axis around which the molecule is capable of rotating, thus forming a quasispherical repulsion envelope.[53] A globular structure is attributed to sulfolane.[76]

Globular substances are very often polymorphic. It is, therefore, expected that more than one transition point will be encountered if the temperature of the substance is lowered. Accordingly, two transition points have been found experimentally for solid sulfolane. The first one is evidenced by dielectric[77] and molar volume[78] measurements, and the latter (as well as the first) by wide-line ^{1}H NMR measurements.[79] About 13°C below the melting point the mesomorphic phase undergoes a transition into a more ordered nonrotational crystalline phase. The enthalpy change associated with this transition is considerably higher than the enthalpy of fusion.[61]

Pure liquid sulfolane has little order.[76] This conclusion is drawn from the insensitivity of the dipole moment and the dielectric constant toward temperature. The dielectric behavior of liquid sulfolane is reasonably well represented by Onsager's formula.[55]

The viscosity behavior of liquid sulfolane as a function of temperature is described by Eq. 2.[80]

$$\log \eta = A + B/T \tag{2}$$

where η is the viscosity in poise, T is the absolute temperature, and A and B are constants obtained from the plot of $\log \eta$ vs. $1/T$. The values of the parameters A and B are reported to be -2.3892 and 1031.3, respectively. From the value of B the activation energy of viscous flow can be calculated.

According to Vaughn and Hawkins,[80] the density data for liquid sulfolane can be expressed by Eq. 3, where d is the density

$$d^t = d^{t_0} - k(t - t_0) \tag{3}$$

TABLE I

PHYSICAL PROPERTIES OF SULFOLANE

Molecular weight	120.16
Melting point (°C)	28.86 (extrapolated),[21] 28.62,[45] 28.5 ± 0.1,[34,37] 28.45[35,36,53]
First transition point (°C)	15.45,[83] 15.43,[53] 15[61,79]
Second transition point (°C)	−63[79]
Vapor pressure (torr)	$\log p = 28.6824 - 4350.7/T(°\mathrm{K}) - 6.5633 \log T(°\mathrm{K})$[1]
Dielectric constant	
2Mc	
6.9°–15.5°–28.5°C	3.84–44.03–41.55[77]
30.0°–73.9°C	41.36–36.20[77]
30°C	43.4,[84] 43.38,[85] 43.36[86]
35°C	42.71,[86] 42.7[84]
40°C	42.12,[85] 42.1,[84] 42.08[86]
45°C	41.48[86]
50°C	40.88,[86] 40.71[85]
60°C	39.71[86]
70°C	38.67[86]
75°C	38.00[86]
80°C	37.67[86]
10 Mc	
30°C	43.3[80]
35°C	42.4[80]
40°C	41.8[80]
45°C	41.4[80]
50°C	40.7[80]
Viscosity (cP)	
30°C	10.3,[33] 10.295,[85,87] 10.29,[55] 10.286,[80] 10.27,[86] 9.876[1]
35°C	9.033[80,87]
40°C	8.007,[85,87] 7.974,[80] 7.959[86]
45°C	7.116[80]
50°C	6.392,[86] 6.346,[85] 6.312[80]
100°C	2.5[33]
200°C	0.97[33]
Density (g ml^{-1})	
7.00–15.35°C	1.3093–1.2969[78]
19.98–28.44°C	1.28924–1.26905[78]
30°C	1.2626,[41] 1.2625,[39] 1.2623,[55] 1.2618,[88] 1.217,[84] 1.2615,[61] 1.2614[80]
35°C	1.2568[80]
40°C	1.2532,[84] 1.2525[80]
45°C	1.2494[80]
50°C	1.2447,[80] 1.2446[84]
60°C	1.2360[84]
100°C	1.201[33]
200°C	1.116[33]

(continued)

TABLE I—*continued*

Refractive index, n_D	
25°C	1.4830[89]
29°C	1.48210[86]
30°C	1.4820,[80] 1.48181,[86] 1.4817,[84] 1.4813[41]
35°C	1.4798[80]
40°C	1.4785,[84] 1.4783[80]
45°C	1.4765[80]
50°C	1.4751,[84] 1.4747[80]
60°C	1.4717[84]
Dipole moment (*D*)	
25°C	
In benzene	4.81,[90] 4.80,[81] 4.71,[91] 4.69[82]
In dioxane	4.91[81]
In CCl_4	4.68[81]
30°C (calculated)	4.90[80]
30.0°–73.9°C	4.65–4.76[77]
Flash point (°C)	177[92]
Cryoscopic constant (deg kg $mole^{-1}$)	64.1,[36] 64.8,[62,93] 65,[53] 66.1,[93] 66.2[61]
15°C	7.57[53]
Heat of fusion (cal/g)	2.84,[36] 2.73[61]
15°C	10.6[61]
Gutmann donor number, DN_{SbCl_5} (25°C)	14.8[94]
Molecular volume (Å)	158 (estimated)[88]
Proton basicity, $-pK_{BH^+}$	12.88[95]
Acidity, pK_a (25°C)	>31 (in DMSO)[96]
Autoprotolysis constant, pK_s	25.45[97]
Molar volume (ml)	
30°C	95.23,[84] 95.20[41,86]
40°C	95.89[84]
50°C	96.55[84]
60°C	97.22[84]
Molar heat capacity, C_p (cal/deg)	
30°C	37.5,[86] 38.5[33]
100°C	42.1[33]
200°C	48.1[33]
Heat of formation (kcal/mole)	−88 (calculated)[70]
Z value (kcal/mole)	77.5[98]
Optical density at 250 nm (5-mm cell)	<0.30[99]
Molar Kerr constant (25°C)	+121[91]
Surface tension (dyn/cm)	
30°C	53.3,[35] 50.18[86]
35°C	49.1[86]
40°C	48.79[86]
50°C	47.79[86]
60°C	46.80[86]
Heat of vaporization (kcal/mole)	
20°C	17.5[21]
100°C	15[33]
200°C	14.7[33]
Specific conductance (Ω^{-1} cm^{-1})	1×10^{-8},[37] 2×10^{-9},[39]

in g/ml, t any temperature in the range 30°–50°C, t_0 the reference temperature (30°C), and k the least square value for the temperature gradient of density in g/ml deg. The measured value for k is reported to be 8.16×10^{-4}. The corresponding equation (Eq. 4) is reported by Lamanna *et al.*[81] for the temperature range 28.50°–74.33°C. The densities at temperatures from

$$d^t = 1.26308 - 0.00086(t - 28.50°\text{C}) \tag{4}$$

15.43° to 28.50°C are obtained from Eq. 5, and those in the range 7.00° 15.35°C are obtained from Eq. 6.[78]

$$d_m = 1.29243 - 0.00070(t - 15.43°\text{C}) \tag{5}$$

$$d_s = 1.31080 - 0.00021t \tag{6}$$

Further, the interpolation equation for the dielectric constant is given by Eq. 7, where temperature values in the range 30.00°–73.94°C are given for t.[77]

$$\varepsilon = 41.25 - 0.1182(t - 30.00°\text{C}) \tag{7}$$

The refractive index is obtained from Eq. 8, where t is an arbitrary temperature in the range 29.00°–74.40°C.[81]

$$n_D = 1.48209 - 3.258 \times 10^{-4}(t - 29.00°\text{C}) - 2.99 \times 10^{-7}(t - 29.00°\text{C})^2 \tag{8}$$

The high dipole moment of sulfolane arises from the contributions of the sulfur–oxygen bonds.[35] The ring formation in sulfolane has not greatly enhanced the dipole moment in comparison to related sulfones,[82] but has caused only the negative end of the dipole to be exposed.

V. Spectral Studies and Structure of Sulfolane

At present, the crystal structure of sulfolane is not known. The polymorphic character of sulfolane was discussed in Section IV. It also has been shown that sulfolane is never self-associated, even when its molecules are closely packed in the plastic crystal.[78]

The molecular structure is not known with certainty. Rérat *et al.*,[100,101] however, report an X-ray study in which the crystal structure of the 1:1 complex of 1,3,5-trinitrohexahydrotriazine and sulfolane has been measured. The crystal structures of 3-sulfolene[102] and 3,4-epoxysulfolane[103] have likewise been determined by X-ray diffraction. The positional parameters are reported in all three cases. Some of the intramolecular distances from the results of these studies are collected in Table II for comparison.

TABLE II

COMPARISON OF X-RAY DATA

	Butadiene sulfone[a]	3,4-Epoxysulfolane[b]	Sulfolane as a 1:1 complex with 1,3,5-trinitro-hexahydrotriazine[c]
Distances (Å)			
S–O	1.440	1.444	1.435
S–C	1.749	1.795	1.80
C-2–C-3, C-4–C-5	1.479	1.501	1.525
C-3–C-4	1.299	1.450	1.56
Angles (degrees)			
O–S–O	117.0	117.69	—
O–S–C	110.2	109.91	—
C–S–C	97.0	97.72	98

[a] From Sands and Day.[102]
[b] From Sands.[103]
[c] From Rérat *et al.*[100]

The length of the S–O bond (1.44 Å) is in agreement with the value obtained in other sulfoxo compounds, being indicative of π bonding involving the 3*d* orbitals of sulfur. The sulfur atom is tetrahedrally coordinated.[100,103] This is evidenced by the weighted mean (109.20°) of the six angles associated with the sulfur atom in the case of 3,4-epoxysulfolane. The corresponding values for 3-sulfolene and for a regular tetrahedron are 109.17° and 109.47°, respectively. Thus, the epoxidation has caused only slight changes in the bond angles at the sulfur atom.[103] On the other hand, the C-3–C-4 distance has increased from 1.30 Å in 3-sulfolene to 1.45 Å in 3,4-epoxysulfolane to a single bond value in the complexed sulfolane. Because the angles formed by the sulfur atom are practically unchanged in the two latter molecules, folding of the five-membered ring is required. This result is in accord with IR and Raman spectroscopic studies,[89,104] which also indicate a puckered ring, in contrast to dipole moment calculations,[91] which support a planar ring structure. In the 3,4-epoxysulfolane molecule the planes C-2–S-1–C-5 and C-2–C-3–C-4–C-5 form an angle of 167.3°. It is expected that this angle is even smaller in the case of sulfolane.

Electron diffraction studies and theoretical calculations based on two types of puckered models and one planar-type model did not show with certainty the real structure of the five-membered ring, as reported by Naumov *et al.*[105] The results are, however, in reasonable accordance with those for the complexed sulfolane in Table II.

The sulfone grouping is an example of a highly polar group that is tightly bound. Consequently, stretching vibrations are well localized and result in good group frequencies.[104,106] The S–O stretching force constant has been reported as 6.54×10^5 dyn cm^{-1} in sulfoxides and 9.86×10^5 dyn cm^{-1} in sulfones.[107] Katon and Feairheller have discussed the structure of sulfolane on the basis of infrared data and made band assignments.[104] They report the SO_2 stretching vibrations to absorb at 1143 and 1298 cm^{-1}, in addition to which the major absorption frequencies for sulfolane in the 700–250 cm^{-1} region occur at 673(m), 567(s), 517(m), 445(s), 438(s), 386(m), and 287(m) cm^{-1}. The absorptions at 567 and 517 cm^{-1} (scissoring and wagging vibrations, respectively), are in accord with the fact that all sulfones have a medium to strong band at 545–610 cm^{-1}. The Raman band observed at 570 cm^{-1} (corresponding to the strong IR band at 567 cm^{-1}) is polarized, indicating that it too is a SO_2 scissoring vibration. The authors suggest further[89] that the four methylene groups of sulfolane are nonequivalent, since four bands are noted in the IR spectrum at 1461, 1448, 1418, and 1412 cm^{-1}. Additional support for this assumption is based on Raman polarization data. There are almost no Raman bands of reasonable intensity to which an infrared band does not coincide. This excludes the planar ring model of sulfolane from consideration and one concludes that the methylene groups are twisted in the sulfolane molecule.

According to Hanley and Iwamoto,[108] the IR stretching frequencies (in cm^{-1}) of the S–O bond in sulfolane are: (ν_{sym}) 1105, 1145; (ν_{asym}) 1295, which is the major band, two less intense adjacent bands at 1255 and 1272 cm^{-1}; (ν_{S-O}) 1210, which equals (average ν_{sym} + average ν_{asym})/2. Buxton and Caruso[109] report a strong band below 100 cm^{-1} and assume it to be due to the dipole-dipole interactions. A second, extremely weak, band was observed with the center at about 140 cm^{-1} and was partly overlapped by the dipole-dipole band. This absorption is assigned to C–S–C ring puckering. A consideration of models shows the ring to be highly strained and this may raise the frequency of the ring puckering vibration in which the sulfur atom is involved.[89]

A transfer from $>$S through $>$SO to $>SO_2$ is associated with a decrease of the UV extinction. This shift of the absorption is a consequence of the successive filling of the S octet. Thus, sulfolane is very transparent in UV light, as shown by Schurz and Stübchen.[46] In this respect sulfolane exceeds the properties of most of the other solvents. It has been reported[34] that the shortest wavelengths at which it is possible to make measurements on sulfolane solutions, which are $\sim 6\ M$ in nitric acid, with cell thicknesses of 10, 1, and 0.1 mm, are 350–360, 315–320, and 250 nm, respectively.

Nuclear magnetic resonance (NMR) measurements of sulfolane are few.[76,99,110] The protons at α positions to the sulfone group show a triplet at 3.26 ppm with a coupling constant of 7 Hz, as measured in a SO_2ClF solution at $-60°C$, with tetramethylsilane as external standard. The corresponding absorption for the β protons is a multiplet centering at 2.43 ppm. The chemical shifts of protons of pure liquid sulfolane exhibit only small temperature dependence. The following chemical shift displacements have been measured at higher temperatures for the α and β protons, with respect to values at 29°C: (31°C) -0.2, -0.2; (48°C) -0.8, -0.3; (64°C) -1.3, -0.3; (98°C) -2.9, -0.7 Hz.[76]

VI. Solubility Characteristics

A. Inorganic Compounds

Sulfolane is miscible with water in all proportions because of its great hygroscopicity. Many inorganic compounds are also found to be soluble in sulfolane. However, only limited quantitative solubility data are presently available. In the most extensive study, made by Starkovich and Janghorbani,[111] the solubilities of some chloride and perchlorate salts were determined by instrumental neutron activation analysis (INAA) at 40°C.

As far as the alkali metal chlorides are concerned, LiCl exhibits the best solubility in sulfolane (361 mmoles/liter).[111] It has been shown that LiCl is highly associated in sulfolane.[55,112] Consequently, very viscous solutions of this solute will be obtained. Apparent polymerization numbers for the LiCl solutions have been reported.[36] Somewhat viscous solutions are reported to form also with $LiBF_4$ as the solute.[113]

The solubilities of other alkali metal chlorides are found to be considerably lower, as shown by the values 0.88, 0.53, 0.84, and 1.1 mmoles/liter for saturated solutions of Na, K, Rb, and Cs salts at 40°C, respectively.[111] This solubility trend with the minimum at KCl parallels that in water. Most of the other alkali halogenides appear to be soluble in sulfolane in amounts sufficient for electrochemical measurements. It seems that good solubility in sulfolane correlates with increasing cation size in the case of chlorine salts.[111]

Sulfolane also dissolves perchlorates to a reasonable extent. The alkali perchlorates generally exhibit better solubilities in sulfolane than the corresponding halogenides. For example, the saturation concentrations of $NaClO_4$ and $KClO_4$ are 981 and 89.5 mmoles/liter at 40°C, respectively.[111] The solubility of NH_4ClO_4 is a 600-fold increase compared to that of NH_4Cl, 0.5 mmole/liter, at 40°C.[111] The exchange of the ammonium hydrogens by alkyl groups leads to higher solubilities in the case of chlorides, but to a lowering of the molar solubility for $(CH_4)_4NClO_4$ (62 mmoles/liter) and $(C_3H_7)_4NClO_4$ (269 mmoles/liter).[111] The solubilities of tetraalkylammonium salts increase strongly on heating.[72] Most of the alkaline earth metal chlorides exhibit very low solubilities in sulfolane. $MgCl_2$, however, is soluble up to 24 mmoles/liter, whereas the corresponding perchlorate, $Mg(ClO_4)_2$, is soluble up to 363 mmoles/liter at 40°C.[111]

Other compounds reported to be sufficiently soluble in sulfolane to allow measurements of different kinds are: $NaBPh_4$,[44] NaSCN,[44] and Ph_4C, Ph_4AsI, $AgBPh_4$, Ph_4AsBPh_4, AgN_3, AgBr, and AgI.[114] Hydroxides, especially alkali hydroxides, are generally only very slightly soluble in sulfolane. However, inorganic complexes insoluble in most other solvents frequently dissolve in sulfolane.[69] In addition to being a good solvent for many sulfurous compounds, sulfolane also dissolves elementary sulfur.[12]

Sulfolane is a good solvent for nitronium salts such as NO_2PF_4, NO_2SbF_6, NO_2PF_6, NO_2AsF_6, NO_2ClO_4, and $NO_2HS_2O_7$, employed as nitrating agents.[115] NO_2BF_4 dissolves up to 7% at 25°C, which corresponds to a 0.5 *M* solution. This is considerably more than in the case of nitromethane.[38,73] The other nitronium salts exhibit greater solubilities, which has been partly explained in terms of the higher association of these salts in sulfolane.[38]

The solubilities of NO and NO_2 are much higher in sulfolane than in water.[71] The good gas absorption power of sulfolane is further demonstrated in the so-called Sulfinol process.[116,117] In this petrochemical application gases such as H_2S, CO_2, COS, and mercaptans are removed from acid–gas mixtures with a solvent system consisting of an alkanolamine mixed with sulfolane.

Heats of solution for some compounds in anhydrous sulfolane are presented in Table III.[42,118–122] Vapor pressure data for O_2[118] and HCl[119] solutions

TABLE III

HEATS OF SOLUTION OF SOME SOLUTES IN SULFOLANE[a]

Compound	Heat of solution (kcal/mole)	Ref.	Compound	Heat of solution (kcal/mole)	Ref.
$NaClO_4$	−5.40	42	Et_4NCl	3.94	42
NaI	−7.41	42	Et_4NBr	4.99	42
$NaBPh_4$	−10.90	42	Et_4NI	5.53	42
$LiClO_4$	−5.62	42	$AsPh_4I$	3.87	42
$KClO_4$	1.10	42	O_2	1.8	118
$RbClO_4$	2.02	42	HCl	−5.8	119
$CsClO_4$	2.21	42	$HClO_4$	−8.0	120
Et_4NClO_4	3.63	42	SO_2	6.7	121
Butadiene	4.9	122			

[a] Temperature, 30°C.

in sulfolane at different temperatures are also available. Louis and Benoit[118] have indicated the validity of Henry's law when the partial pressure of oxygen is varied by the concentration of oxygen in sulfolane at 30°C.

B. Organic Compounds

Sulfolane is miscible with most of the common protic as well as aprotic solvents, such as alcohols, esters, ketones, sulfoxides, sulfones, halogenated and lower aromatic hydrocarbons, pyridine, tetrahydrofuran, and dioxane. It is also a good solvent for numerous indicators used in titration studies.[62] According to Parker,[69] some nitroaromatic disulfides and diselenides dissolve readily in sulfolane. The solubilities of common organic acids and bases are normally good, whereas some dicarboxylic acids such as phthalic and succinic acid require heating to achieve solution.[72] Paraffins, saturated cyclic hydrocarbons, olefins, naphthenes, and long-chain alcohols are only very slightly soluble in sulfolane.[69] *n*-Heptane and cyclohexane are, however, sufficiently soluble to allow cryoscopic measurements to be made in dilute solutions.[53]

Sulfolane has proved to have superior properties for the extraction of aromatics from aliphatic hydrocarbons in pilot-plant studies.[33]

Sulfolane is also a solvent for organic mercury derivatives and many polymers such as nitrocellulose, cellulose acetate, and wood extracts.[69] Polyacrylonitrile (PAN) is one of the most important polymers soluble in sulfolane. The high-melting nature of linear PAN is believed to be due to the strong hydrogen bonds between the α-hydrogens and the nitrile nitrogens of an adjacent chain. A solvent for PAN must, therefore, be able to break these hydrogen bonds.[123] The solid 1,3,5,7-tetracyanoheptane has been used as a model compound in studies of the PAN dissolution process because this substance reflects the relative solvent power of liquids for the polymer.[124] As a result of these investigations, it seems likely that the high dipole moment of a solvent contributes to the solution of PAN and lower nitriles. In addition, no appreciable hydrogen bonding between solvent molecules is permitted. Because sulfolane fulfills these requirements particularly well, it is not surprising that it is found to be a good solvent for PAN. Methacrylate, styrene, and vinylidene chloride polymers are insoluble in sulfolane.[125]

C. Cryoscopic Behavior

The exceptionally high cryoscopic constant for sulfolane has received attention as a possible basis for precise molecular weight determinations. It has been shown, however, by Garnsey and Prue[36] that the advantages of a high cryoscopic constant for precise work are largely illusory. This is partly due to the poor thermal buffering capability of sulfolane and partly to the growing inaccuracy in the freezing point determination with increasing solute concentration. It has also been pointed out that sulfolane is able to dissolve solutes and impurities in its solid phase which is assumed to be closely similar to the liquid phase. This conclusion has been drawn from the anomalous cryoscopic behavior of globular solutes such as CCl_4 and cyclohexane.[53] It has been shown[53] that the steepness of the slope of the liquid–solid equilibrium temperature vs. molality plot is greater as the mutual size and shape difference of the solute and solvent molecules increases, respectively. Also pyridine is suspected to form plastic mixed crystals with sulfolane.[53]

Freezing point depression constants for several solutes in sulfolane are given in Table IV.[36,53,61,62,126,127] The value 65 deg kg mole^{-1} is considered as the true cryoscopic constant of sulfolane. For example, H_2SO_4 exhibits nearly unit *i* factors in sulfolane solutions. This behavior has been assumed to indicate that H_2SO_4 is dissolved in molecular units or it has protonated the solvent to give very tight ion pairs of sulfolanium bisulfate, $C_4H_8SO_2H^+HSO_2^-$.[62] Studies with indicator acids show parallel results.[128] Also acetic[61] and hydrochloric acids[129] behave as monomers. In fact, it has been reported that practically all acids are not dissociated in sulfolane.[53] Exceptions are evidently the strong acids $HSbCl_6$[120] and HBF_4.[128] The cryoscopic constants for the perchlorates of Li, Na, K, and Rb exhibit a strong electrolyte (dimer) behavior, as can be deduced from the data of Garnsey and Prue.[36] Data concerning the osmotic coefficients of these salts are also available.[36]

Water behaves as a dimer at least over the concentration range from 0.01

TABLE IV

CRYOSCOPIC CONSTANTS $-K_f$, FOR SOME SOLUTES IN SULFOLANE

Solute	$-K_f$ (deg kg mole^{-1})	Ref.
Acetanilide	66.2 ± 0.6	61
Benzothiophene sulfone	66.2 ± 0.6	61
Naphthalene	65	53
Diphenyl methane	65	53
n-Heptane	65	53
Benzoic acid	65, 64.1 ± 0.2	36, 53
H_2SO_4	64.8	62
Tributylphosphine oxide	59.3	126
Dioxane	54	127
Trifluoroacetic acid	53.2	126
Dimethyl sulfoxide	47.3	126
CCl_4	35	36, 53
Cyclohexane	17	36, 53

to 0.1 *m*.[61] Methanol is reported to show some association and tetraethylammonium iodide is extensively dissociated.[61]

Della Monica *et al.*[53] have shown that the cryoscopic plots in most cases exhibit a sharp inflection at about the same temperature where pure sulfolane has its phase transition (15.43°C).

D. Solute–Solvent Interactions

If an inorganic solute dissolves in a solvent, solution is nearly always accompanied by solvent coordination and alterations in, for example, the IR spectrum can be expected to occur.[98,130] The effect of a variety of dissolved salts on the ^{1}H NMR chemical shifts of sulfolane have been extensively studied.[76,131] For the majority of these salts, the shifts for α protons exhibit greater absolute values than those for the β protons, whereas the molar shifts are independent of concentration.[76] The ^{23}Na chemical shifts in sulfolane show a marked concentration dependence for NaI, a slight dependence for $NaClO_4$, and no dependence for $NaBPh_4$. This downfield shift occurring with increasing concentration is assumed to reflect the formation of contact ion-pairs.[44] Similar conclusions are also drawn on the basis of ^{133}Cs shift measurements.[132–134]

It has been reported that IR and Raman spectra do not reveal any differences between the dissolved nitronium salts in sulfolane as compared with those of the salts as mulls. This is understandable if one thinks of the solutes as existing in the form of closely associated ion-pairs.[115]

Based on the lowering of the IR SO_2 stretching frequencies, it has been assumed that sulfolane coordinates via the lone electron pair of one of the SO_2 oxygens.[130] The sulfolane molecule has a rather low donor ability toward Lewis acids and metal ions, but it still yields coordination compounds with moderate stability with many inorganic compounds.[130] The coordination ability of some sulfoxo solvents toward BF_3, based on the ^{19}F NMR data,[108] decreases in the order: DMSO $\gg$ diethyl sulfite $>$ sulfolane $>$ methyl methanesulfonate $>$ diethyl sulfate. The donor strengths of these solvents, as established by ^{1}H NMR measurements, follow the same order. The donor strength of sulfolane toward iodine is comparable to that of benzene.[98,107]

The donor number of Gutmann, DN_{SbCl_5}, is a measure of nucleophility, i.e., the Lewis basicity of the solvent.[87,94,135] The *DN* value of aprotic solvents allows them to be arranged in a practical sequence. A high value of *DN*, together with a high dielectric constant, means that the solvent has a high ionizing power. Thus, sulfolane ($DN = 14.8$; $\varepsilon = 43.2$) is a better ionizing solvent than, say, acetonitrile ($DN = 14.1$; $\varepsilon = 38.0$).[135]

The concept basicity is employed to indicate the relative position of a donor–acceptor equilibrium in a solvent.[98] In spite of the high dipole moment, sulfolane does not exhibit a highly basic or acidic character. This may be due to the greater strength of the S–O bonds in sulfones as compared to the S–O bonds in sulfoxides, which in turn are known to be of greater polar character.[95] For some common dipolar aprotic solvents with different donor groups the following order of increasing basicities has been proposed[136]: sulfolane < acetonitrile < propylene carbonate ≪ DMSO. It has also been found that for certain sulfones the order of decreasing basicity is[137]: dimethyl sulfone > diethyl sulfone > di-*n*-propyl sulfone > di-*n*-butyl sulfone > sulfolane.

In an extensive study by Reedijk *et al.*,[130] sulfolane is shown to form a number of adducts and complex solvates with metal halogenides, and a description is given concerning their synthesis and characteristics. The compounds have the general formula MX_n(sulfolane)$_m$ in which $X = Cl^-$, Br^-, $SbCl_6^-$, $InBr_4^-$, $InCl_4^-$, $FeCl_4^-$, $AlCl_4^-$ and $n = 2$–5, $m = 1$–6. These complexes are, in general, highly unstable upon hydrolysis. The adducts are formed when the metal halides are dissolved in excess sulfolane at 30°C.

Based on the investigations of Reedijk and co-workers,[130] the complexes of sulfolane are known to have a lower stability than those with ligands such as tetramethylene sulfoxide or water, but they are slightly stronger than complexes with the ligand $POCl_3$. The ions M(sulfolane)$_6^{2+}$ (where M = Fe, Co, Ni, or Cu) are shown to be octahedrally coordinated on the basis of ligand-field spectra. The number of sulfolane molecules in these adducts seems to correlate with the size of the anion. No hexasolvates can be obtained if the anion size is too small, i.e., smaller than the size of chloro- or bromoanions.[130] Thus, for the ClO_4^- ion, Langford and Langford[138] found that the dissolution of $Co(ClO_4)_2$ in sulfolane gives the complex Co(sulfolane)$_3(ClO_4)_2$. The following metal halogenides do not give any solid adduct with sulfolane: $FeCl_2$ $NiCl_2$, $SnCl_2$, $AlCl_3$, $CrBr_3$, $CrCl_3$, $FeBr_3$, $FeCl_3$, $InBr_3$, and $InCl_3$.[130]

Puchkova *et al.*[139] reported on the adducts sulfolane·$AlBr_3$ and sulfolane·$2AlBr_3$. The adduct $C_3H_6N_6O_6$·sulfolane was already mentioned in Section V. Sulfolane and the gaseous BF_3 combine together in a 1:1 mole ratio. The complex, which is very hygroscopic, can be isolated in the form of square, white plates.[51] The corresponding complex with PF_5 seems to be difficult to isolate because PF_5 is a weaker Lewis acid than BF_3. The interaction of BF_3 with sulfolane has also been studied by the ^{19}F NMR method.[108] The resonance signal of the complex is a singlet and the upfield shift caused by solvation is 143.5 ppm relative to CCl_3F. Decaboranes $B_{10}H_{10}^{2-}$ and dodecaboranes $B_{12}H_{12}^{2-}$ form stable, but hygroscopic, complexes with sulfolane. The melting points of these adducts are, in general, near 200°C.[140,141]

In the superacid system, FSO_3H–SbF_5–SO_2ClF at $-80°C$, sulfolane is easily protonated.[110] The protonation occurs at one of the two oxygens. As established by ^1H NMR spectroscopy, the deshielding of the α protons in the sulfolane ring is about 1.15 ppm lowfield from the value in the parent compound at the same temperature. The corresponding deshielding for β protons is measured as 0.67 ppm. No significant changes could be detected in the ^1H NMR spectrum of the protonated sulfolane in the temperature range from $-60°C$ to room temperature. This is an indication of the stability of the protonated specimen in the superacid system.

A common feature of dipolar aprotic solvents is that hydrogen bonds do not exist between the solvent molecules. The low hydrogen self-bonding capability makes sulfolane useful in homoconjugation studies.[72] Although sulfolane has a moderately high dielectric constant, it still has very little tendency to form hydrogen bonds with other protic solutes. In this respect sulfolane differs clearly from DMSO which forms much stronger hydrogen bonds.[95] With phenol both oxygen atoms coordinate to the hydrogen and a greater donor strength is observed toward this acid.[98] When a hydrogen bonding interaction occurs, the ^1H NMR chemical shift of the proton involved usually moves to lower field.[142] In CCl_4 solution the observed ^1H NMR chemical shift δ_{obs}, of the phenolic hydrogen, in the presence of varying proportional amounts of sulfolane, has the limiting value of 5.95 ppm downfield from the value without sulfolane at 38°C.[108]

Similar structures are also assumed to form with 1,1,1,3,3,3-hexafluoro-2-propanol and 2,2,2-trifluoroethanol.[143–145]

Using IR spectra Odinokov and co-workers[146,147] have studied the ability of some hydrogen-bond acceptors to form hydrogen bonds with cyclohexanol, phenol, benzoic acid, and salicylic acid in CCl_4 solutions. On the basis of these results, the acceptors were arranged in the following order of decreasing strength of associated hydrogen bonds: DMSO $>$ dioxane $>$ acetone $>$ sulfolane.[146]

Bis(fluorodinitromethyl)mercury[148] and bis(trinitromethyl)mercury[149] are reported to form direct 1:1 complexes with sulfolane. The melting point of the latter complex is 92°–93°C.

VII. Electrochemical Studies

A. Polarographic and Voltammetric Studies

Sulfolane was first introduced in polarographic studies by Hills.[150] Since then, only a limited number of papers concerning polarography in sulfolane have been published. Most of these studies have been undertaken to examine the properties of inorganic species. There are considerably fewer studies dealing with organic depolarizators. Polarography is also suitable for sulfolane purity control.[45,48] Impurity concentrations as low as 1×10^{-6} equivalents/liter can be detected by this method.[39]

Owing to its high dielectric constant sulfolane is a very suitable solvent for polarographic studies, but its high freezing point is a restriction for its use at temperatures much below normal room temperature. The lowest working temperature depends mainly on the concentration of the supporting electrolyte which effectively depresses the freezing point of the system. $LiClO_4$, $NaClO_4$, $(Et)_4NClO_4$ (TEAP), $(Bu)_4NClO_4$ (TBAP), $(Pr)_4NI$, $(Bu)_4NI$, $AgClO_4$, NH_4PF_6, and KPF_6 have been used as supporting electrolytes, usually in 0.1 *M* concentrations.

Several types of reference electrodes suitable for work in nonaqueous sulfolane have been presented. Louis and Benoit[118] used an I^-/I_3^- reference electrode which consisted of a platinum wire in a sulfolane solution containing 0.015 *M* of both $(Et)_4NI$ and $(Et)_4NI_3$. The potential of this electrode is reported to be +0.17 V vs. a saturated calomel electrode (SCE). A bridge of sulfolane (0.1 *M* in TEAP) can be employed to make the connection with the sample compartment.

A sheet of silver (5 × 2 cm) coated with AgCl and immersed in sulfolane saturated by AgCl and $(Et)_4NCl$ was used by Headridge *et al.*[54] as the reference electrode. With this electrode, the polarographic range is 2.5 V (from +1.2 to −1.3 V) in a 0.1 *M* $NaClO_4$ solution and 3.5 V (from +1.2 to −2.3 V) in a 0.1 *M* TEAP solution at 40°C. Using a rotating platinum electrode (RPE), the accessible potential range for $NaClO_4$ as supporting electrolyte is from +3.3 to −1.3 V vs. Ag/AgCl electrode.[54] Desbarres *et al.*[48] report an $Ag/AgClO_4$ (0.01 *M* in sulfolane) reference electrode with an emf of 0.64 V vs. SCE at 25°C. According to Coetzee *et al.*[39] the range between the anodic oxidation and the cathodic reduction of TEAP in sulfolane is 3.54 V (from +0.04 to −3.50 V) vs. an $Ag/AgClO_4$ (0.1 *M* in sulfolane) reference electrode (AgRE) at 30°C. The corresponding range vs. SCE is the same, but the limits of electrolyte discharge are shifted to more positive values (+0.70 V). For TEAP the corresponding voltammetric range covers the potentials from +2.3 to −2.9 V vs. the AgRE reference electrode.[39] TBAP has been reported to have a voltammetric discharge range from +0.975 to −2.8 V vs. an $Ag/AgClO_4$ reference couple at 50°C.[49] The discharge ranges of electrolytes such as $LiClO_4$, $NaClO_4$, TEAP, and NH_4PF_6 at Pt and Hg electrodes have also been determined.[48]

Owing to the appreciably high viscosity of sulfolane solutions at ordinary temperatures, the resistance of the polarographic system will normally be very high. For the same reason, *iR* corrections in two-electrode measurements are large,[151] and the diffusion coefficients are low.[39,54] Thus, concentrations of about 2–3 $\times 10^{-3}$ M of the depolarizators are required. However, the residual currents are generally very low; values as low as 0.2 μA have been reported.[39]

Polarographic and voltammetric measurements for 22 inorganic substances in anhydrous sulfolane have been carried out by Headridge *et al.*[54] In these measurements both dropping mercury electrode (DME) and RPE have been employed as indicator electrodes. As was specially demonstrated by the Tl(I) ion, the diffusion currents were proportional to the concentrations for most of the inorganic depolarizators in sulfolane. This is also verified by the sodium ion.[39] Small water concentrations (<0.01 M) do not cause any detectable shift in half-wave potentials for the majority of inorganic ions.[151]

Owing to the high viscosity of sulfolane, the polarograms obtained with DME exhibit a pronounced widening of the current range between drop maximum and drop fall until the diffusion current level is reached.[39,45] This phenomenon requires one to make a choice between the maximum, mean, or minimum current values when evaluating the slope of the E vs. $\log[(i_d - i)/i]$ plot of a polarogram. Coetzee *et al.*[39] have employed the maximum value and this, indeed, seems to be the best choice.

The Sm(III) ion exhibits two waves in sulfolane.[54] These appear at $E_{1/2} = -0.51$ and -1.22 V vs. an Ag/AgCl electrode. The corresponding waves in water occur at $E_{1/2} = -1.80$ and -1.96 V, respectively. The pronounced difference in the polarographic behavior of the Sm(III) ion in these two systems indicates the greater facility of Sm(II) ion formation in sulfolane than in water. Hanley and Iwamoto[108] have examined the electrochemical behavior of Cu(II) and Cu(I) ions in a series of sulfoxo solvents, including sulfolane. For solutions of Cu(II) ion in sulfolane, two-step reduction waves are established with RPE, but only one-step reduction waves with DME. This is due to the closeness of the potentials of the Cu(II), Cu(I) and Cu(I), Cu(Hg) couples to each other in sulfolane and in the other sulfoxo solvents. Further data are given for the reduction of several metal ions in sulfolane and are compared with those found in acetonitrile and water.[48]

A polarographic run of oxygen dissolved in sulfolane containing 0.1 M TEAP as supporting electrolyte shows two reduction waves at about -0.90 and -1.9 V vs. SCE.[118] The wave at -1.9 V is reported to be disturbed by irregular drop fall, a phenomenon frequently encountered at very negative potentials when one is working with a DME in anhydrous sulfolane. An increase in the water concentration up to 0.12 M shifts the half-wave potential of the first wave to -0.82 V, whereas only a slight increase occurs in the wave height. The diffusion coefficient of oxygen in sulfolane (0.1 M in TEAP) is reported[118] to be 1.2×10^{-1} cm^2/sec, which is in accordance with the values obtained with DMSO and propylene carbonate as solvents, but less than that in dimethylformamide. Similar results are obtained with TBAP as supporting electrolyte, whereas with $KClO_4$ three waves occur with half-wave potentials at -0.74, -1.07, and -1.8 V. In the presence of a proton donor such as phenol, two drawn-out waves appear with $E_{1/2}$ values at -0.58 and -1.24 V vs. SCE, respectively. This result parallels that obtained in water. The first wave at -0.90 V vs. SCE is found to represent a quasi-reversible one-electron reduction of the oxygen molecule

$$O_2 + e^- \rightleftharpoons O_2^- \tag{9}$$

with further decomposition of the O_2^- ion to give O_2, OH^-, and peroxide.

In a voltammetric study by Starkovich,[152] reasonable well-defined peaks with characteristic peak potentials were observed for sulfur compounds such as *n*-butyl mercaptan, sulfide, disulfide, and phenyl disulfide using sulfolane as solvent. Gold, glassy carbon, and platinum electrodes give anodic waves for the mercaptan and sulfide and cathodic waves for the disulfide. All the aliphatic sulfides undergo irreversible reductions, but phenyl disulfide is reversibly reduced on mercury.

Some aromatic nitro compounds are easily reduced at DME in sulfolane.[45] In the special case of nitrobenzene (NB), it is shown that two reduction waves occur at half-wave potentials of -1.1 and -1.9 V vs. aqueous SCE at room temperature. The former wave is due to a reversible one-electron reduction of NB to give a radical anion (Eq. 10). The formation of the radical anion

$$NB + e^- \rightleftharpoons NB^{\cdot -} \tag{10}$$

can be verified by the polarographic oxidation current, the visible spectrum, and the electron spin resonance spectrum.

Voltammetric investigations have shown that para-substituted benzaldehydes undergo one-electron reductions in sulfolane.[49] The radical anion formed undergoes a subsequent protonation followed by a dimerization. The source of protons is proposed to be residual water. This type of secondary reaction is always pronounced when one is working at such dilute depolarizator concentrations as in polarography. The concentration of the depolarizator usually is of the same order as that of the residual water in the "anhydrous" solvent purified by whatever means.

B. Conductivity Measurements

Conductivity parameters, Λ_0, for some 1:1 electrolytes in sulfolane at 30°C are presented in Table V.[55,61,64–66,74,88,112,153–156] The data in Table V, are, in general, much lower than those found in most other solvents even at lower temperatures.

Della Monica *et al.*[64] have determined coulometrically the limiting ionic conductivity, λ_0^-, of the ClO_4^- ion in sulfolane at 30°C to be 6.685 Ω^{-1}

TABLE V

CONDUCTIVITY PARAMETERS FOR 1:1 ELECTROLYTES IN SULFOLANE[a]

Electrolyte	Conductivity ($\Lambda_0\ \Omega^{-1}\ cm^2\ mole^{-1}$)	Ref.
$HClO_4$	15.49 ± 0.26	74, 153
LiCl	13.629 ± 0.016	55
LiCl[b]	15.919	112
LiBr	13.250 ± 0.010	55
LiI	11.528 ± 0.009	55
$LiClO_4$	11.012	154
$LiNO_3$	~10.6	61
NaI	10.865 ± 0.003	55
$NaClO_4$	10.293	154
NaSCN	13.200	155
KI	11.253 ± 0.005	55
$KClO_4$	10.737	154
KSCN	13.633	155
KPF_6	9.995 ± 0.004	55
Kpicrate	9.36 ± 0.03	65
$RbClO_4$	10.840	154
RbSCN	13.866	155
$CsClO_4$	10.955	154
CsSCN	13.963	155
NH_4ClO_4	11.653	154
$AgClO_4$	~11.5	64
Et_4NI[c]	11.201 ± 0.008	55
Me_4NClO_4[c]	10.994	88
Et_4NClO_4[c]	10.632	88
Pr_4NClO_4[c]	9.912	88
Bu_4NClO_4[c]	9.486	88
Ph_4AsCl[c]	~10.5	61
n-Am_4NBr[c]	11.416 ± 0.002	65
i-Am_3BuNpicrate[c]	6.282 ± 0.006	66
i-Am_3BuNI[c]	9.801 ± 0.036	66
i-$Am_3BuNBPh$[c]	4.900 ± 0.038	66
Dimethylmorpholinium iodide	~12.5	61
Pyridinium perchlorate	11.1	156
o-Anisidinium perchlorate	10.3	156

[a] At 30°C.

[b] At 35°C.

[c] Am, amyl; Ph, phenyl; Bu, butyl; Pr, propyl; Et, ethyl; Me, methyl.

$cm^2\ mole^{-1}$. Using this result, the limiting ionic conductivities for other ions shown in Table V can easily be calculated. An alternative basis for these calculations is proposed by Zipp,[66] who made use of the conductivity equivalence of the ions $i\text{-}Am_3BuN^+$ and BPh_4^-, which has been shown to apply in the case of other nonaqueous solvents. The λ_0^- value found this way for the I^- ion in sulfolane is $7.33 \pm 0.04\ \Omega^{-1}\ cm^2\ mole^{-1}$, which is in reasonable agreement with the value 7.25 ± 0.02 established by the method described above.

Limiting ionic conductances of ions in water and in a number of dipolar aprotic solvents, including sulfolane, and the associated Walden products are presented and comparisons made with other solvents in a comprehensive article by Cox.[157] Both cations and anions have considerably higher Walden products in sulfolane than in other dipolar aprotic solvents. Owing to the high viscosity of sulfolane, this may not, however, have any physical significance.[157]

Conductivity measurements of electrolytes have been made not only in the liquid phase of sulfolane, but also in the plastic phase, as demonstrated by Della Monica.[83] As deduced from the resistance measurements of a solution (1.2×10^{-4} *M* in$KClO_4$), a sharp change in conductivity occurs at 15.45°C on the heating curve. This result appeared to be reproducible and the transition point thus obtained is in excellent agreement with the corresponding values obtained with cryoscopic[53,158] and dielectric constant[77] measurements. In the cooling curve, the change occurs far below the transition point indicating that a metastable state on cooling is possible. In the temperature range 27.5°–28.5°C another peculiarity is found. Beginning at 27.5°C the conductivity decreases until the value at 28.40°C is reached, at which temperature again a sudden increase in the curve takes place. This phenomenon is explained by the increased electrolyte concentration in small liquid zones dispersed within the bulk of the solid solvent.[83]

C. Potentiometric and Conductometric Titrations

Sulfolane was apparently first introduced for potentiometric titrations by Morman and Harlow.[72] The low acidity and basicity of sulfolane becomes apparent in that it has one of the largest potential ranges (from −800 to +1200 mV) available in any titration solvent. Moreover, it is a very suitable solvent for titration of very weak acids or bases, being a nonleveling solvent for strong acids and bases. For example, using an 1.2 *N* solution of tetra-*n*-butylammonium hydroxide (TBAH) in isopropyl alcohol as titrant and a general purpose glass and calomel electrode pair, the following mixture of monobasic acids is well resolved in sulfolane[72]: $HClO_4$, HNO_3, 2,4-dinitrophenol, *o*-nitrophenol, and 2,6-di-*tert*-butyl-4-methylphenol. Perchloric and hydrochloric acids are also well-differentiated from each other, whereas perchloric and sulfuric acids are resolved only partially. Dibasic acids such as oxalic, phthalic, and succinic, when titrated with a 0.2 N TBAH titrant, have half-neutralization potentials for the first equivalent at −157, −180, and +30 mV, respectively.[72]

Although phenol is a monofunctional acid, it still shows two inflections when titrated with 1.2 N TBAH. This is explained on the basis of homoconjugation, i.e., on the basis of an acid–anion complex as indicated by Eq. 11. This type of solute behavior is expected if the solvent exhibits a moderately high

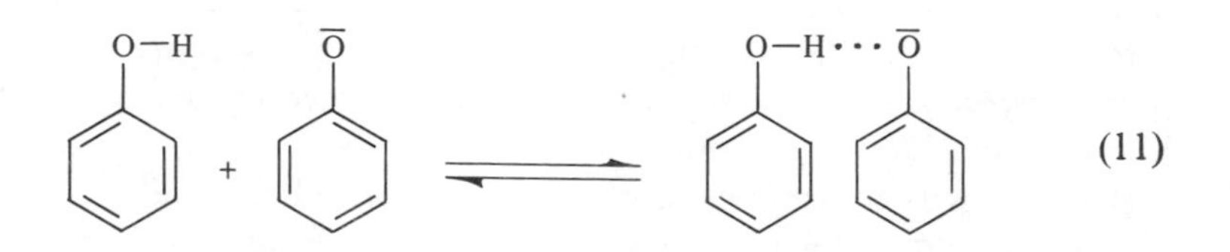

dielectric constant, a low hydrogen-bond capability, and a poor tendency to solvate ions.[57,72] A homoconjugation study of 2,6-dihydroxybenzoic acid in sulfolane has also been reported.[159]

A mixture of bases such as guanidine, piperidine, pyridine, and *o*-chloroaniline is easily differentiated using 0.2 N $HClO_4$ in dioxane as titrant.[72] As reported by Morman and Harlow,[72] it is also possible to titrate caffeine using sulfolane as solvent.

Potentiometric titration in sulfolane can be employed to determine the solubility product constants of the AgX salts (where X = Br,I, and SCN) and the stability of the $Ag_nX_m^{(m-n)-}$ complexes. A spongy silver wire can be used as the indicator electrode and the potentials measured against aqueous SCE. In the actual measurements silver perchlorate solutions in sulfolane are titrated with Et_4NBr, KI, and KSCN solutions.[67]

Coetzee and Bertozzi[160] have found that perchloric acid is extremely acidic in sulfolane. As supported by measurements with the hydrogen electrode in sulfolane, the hydrogen ion activity of a 10^{-2} M perchloric acid solution is ca. 10^8 times greater in sulfolane than in water. The corresponding activity is about $10^{6.5}$ times greater in a 10^{-4} M perchloric acid solution.[160] Furthermore, $HClO_4$ exhibits a remarkable chemical stability in sulfolane solutions.[159] Although $HClO_4$ is not completely dissociated in sulfolane (pK_a −2.7[120]), it still is a recommended titrant for nonaqueous titrations in inert solvents.

Four general types of titration curves were established by Coetzee and Bertozzi in a comprehensive conductometric study of the protonation of the following weak bases in sulfolane: water, alcohols, ketones, amines, and some dipolar aprotic solvents such as acetonitrile, nitrobenzene, dimethylformamide, and DMSO.[159,161]

$HSbCl_6$ is found to be essentially completely dissociated in sulfolane.[120] Its chemical stability in this solvent is, however, less than that of $HClO_4$.[159] Based on ^1H NMR and conductometric investigations[120] the following order of acid strengths in sulfolane has been established: $HClO_4 > HSO_3F > H_2S_2O_7$.

Sulfolane is also a suitable solvent for potentiometric titration of barbiturates and sulfa drugs such as barbituric acid, phenobarbital, amobarbital, barbital, secobarbital, sulfapyridine, sulfadiazine, and sulfamerazine.[57,58] A useful electrode system for these titrations consists of a glass indicator electrode and a fiber-type calomel reference electrode. $HClO_4$ is the recommended titrant.[58] The titration of barbituric acid is also shown to be feasible with various other electrode systems.[57]

In a study by Zipp[60] it has been shown that in sulfolane potentiometric titration is more accurate than titrations with most of the common acid–base indicators. This arises from the fact that these indicators do not exhibit color changes sufficient for satisfactory end-point detection especially when weak acids such as phenol or *p*-toluidine are titrated.

Some difficulties in the electrochemical investigations discussed in this section are reported to arise from the high hygroscopicity of sulfolane and especially of its $HClO_4$ solutions, as well as from the high viscosity of those solutions.[74,159]

D. Miscellaneous Electrochemical Measurements

To determine the autoprotolysis constant pK_s of sulfolane, the emf's of the following electrochemical cells were measured at 25°C by Kreshkov *et al.*[97]: glass electrode/HCl,sulfolane/AgCl,Ag and glass electrode/Bu_4NOH, sulfolane/AgCl,Ag. From the data obtained with different dilute solutions of HCl and Bu_4NOH, the standard emf's were calculated and the autoprotolysis constant determined. The value of pK_s for sulfolane was found to be 25.45.

In a report by Brenner[113] a single cell with a back emf as high as 5.2 V has been described. The cell consists of a 10% solution of $LiBF_4$ in sulfolane and a graphite rod. On making the graphite the anode, the high back emf can be observed after several minutes of electrolysis with a reproducibility of about 0.1 V. No satisfactory explanation could, however, be given for the cause of this high voltage, but it was believed to originate from some kind of reactive material formed when the graphite was made into an anode. The replacement of any one of the three components of the cell resulted in a lower emf.

Capacity and electrocapillary investigations of the electrical double layer at the mercury–sulfolane interface have indicated that KPF_6 is weakly adsorbed from sulfolane solutions.[36,162]

The standard (reduction) potential of the hydrogen electrode in perchloric acid solution is 0.075 V vs. AgRE.[74] A value 0.18 V less positive is reported by Benoit and Pichet.[156] The hydrogen electrode vs. Ag/AgCl reference electrode is found to be a good indicator electrode for the study of acids such as $HClO_4$, *p*-toluenesulfonic acid, tetrazole, barbituric acid, benzoic acid, and phenol in sulfolane.[57]

VIII. Reactions in Sulfolane

A. Chemical Stability of Sulfolane

Although sulfolane is known as a chemically "inert" solvent, being, for example, stable toward strong acids and bases,[12] it still undergoes reactions under appropriate conditions. These reactions may be divided into two categories depending on whether cleavage of the five-membered ring occurs.

To achieve an opening of the sulfolane heterocycle, relatively drastic conditions must be employed. For example, sulfolane undergoes an exothermic reaction with sodium and potassium metals at elevated temperatures to give dimeric sodium or potassium bis-1,8-octanedisulfinate in 5–6% yields, but it is unaffected by lithium.[163] Dimerization occurs unless the reaction product is detached from the metal surface during the reaction. The corresponding reaction with sodium amide leads obviously to 3-butenesulfinate.

Wallace *et al.*[164] have shown that the base-catalyzed elimination reaction of sulfolane at 55°C using DMSO as the solvent and potassium *tert*-butoxide as the base leads to a complete conversion of sulfolane to an unidentified product, in contrast to tetramethylene sulfide and sulfoxide, which yield mainly butadiene. It is assumed in the case of sulfolane that at first a β-proton abstraction occurs after which the S–C bond splits to give a sulfinate anion.[164] $CH_2{=}CHCH{=}CHSO_2Na$, butadiene, and isomeric octadienes are formed when sulfolane is heated with different proportions of sodium ethylate.[165] With sodium nonylate and phenylate much the same products are obtained.[166]

The photochemically generated hydroxide radical ·OH is shown[52] to attack sulfolane by hydrogen abstraction. The resulting sulfolane radical can be trapped by using a radical scavenger such as *tert*-nitrosobutane to give a nitroxyl radical stable enough for electron spin resonance (ESR) measurement. On the basis of the ESR spectrum, it is concluded that a cleavage of the C–S bond occurs giving rise to a radical of the type $\cdot CH_2{-}CH_2{-}R$.[52] The reaction of atomic hydrogen with sulfolane at −140° to −160°C also proceeds via hydrogen abstraction from the α position with respect to the sulfone group.[167] On pyrolysis at 490°–500°C, sulfolane dissolved in benzene gives

SO_2, saturated hydrocarbons, hydrogen, and ethylene.[168] Darkening of the solvent has been reported to occur within 0.5 to 2 hr when treated with concentrated H_2SO_4 or NaOH solutions at elevated (reflux) temperatures.[12]

An energy-rich radiation also brings about the ring cleavage. For example, the mercury-photosensitized reaction of sulfolane in the vapor phase and in the temperature range 70°–130°C, as reported by Honda and co-workers,[70] yields SO_2, H_2, C_2H_4, C_2H_2, and c-C_4H_8 as the main cleavage products, and CH_4, C_2H_6, C_3H_8, and C_3H_6 in minor amounts. In this investigation the 253.7 nm mercury resonance line was used. A direct photolysis of sulfolane for several hours at 110°C does not yield any decomposition products in detectable amounts. Sulfolane is found to be stable toward benzene photosensitization also.[70] The radiolytic decomposition of sulfolane at low temperatures has been reported to give radicals, detectable by ESR spectroscopy, and hydrogen.[169]

The major peaks with intensities greater than or equal to 10% of the base peak in the mass spectrum of sulfolane appear at m/e values[170,171] of 120 (35%), 56 (82%), 55 (68%), 41 (100%), 29 (10%), 28 (50%), and 27 (17%).

Of those reactions of sulfolane in which ring cleavage does not occur, halogenation seems to be the most widely studied. In the presence of sulfuryl chloride, sulfolane undergoes a selective β-chlorination at 60°C (Eq. 12).[172,173] A 10-fold molar excess of SO_2Cl_2 in relation to sulfolane and a

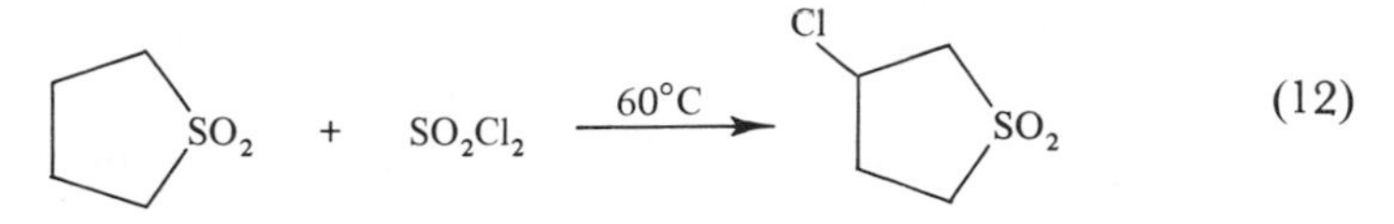

(12)

reaction time of 120 hr are required to obtain 3-chlorosulfolane in 46% conversion. No 2-chlorinated product is reported to occur in this reaction.

Boiling sulfolane (0.05 mole) in dry CCl_4 for 10 hr in the presence of a two to fourfold excess of chlorine under simultaneous irradiation with UV light yields 3-chloro-, 3,4-dichloro-, and 3,3,4-trichlorosulfolane.[174,175] With amounts as high as 1.17 moles of chlorine, yields up to 77% 3,3,4,4-tetrachlorosulfolane are achieved.[176] Bromination of sulfolane by BrCl in boiling CCl_4 under irradiation gives 2-bromosulfolane, which reacts further to give *cis*-2,5-dibromosulfolane. With prolonged irradiation other dibromo isomers are formed.[177] As demonstrated by Tidswell and Doughty,[50] the fluorine radical F· formed by anodic oxidation of the borofluoride ion BF_4^- in sulfolane attacks the 2- position to give 2-fluorosulfolane.

Sulfolane can be converted to thiophene in the presence of Al–Cr, Al–Mo, Al–Pt, or Al–Co–Mo catalysts at atmospheric pressure in elevated temperatures as high as 300°–500°C. The best yield (46%) is achieved with the Al–Cr catalyst.[178]

Diisobutylaluminum hydride is shown to reduce sulfolane effectively to the corresponding sulfide.[179] A threefold excess of the hydride with respect to

$$>SO_2 + 2(C_4H_9)_2AlH \rightarrow >S + 2(C_4H_9)_2AlOH \quad (13)$$

sulfolane yields 73% tetramethylene sulfide in 72 hr at 20°–25°C when the reaction is carried out in mineral oil. The action of the Grignard reagent EtMgBr, on sulfolane in diethyl ether, benzene, or tetrahydrofuran leads to the formation of 2-mono- and 2,5-di-MgBr derivatives.[180] This reaction is very slow at −20°C, 20°C is the optimum temperature, and at 80°C side reactions such as ring splitting occur. Tributylstannylamines react with sulfolane to give poisonous 2,5-bis(tributylstannyl)sulfolane.[181]

B. Sulfolane as an Inert Solvent

Many syntheses, especially those involving anions, can be performed with advantage in dipolar aprotic solvents because of the poor anion solvation in these media.[67,182,183] Reactions involving anions are, in general, facilitated when sulfolane is used as the reaction medium, which usually is reflected in enhanced reaction rate and possibly yield.[184]

As reported by Tabushi *et al.*,[173] the use of sulfolane as a solvent in the ionic chlorination of aliphatic compounds with sulfuryl chloride causes a remarkable acceleration of the reaction as compared to the corresponding chlorination in the absence of sulfolane. Adamantane, norbornane, cyclohexane, isooctane, and *n*-hexane are chlorinated with great facility and regioselectivity with nearly complete retention of the skeletal structure. The excellent solvent properties of sulfolane in the chlorination of alkylamines have also been reported.[185]

In the preparation of highly fluorinated aromatic compounds with potassium fluoride from corresponding chlorinated or brominated compounds, sulfolane has proved to be an excellent solvent.[75,186] Another desirable solvent for these reactions is *N*-methylpyrrolidone, but the yields are reported to be slightly higher and somewhat less complex products are obtained when sulfolane is used as solvent.[186] The same is true with respect to dimethylformamide. The greater degree of fluorination may be due in part to a temperature effect because sulfolane permits higher reaction temperatures.[75] The reaction of hexachlorobenzene with KF in sulfolane at 230°–240°C for 18 hr yields the following products: C_6F_6 (0.4%), C_6ClF_5 (25%), $C_6Cl_2F_4$ (24%), and $C_6Cl_3F_3$ (30%).[186] In general, KF appears to be a much more active fluorinating agent in sulfolane than in other aprotic solvents. The more powerful fluorination agent, CsF, gives hexafluorobenzene in 42% yield at 160°–190°C with a reaction time of 18 hr.[186] The strong fluorinating agent, SbF_5, is reported to be unsuitable for use in sulfolane because it also attacks the solvent.[38]

As demonstrated by Fitzgerald *et al.*,[187,188] sulfolane is preferable to other solvents such as DMSO, dimethyl acetamide, or methanol in the study of the isomerization of *cis-* and *trans-*$[CoCl_2(ethylenediamine)_2]^+$ ions. No solvent-containing complexes occur with sulfolane as with DMSO. In dimethylacetamide subsequent reduction reactions occur, and in methanol the measurements are not practicable. Sulfolane was found ideal for the study of isomerization kinetics and equilibria.

Sulfolane is also an excellent reaction medium for Friedel–Crafts type nitration of aromatics by nitryl halides, dinitrogen tetroxide, dinitrogen pentoxide, and stable nitronium salts.[38] By using sulfolane as the solvent, the amount of chlorinated by-products in the nitration by nitryl chloride will be lower than in the CCl_4 medium.

Olah and co-workers[115,189–191] have shown that sulfolane is a suitable solvent for investigations of aromatic nitration by nitronium salts such as NO_2BF_4. Similar investigations have also been reported by other authors.[73,192,193] The need for high-purity solvents such as sulfolane in these nitrations, especially in dilute systems, has convincingly been demonstrated.[37] Homogeneous nitrations of benzene and alkylbenzenes can be conventiently carried out in sulfolane also by using mixed acid ($HNO_3 + H_2SO_4$) and nitric acid.[93]

Sulfolane is a suitable medium for sodium borohydride reductions.[194,195] It is possible to displace halogens and tosylates selectively with hydrogen in certain primary, secondary, and tertiary alkyl and benzylic halides and tosylates, in spite of the presence of other groups including carboxylic acid, ester, and nitro groups. Among the halides and tosylates which undergo such reduction are: 1-iododecane, 1-bromododecane, *n*-dodecyl tosylate, 1-chlorododecane, α,2,6-trichlorotoluene, 2-iodooctane, 2-bromododecane, α-phenylethyl bromide, styrene dibromide, benzhydryl chloride, benzhydryl bromide, and cumyl chloride. At 100°C the reductions give yields of hydrogenolysis products from 53% to 98%.[194] Moreover, the aromatic nitro group is reduced readily by sodium borohydride at 85°C in sulfolane to give initially azoxy compounds which may further reduce to mixtures of the corresponding azo derivatives and amines.[195] The use of sulfolane affords greater yields of amines, whereas the reaction time is longer in comparison to DMSO as solvent. This reduction can also be carried out in the presence of cyano and amido groups.

On treatment with a solution of HBF_4 in sulfolane, phenylhydroxylamine condenses with aromatic compounds of moderate reactivity to give substituted 2- and 4-aminobiphenyls and diphenylamines.[196]

Owing to its low basicity sulfolane is a suitable solvent for the calorimetric study of reactions of the type

$$X^-(\text{solv}) + HY(\text{solv}) \rightleftharpoons HYX^-(\text{solv}) \tag{14}$$

where X = Cl, Br, or I, and Y = Cl.[119,197] The formation of the hydrogen dihalide ions is generally favored in dipolar aprotic solvents because the differences in solvation energies for small X^- and large HYX^- anions are considerably smaller in these solvents than in protic solvents. Similarly, the I_3^- ion is more stable than the I^- ion in sulfolane.[198,199] This may be due to the much weaker solvation of I^- in the aprotic medium. Parallel results are also found for the Br_3^- and Cl_3^- ions.[200–202]

As seen by polarography, nitrobenzene forms radical anions via one-electron reduction at mercury in anhydrous sulfolane.[45] Using mercury pools, a prolonged electrolysis of a nitrobenzene–sulfolane solution (0.1 *M* in tetrabutylammonium iodide) yields appreciable concentrations of the radical anion at the cathode (Eq. 10). On adding such an electrolyzed solution to deaerated acrylonitrile, polymerization occurs.[203] The structure of the PAN polymer, yellowish in color, deviates from that of linear PAN by having a great content of cyanoethyl groups.[204] The polymerization mechanism is evidently anionic. Tidswell and Doughty[50] found that the electro-initiated polymerization of styrene is easily performed in sulfolane. On electrolysis of a sulfolane–$NaBF_4$–styrene system, cationic polymerization of styrene occurs in the anode compartment. Small amounts of water have a catalytic effect on this polymerization. Martinmaa *et al.*[205] employed a pulse electrolysis technique in polymerization of styrene, methyl methacrylate, β-methylstyrene, β-nitrostyrene, vinyl isobutyl ether, and eugenol in sulfolane–$NaBF_4$ solutions at Pt electrodes. The results substantiate those of Tidswell and Doughty with respect to yield, reaction mechanism, and molecular weight of the resulting polymers. Only methyl methacrylate polymerizes in considerable yields at the Pt cathode. A comparison with results obtained in solvents other than sulfolane suggests that the general initiation mechanism in sulfolane is the same as in dimethylformamide, dimethylacetamide, and tetrahydrofuran.

Orthophosphoric acid, which is stable during electrolysis in aqueous solution, can be cathodically reduced to phosphorous acid in sulfolane.[206]

Sulfolane has proved to be a better solvent than tetraglyme for fluoride ion-catalyzed polymerization of hexafluorobut-2-yne.[207] Sulfolane has also been reported to be an useful reaction medium for the solution polymerization of diamine and ureas. Increased polymerization rates and uncontaminated products are obtained.[208] Sulfolane is also a good medium for preparation of polycondensates from benzyl derivatives and certain ferrocene derivatives.[209] The use of sulfolane as reaction environment in these polymerizations is advantageous because of the low basicity and the poor cation-solvating power of this solvent. Additional advantages, in comparison to the melt polycondensation methods, include better conversions, lower reaction temperatures, fewer side reactions, simpler polymerization apparatus, and better molecular weight control.[209]

The use of sulfolane as a reaction solvent in the copolymerization of styrene with 1-bicyclobutanecarbonitrile or methyl 1-bicyclobutanecarboxylate in the presence of $ZnCl_2$ is reported to lead to copolymers soluble in both sulfolane and dimethylformamide.[210] The corresponding polymerization in bulk yields crosslinked and intractable copolymers.

Sulfolane has also been employed as a catalyst or cocatalyst in several other polymerization reactions such as in the preparation of polysiloxanes,[211] in trioxane[212] and alkene oxide[213] polymerization, in ethylene dimerization to *n*-butenes,[214] as well as in the curing of epoxy resins.[215]

The phosphorylation of riboflavine by $POCl_3$, $O{=}P(OH)_2Cl$, $O{=}P(OH)Cl_2$, or a mixture of $POCl_3$ and H_3PO_4[216] also employs sulfolane as the solvent. The Rosenmund–von Braun synthesis for preparation of aromatic nitriles from a *p*-halophenol is conveniently carried out by using sulfolane as the solvent.[217] When 1,4-cyclohexanedisulfonamide is treated with phosgene in sulfolane, the corresponding isocyanate is obtained. The reaction is rapid in sulfolane and high yields of a pure product will result.[218] Sulfolane has also been used successfully as a solvent in the preparation terephthalonitrile in yields of 75%, by heating diammonium terephthalate.[219]

In a recent paper by Matawowski some interesting effects of sulfolane on linalool dehydration are reported.[220] As compared with two other dipolar aprotic solvents, DMSO and dimethylformamide (DMF), sulfolane exhibited the greatest dehydration power in the order: sulfolane > DMSO > DMF, at 180°–184°C. The catalytical behavior of sulfolane in this reaction has been explained in terms of the formation of a carbocation from the parent linalool molecule and in that sulfolane favors the formation of this cation owing to its acidic α-hydrogen atoms.[220]

Other reactions in which sulfolane has proved to be a suitable solvent are the nitration of pyridine by NO_2BF_4,[221] the reaction of aryl isocyanates with aromatic compounds to give anilides,[222] and the quaternization of pyridine derivatives.[223,224]

IX. Other Studies

A. Toxicology

The acute oral LD_{50} value for rats is about 1.5–2.5 g sulfolane/kg. The corresponding value for mice is 1.6–2.5 g/kg.[225–227] Death occurs within 24 hr with no specific pathological lesion, and seems to be caused by anoxia.[227] Chronic toxicity tests for 4 months with rats, when a total dose of 5.1 g/kg was given, revealed no weight changes, no dyscrasias (a depraved state of the humors), and no blood pressure changes; a normal liver function was established.[226] No apparent systemic effects were found in rats when sulfolane was applied to the skin in amounts of 3.8 g/kg.[227]

Based on experiments with rabbits, sulfolane seems to have no skin-irritating nor skin-sensitizing properties and it is only a very mild eye irritant.[227] The single skin penetration LD_{50} for rabbits is reported to be within the limits of 2.3–4.3 ml/kg.[225]

The following threshold concentrations of sulfolane have been reported[228]: organoleptic taste, 0.58 mg/liter; smell, 0.69 mg/liter; and for public health norms, 1 mg/liter. Sulfolane has a positive inotropic activity on guinea pig left atria[229] and it induces the absorption of heparin from the rabbit intestine.[230]

B. Uses in Chromatography and Other Separation Schemes

Sulfolane is shown to be a very suitable stationary liquid for the gas chromatographic analysis of complex C_2–C_6 hydrocarbon mixtures.[231] A column charged with firebrick impregnated with sulfolane exhibits high separation efficiency and short time of analysis at 30°C with hydrogen carrier gas. The following mixtures of hydrocarbons are completely separated: ethane–ethylene–nitrous oxide–carbon dioxide–acetylene and isobutene–1-butene.

The property of sulfolane to extract aromatic hydrocarbons from aliphatic ones is widely utilized in the petrochemical industry and numerous patents dealing with this subject have been published during the last three decades. The specific gravity, vicosity, and interfacial tensions of sulfolane are within the ranges which make for high efficiency in extraction equipment.[33] The extraction properties are fairly insensitive toward water addition up to 5% (w/w), which is of some practical value. This and other extraction applications covering the earlier literature are reviewed by Morrow.[1]

Sulfolane has been utilized in the separation of various mixtures and azeotropes by solvent extraction. Examples include the separation of *o*- and *p*-chlorotoluene,[232] the separation of thiophene from benzene,[233,234] and the extraction of HF from mixtures of HF, HCl, and $F_2C{=}CF_2$.[235] Sulfolane is also a suitable solvent for separating chloroethane/chloroethylene azeotropes by extractive distillation. An example is the azeotrope of $ClCH_2CH_2Cl$ with $Cl_2C{=}CHCl$.[236]

Parish *et al.*[237] have found that sulfolane is a very good alternative for sucrose in density gradient centrifugation of nucleic acids. It is also established that sulfolane may reduce the intermolecular interactions of RNA and that the Dische test for DNA operates well in this solvent.

C. Binary Mixtures

Densities, viscosities, surface tensions, dielectric constants, vapor pressures, activities, partial molar volumes, and heats of mixing of binary mixtures of sulfolane with water, methanol, and ethanol have been extensively measured at various temperatures by Tommila *et al.*[86] and also by other authors.[84, 238] In sulfolane–water mixtures the partial molar volumes of both constituents do not vary appreciably over the whole composition range at temperatures of 30°, 40°, 50°, or 60°C. The dielectric constant vs. volume–composition plot shows positive deviation from linearity, while the molar refraction of the mixtures is a linear function of composition indicating validity of the additivity rule.[84] Thus, the structure of water is only slightly affected by the presence of sulfolane.

The heats of mixing, ΔH^{M}, are larger for sulfolane–methanol mixtures than for sulfolane–water mixtures, and still larger for sulfolane–ethanol mixtures. They are positive in all cases. This behavior is different from that found for mixtures containing DMSO rather than sulfolane. In addition, the entropy of mixing for sulfolane–water systems shows nearly ideal behavior.[86]

The viscosities of the sulfolane–water mixtures show a steady increase from pure water to pure sulfolane.[86] The same is also true for sulfolane–methanol and sulfolane–ethanol mixtures.[86, 87] However, the mixtures of sulfolane with higher alcohols such as propanol, isopropanol, *n*-butanol, isobutanol, *sec*-butanol, *tert*-butanol, and *n*-pentanol show viscosity curves with a minimum when the mole fraction of sulfolane is less than 0.2.[87] This indicates that sulfolane will act as an alcohol structure breaker in these mixtures.[87, 239]

Data on standard electrode potentials of the silver/silver chloride electrode in sulfolane–water mixtures are also available.[128] It has been reported that sulfolane–water mixtures behave as an ideal conducting medium.[240]

Hydroxides are known to be only very slightly soluble in pure sulfolane. The solubility, however, increases distinctly by the addition of water and also by increasing the size of the cation. Because of the poor solvating ability of sulfolane, the activity of the hydroxide ion is remarkably high and, consequently, very strong basicities can be established. For example, a 0.01 *M* solution of phenyltrimethylammonium hydroxide in sulfolane containing 5 mole % water is reported to be over 10^6 times more basic than the corresponding aqueous solution.[241] These strongly basic systems have been employed to determine pK_a values of various indicators on the basis of the Hammett acidity function concept.[63, 241–243] The rate enhancement of alkaline hydrolysis of ethyl benzoates at low water concentrations in sulfolane is also attributed to an increase in the activity of the hydroxide ion.[244] Data on kinetics of acid-catalyzed hydrolysis of phenylacetohydroxamic acid in aqueous sulfolane are also available.[245]

Investigations concerning mixtures of sulfolane with DMSO,[246] benzene,[247,248] CCl_4,[247] dioxane,[247] and a series of hydrocarbons[249,250] have also been reported.

D. Miscellaneous Applications

Sulfolane is found to be a good color-enhancing agent. For example, addition of about 20–30% (v/v) sulfolane to a solution containing silicomolybdic acid causes a pronounced deepening of the color.[251] Similar color enhancement has also been reported to occur in the case of several metal salts and some metal complexes when sulfolane is added to their water solutions.[252] As demonstrated by the adsorbances of the so-called Vogel Blue ($Co^{2+} + 2SCN^-$), sulfolane is more than 2.5 times better an enhancing agent than *tert*-butyl alcohol and more than 2 times better than acetone.[252]

Owing to the low volatility of sulfolane at ordinary temperatures, it is used as a plasticizer for many polymers, such as vinylidene fluoride polymers[253] and polylactams,[254] and it is shown to be a suitable solvent for spinning fibers from its PAN solutions.[123] Sulfolane has also found use in the preparation of permeable membranes from cellulose acetate,[255] in the preparation of photographic emulsions,[256,257] and in the preparation of metallized films and fibers.[258]

Aqueous sulfolane solutions are efficient delignifying media for aspenwood.[259,260] Acidified aqueous solutions of sulfolane are also effective for this purpose.[261]

E. Sulfolane Derivatives

Numerous ring-substituted derivatives of sulfolane are known. The methyl-substituted ones appear to have some significance as nonaqueous solvents, The introduction of one methyl group into the 3-position of sulfolane leads to the following physical properties: boiling point, 276°C[72]; freezing point, −1.9°C[93]; density (30°C), 1.1838 g/ml[66]; viscosity (30°C), 0.1013 cP[66]; dielectric constant (30°C), 29.5[66]; and cryoscopic constant, 10.1 ± 0.3 deg kg $mole^{-1}$.[93] 3-Methylsulfolane is a poorer solvent for nitronium salts than sulfolane.[93]

The 2,4-disubstituted methyl derivative of sulfolane has the following physical properties[72]: boiling point, 281°C; freezing point, −3.3°C; density, 1.14 g/ml; and viscosity (30°C), 7.9 cP. 2,4-Dimethylsulfolane is not a solvent for polyacrylonitrile.[123]

Both of the methyl-substituted sulfolanes are as suitable as solvents for potentiometric titrations as sulfolane itself.[72] This seems to be true also for conductometry.[262] Sulfolane containing about 5% 3-methylsulfolane has a freezing point near 13°C and has proved to be an useful medium for potentiometric studies.[72]

References

1. G. S. Morrow, *Encycl. Chem. Technol.* **19**, 250 (1969).
2. J. von Braun and A. Trümpler, *Chem. Ber.* **43**, 550 (1910).
3. E. Grischkevitsch-Trochimovsky, *J. Chem. Soc., London* **112**, 154 (abstr.) (1917).
4. H. Staudinger and B. Ritzenthaler, *Chem. Ber.* **68**, 455 (1935).
5. L. R. Drake, S. C. Stowe, and A. M. Partansky, *J. Amer. Chem. Soc.* **68**, 2521 (1946).
6. Y. Minoura and S. Nakajima, *J. Polym. Sci., Part A-1* **4**, 2929 (1966).
7. O. Grummit, A. E. Ardis, and J. Fick, *J. Amer. Chem. Soc.* **72**, 5167 (1950).
8. E. de Roy van Zuydewijn, *Rec. Trav. Chim. Pays-Bas* **56**, 1047 (1937).
9. E. de Roy van Zuydewijn, *Rec. Trav. Chim. Pays-Bas* **57**, 445 (1938).
10. H. J. Backer and C. C. Bolt, *Rec. Trav. Chim. Pays-Bas* **54**, 538 (1935).
11. L. Bateman and F. W. Shipley, *J. Chem. Soc., London* p. 2888 (1958).
12. T. E. Jordan and F. Kipnis, *Ind. Eng. Chem.* **41**, 2635 (1949).
13. A. V. Mashkina and Yu. A. Savostin, French Patent 2,061,890 (1971).
14. Shell Int. Res. Maatsch. N. V., Neth. Patent Appl. 6,902,819 (1969).
15. P. F. Warner, U.S. Patent 3,417,103 (1968).
16. Shell Int. Res. Maatsch. N. V., Neth. Patent Appl. 6,600,128 (1967).
17. D. W. Jones, British Patent 1,051,089 (1966).
18. A. H. Turner and T. W. Whitehead, British Patent 1,054,534 (1967).
19. Shell Int. Res. Maatsch. N. V., Neth. Patent Appl. 6,604,436 (1967).
20. A. V. Mashkina and Yu. A. Savostin, British Patent 1,260,037 (1972).
21. S. F. Birch and D. T. McAllan, *J. Chem. Soc., London* p. 2556 (1951).
22. R. W. Bost and M. W. Conn, *Ind. Eng. Chem.* **23**, 93 (1931).
23. L. Horner, H. Schaefer, and W. Ludwig, *Chem. Ber.* **91**, 75 (1958).
24. D. S. Tarbell and C. Weaver, *J. Amer. Chem. Soc.* **63**, 2939 (1941).
25. E. V. Whitehead, R. A. Dean, and F. A. Fidler, *J. Amer. Chem. Soc.* **73**, 3632 (1951).
26. N. W. Connon, *Org. Chem. Bull.* **44**, 1 (1972).
27. L. Hodossy and B. Szabady, *Magy. Kem. Lapja* **26**, 287 (1971); *Chem. Abstr.* **76**, 14224 (1972).
28. H. Szabady, L. Hodossy, B. Szeiler, and E. Haidegger, *Hung. Teljes* p. 521 (1970); *Chem. Abstr.* **74**, 64202 (1971).
29. G. Van Gooswilligen, H. Voetter, and W. Ridderikhoff, British Patent 1,134,582 (1968).
30. H. M. Van Tassel, U.S. Patent 3,396,090 (1968).
31. E. de Roy van Zuydewijn and J. Böeseken, *Rec. Trav. Chim. Pays-Bas* **53**, 673 (1934).
32. V. S. Foldi and W. Sweeny, *Makromol. Chem* **72**, 208 (1964).
33. G. H. Deal, Jr., H. D. Evans, E. D. Oliver, and M. N. Papadopoulos, *Petrol. Refiner* **38**, 185 (1959).
34. J. G. Hoggett, R. B. Moodie, and K. Schofield, *J. Chem. Soc. B p.* 1 (1969).
35. J. Lawrence and R. Parsons, *Trans. Faraday Soc.* **64**, 751 (1968).
36. R. Garnsey and J. E. Prue, *Trans. Faraday Soc.* **64**, 1206 (1968).

124. M. K. Phibbs, *J. Phys. Chem.* **59**, 346 (1955).
125. Shell Chemical Co., Information Bulletin (1962).
126. J. Husar and M. M. Kreevoy, *J. Amer. Chem. Soc.* **94**, 2902 (1972).
127. L. Jannelli, A. Inglese, A. Sacco, and P. Ciani, *Z. Naturforsch. A* **30**, 87 (1975).
128. R. W. Alder, G. R. Chalkley, and M. C. Whiting, *Chem. Commun.* p. 405 (1966).
129. E. Tommila and I. Belinskij, *Suom. Kemistilehti B* **42**, 185 (1969).
130. J. Reedijk, P. Vrijhof, and W. L. Groeneveld, *Inorg. Chim. Acta* **3**, 271 (1969).
131. J. F. Coetzee and W. R. Sharpe, *J. Phys. Chem.* **75**, 3141 (1971).
132. J. D. Halliday, R. E. Richards, and R. R. Sharp, *Proc. Roy. Soc., Ser. A* **313**, 45 (1969).
133. A. K. Covington, I. R. Lantzke, and J. M. Thain, *J. Chem. Soc., Faraday Trans. 1* **70**, 1869 (1974).
134. A. K. Covington and J. M. Thain, *J. Chem. Soc., Faraday Trans. 1* **70**, 1879 (1974).
135. V. Gutmann, *Angew. Chem.* **82**, 858 (1970).
136. R. L. Benoit and S. Y. Lam, *J. Amer. Chem. Soc.* **96**, 7385 (1974).
137. G. Nicklers, ed., "Organic Sulphur Chemistry." Elsevier, Amsterdam, 1968.
138. C. H. Langford and P. O. Langford, *Inorg. Chem.* **1**, 184 (1962).
139. V. V. Puchkova, E. N. Guryanova, R. R. Shifrina, and K. A. Kocheshkov, *Dokl. Akad. Nauk SSSR* **207**, 886 (1972); *Chem. Abstr.* **78**, 66385 (1973).
140. H. C. Miller, W. R. Hertler, E. R. Muetterties, W. H. Knoth, and N. E. Miller, *Inorg. Chem.* **4**, 1216 (1965).
141. B. L. Chamberland and E. L. Muetterties, *Inorg. Chem.* **3**, 1450 (1964).
142. D. P. Eyman and R. S. Drago, *J. Amer. Chem. Soc.* **88**, 1617 (1966).
143. A. Kivinen, J. Murto, and L. Kilpi, *Suom. Kemistilehti B* **40**, 301 (1967).
144. A. Kivinen, J. Murto, and L. Kilpi, *Suom. Kemistilehti B* **42** 19 (1969).
145. A. Kivinen, J. Murto, and M. Lehtonen, *Suom. Kemistilehti B* **41**, 359 (1968).
146. S. E. Odinokov, O. B. Maximov, and A. K. Dzizenko, *Spectrochim. Acta, Part A* **25**, 131 (1969).
147. S. E. Odinokov, A. V. Iogansen, and A. K. Dzizenko, *Zh. Prikl. Spektrosk.* **14**, 418 (1971); *Chem. Abstr.* **75**, 12716 (1971).
148. L. V. Okhlobystina, T. I. Ivanova, and Yu. M. Golub, *Izv. Akad. Nauk SSSR, Ser. Khim.* p. 2533 (1972); *Chem. Abstr.* **78**, 72318 (1973).
149. A. L. Fridman, T. N. Ivshina, V. A. Tartakovskii, and S. S. Novikov, *Izv. Akad. Nauk SSSR, Ser. Khim.* p. 2839 (1968); *Chem. Abstr.* **70**, 78115 (1969).
150. G. J. Hills, *Res. Develop. Ind.* **33**, 18 (1964).
151. J. F. Coetzee and J. M. Simon, *Anal. Chem.* **44**, 1129 (1972).
152. J. A. Starkovich, *Diss. Abstr. Int. B* **34**, 3682 (1974).
153. P. M. P. Eller, *Diss. Abstr. Int. B* **33**, 2976 (1972).
154. M. Della Monica, U. Lamanna, and L. Jannelli, *Gazz. Chim. Ital.* **97**, 367 (1967); *Chem. Abstr.* **67**, 15457 (1967).
155. M. Della Monica and U. Lamanna, *Gazz. Chim. Ital.* **98**, 256 (1968); *Chem. Abstr.* **69**, 66794 (1968).
156. R. L. Benoit and P. Pichet, *Electroanal. Chem.* **43**, 59 (1973).
157. B. G. Cox, *Annu. Rep. Progr. Chem. Sect. A. The Chem. Soc. (London)* **70A**, 249 (1973).
158. L. Jannelli, M. Della Monica, and A. Della Monica, *Gazz. Chim. Ital.* **94**, 552 (1964); *Chem. Abstr.* **61**, 15378 (1964).
159. J. F. Coetzee and R. J. Bertozzi, *Anal. Chem.* **45**, 1064 (1973).
160. J. F. Coetzee and R. J. Bertozzi, *Anal. Chem.* **41**, 860 (1969).
161. R. J. Bertozzi, *Diss. Abstr. Int. B* **33**, 3521 (1973).
162. J. Lawrence and R. Parsons, *J. Phys. Chem.* **73**, 3577 (1969).
163. E. Wellisch, E. Gipstein, and O. J. Sweeting, *Polym. Lett.* **2**, 39 (1964).

164. T. J. Wallace, J. E. Hofmann, and A. Schriesheim, *J. Amer. Chem. Soc.* **85**, 2739 (1963).
165. V. I. Dronov and A. U. Baisheva, *Khim. Seraorg. Soedin., Soderzh. Neftyakh Nefteprod.* **8**, 144 (1968); *Chem. Abstr.* **71**, 80599 (1969).
166. V. I. Dronov, A. U. Baisheva, and L. J. Samigullina, *Khim. Seraorg. Soedin., Soderzh. Neftyakh Nefteprod.* **9**, 225 (1972); *Chem. Abstr.* **79**, 126192 (1973).
167. V. D. Shatrov, L. I. Belen'kii, and I. I. Chkheidze, *Khim. Vys. Energ.* **4**, 235 (1970); *Chem Abstr.* **73**, 55389 (1970).
168. L. A. Dement'eva, A. V. Iogansen, and G. A. Kurkchi, *Opt. Spektrosk.* **29**, 868 (1970); *Chem. Abstr.* **74**, 69700 (1971).
169. V. I. Trofimov, I. I. Chkheidze, and L. I. Belen'kii, *Teor. Eksp. Khim.* **5**, 406 (1969); *Chem. Abstr.* **71**, 65982 (1969).
170. D. S. Weinberg, C. Stafford, and M. W. Scoggins, *Tetrahedron* **24**, 5409 (1968).
171. R. Smakman and T. J. De Boer, *Org. Mass. Spectrom.* **3**, 1561 (1970).
172. I. Tabushi, Y. Tamaru, and Z. Yoshida, *Tetrahedron Lett.* p. 3893 (1971).
173. I. Tabushi, Z. Yoshida, and Y. Tamaru, *Tetrahedron* **29**, 81 (1973).
174. V. I. Dronov and V. A. Snegotskaya, *Khim. Geterotsikl. Soedin.* **3**, 5 (1971); *Chem. Abstr.* **78**, 71806 (1973).
175. V. I. Dronov and V. A. Snegotskaya, *Khim. Seraorg. Soedin., Soderzh. Neftyakh Nefteprod.* **8**, 133 (1968); *Chem. Abstr.* **71**, 81066 (1969).
176. V. I. Dronov, V. A. Snegotskaya, L. P. Ivanova, D. P. Voronchikhina, and N. N. Bannikova, *Khim. Seraorg. Soedin., Soderzh. Neftyakh Nefteprod.* **9**, 218 (1972); *Chem. Abstr.* **79**, 115056 (1973).
177. V. I. Dronov and V. A. Snegotskaya, *Zh. Org. Khim.* **6**, 2029 (1970); *Chem. Abstr.* **74**, 22632 (1971).
178. A. V. Mashkina, *Khim. Seraorg. Soedin., Soderzh. Neftyakh Nefteprod.* **6**, 316 (1964); *Chem. Abstr.* **61**, 6979 (1964).
179. J. N. Gardner, S. Kaiser, A. Krubiner, and H. Lucas, *Can. J. Chem.* **51**, 1419 (1973).
180. T. E. Bezmenova and N. M. Kamakin, *Khim. Seraorg. Soedin., Soderzh. Neftyakh Nefteprod.* **8**, 140 (1968); *Chem. Abstr.* **71**, 81065 (1969).
181. D. J. Peterson, J. F. Ward, and R. A. Damico, German Patent 2,246,939 (1973).
182. A. J. Parker and R. Alexander, *J. Amer. Chem. Soc.* **90**, 3313 (1968).
183. F. Madaule-Aubry, *Bull. Soc. Chim. Fr.* p. 1456 (1966).
184. A. J. Parker, *Pure Appl. Chem.* **25**, 345 (1971).
185. H.-D. Dell, *Naturwissenschaften* **53**, 405 (1966).
186. G. W. Holbrook, L. A. Loree, and O. R. Pierce, *J. Org. Chem.* **31**, 1259 (1966).
187. W. R. Fitzgerald and D. W. Watts, *J. Amer. Chem. Soc.* **89**, 821 (1967).
188. W. R. Fitzgerald, A. J. Parker, and D. W. Watts, *J. Amer. Chem. Soc.* **90**, 5744 (1968).
189. G. A. Olah, S. J. Kuhn, and S. H. Flood, *J. Amer. Chem. Soc.* **83**, 4571 (1961).
190. G. A. Olah, S. J. Kuhn, and S. H. Flood, *J. Amer. Chem. Soc.* **83**, 4581 (1961).
191. G. A. Olah and S. J. Kuhn, *J. Amer. Chem. Soc.* **86**, 1067 (1964).
192. P. F. Christy, J. H. Ridd, and N. D. Stears, *J. Chem. Soc., B* p. 797 (1970).
193. C. D. Ritchie and H. Win, *J. Org. Chem.* **29**, 3093 (1964).
194. R. O. Hutchins, D. Hoke, J. Keogh, and D. Koharski, *Tetrahedron Lett.* **40**, 3495 (1969).
195. R. O. Hutchins, D. W. Lamson, L. Rua, C. Milewski, and B. Maryanoff, *J. Org. Chem.* **36**, 803 (1971).
196. J. H. Parish and M. C. Whiting, *J. Chem. Soc., London* p. 4713 (1964).
197. R. L. Benoit, A. L. Beauchamp, and R. Domain, *Inorg. Nucl. Chem. Lett.* **7**, 557 (1971).
198. R. L. Benoit, *Inorg. Nucl. Chem. Lett.* **4**, 723 (1968).
199. R. L. Benoit and C. Louis, *Inorg. Nucl. Chem. Lett.* **6**, 817 (1970).
200. M. Deneux and R. L. Benoit, *Can. J. Chem.* **48**, 674 (1970).

201. R. L. Benoit, M. Guay, and J. Desbarres, *Can. J. Chem.* **46**, 1261 (1968).
202. R. L. Benoit and M. Guay, *Inorg. Nucl. Chem. Lett.* **4**, 215 (1968).
203. J. Martinmaa and P. Törmälä, *Suom. Kemistilehti B* **43**, 378 (1970).
204. J. Martinmaa and P. Patrakka, *Makromol. Chem.* **175**, 3275 (1974).
205. J. Martinmaa, M. Luoto, B. Malm, and J. J. Lindberg, *Suom. Kemistilehti B* **46**, 83 (1973).
206. M. Baudler and D. Schellenberg, *Z. Anorg. Allg. Chem.* **356**, 140 (1968).
207. J. A. Jackson, *J. Polym. Sci., Chem. Ed.* **10**, 2935 (1972).
208. F. B. Jones, U.S. Patent 3,476,709 (1969).
209. E. W. Neuse and K. Koda, U.S. Patent 3,437,644 (1969).
210. H. K. Hall, Jr. and J. W. Rhoades, *J. Polym. Sci., Part A-1* **10**, 1953 (1972).
211. J. R. Elliott and G. D. Cooper, French Patent 1,354,443 (1964).
212. Yawamata Chem. Ind., British Patent 1,134,934 (1968).
213. C. F. Woffard, U.S. Patent 3,484,388 (1969).
214. H. S. Klein, U.S. Patent 3,354,236 (1967).
215. R. E. Stolton, British Patent 1,061,699 (1967).
216. C. Podesva, Canadian Patent 748,888 (1966).
217. H. E. Harris and H. L. Herzog, U.S. Patent 3,259,646 (1966).
218. R. P. Williams, U.S. Patent 3,689,549 (1972).
219. S. D. Turk and R. P. Williams, U.S. Patent 3,644,470 (1972).
220. A. Matawowski, *Ann. Soc. Chim. Pol.* **49**, 37 (1975).
221. J. Jones and J. Jones, *Tetrahedron Lett.* **31**, 2117 (1964).
222. R. W. Alder, G. R. Chalkley, and M. C. Whiting, *J. Chem. Soc., D* p. 52 (1966).
223. B. D. Coleman and R. M. Fuoss, *J. Amer. Chem. Soc.* **77**, 5472 (1955).
224. C. L. Arcus and W. A. Hall, *J. Chem. Soc., London* p. 5995 (1964).
225. H. F. Smyth, Jr., C. P. Carpenter, C. S. Weil, U. C. Pozzani, J. A. Striegel, and J. S. Nycum, *Amer. Ind. Hyg. Ass., J.* **30**, 470 (1969).
226. Z. Kh. Filippova, *Khim. Seraorg. Soedin., Soderzh. Neftyakh Nefteprod.* **8**, 701 (1968); *Chem. Abstr.* **71**, 111000 (1969).
227. V. K. H. Brown, L. W. Ferrigan, and D. E. Stevenson, *Brit. J. Ind. Med.* **23**, 302 (1966).
228. V. A. Trofimov, F. G. Murzakaev, and Z. V. Latypova, *Tr. Ufim. Nauch.-Issled. Inst. Gig. Prof. Zabol.* p. 61 (1971); *Chem. Abstr.* **78**, 88390 (1973).
229. B. Spilker, *J. Phamacol. Exp. Ther.* **175**, 361 (1970).
230. T. Y. Koh, *Can. J. Biochem.* **47**, 951 (1969).
231. A. Van der Viel, *Nature (London)* **187**, 142 (1960).
232. C. Hanson, A. N. Patel, and D. K. Chang-Kakoti, *J. Appl. Chem.* **18**, 89 (1968).
233. C. Hanson, A. N. Patel, and D. K. Chang-Kakoti, *J. Appl. Chem.* **19**, 320 (1969).
234. C. Hanson, A. N. Patel, and D. K. Chang-Kakoti, *J. Appl. Chem.* **20**, 42 (1970).
235. W. M. Hutchinson, U.S. Patent 3,488,920 (1970).
236. J. Becuwe, German Patent 2,404,131 (1974).
237. J. H. Parish, J. R. B. Hastings, and K. S. Kirby, *Biochem. J.* **99**, 19P (1966).
238. R. L. Benoit and G. Choux, *Can. J. Chem.* **46**, 3215 (1968).
239. A. Sacco and A. K. Rakshit, *J. Chem. Thermodyn.* **7**, 257 (1975).
240. M. Castagnolo, L. Jannelli, G. Petrella, and A. Sacco, *Z. Naturforsch. A* **26**, 755 (1971).
241. C. H. Langford and R. L. Burwell, *J. Amer. Chem. Soc.* **82**, 1503 (1960).
242. R. Stewart and J. P. O'Donnell, *J. Amer. Chem. Soc.* **84**, 493 (1962).
243. K. Bowden, A. Buckley, and R. Stewart, *J. Amer. Chem. Soc.* **88**, 947 (1966).
244. E. Tommila and J. Martinmaa, *Suom. Kemistilehti B* **40**, 216 (1967).
245. D. C. Berndt. *J. Org. Chem.* **39**, 840 (1974).
246. P. Bruno and M. Della Monica, *Chim. Ind. (Milan)* **54**, 878 (1972).
247. A. Sacco and L. Jannelli, *J. Chem. Thermodyn.* **4**, 191 (1972).

248. J. W. Powell and M. C. Whiting, *Proc. Chem. Soc., London* p. 412 (1960).
249. J. Dojcansky and J. Surovy, *Chem. Zvesti* **23**, 254 (1969); *Chem. Abstr.* **71**, 116984 (1969).
250. D. Grigoriu and M. Bogdan, *Rev. Chim. (Bucharest)* **21**, 627 (1970); *Chem. Abstr.* **74**, 57948 (1971).
251. H. Flaschka and J. J. Tice, IV, *Talanta* **20**, 423 (1973).
252. H. Flaschka and R. Barnes, *Anal. Chim. Acta* **63**, 489 (1973).
253. D. R. Seibel and F. P. McCandless, *Ind. Eng. Chem., Process Des. Develop.* **13**, 76 (1974).
254. W. E. Garrison and T. J. Hyde, U.S. Patent 3,361,697 (1968).
255. W. E. Skiens, German Patent 1,923,187 (1970).
256. Gevaert-Agfa N.V., Neth. Patent Appl., 6,516,423 (1966).
257. K. J. Jacobson, British Patent 1,038,029 (1966).
258. J. Patschorke, German Patent 2,022,109 (1971).
259. L. P. Clermont, *Tappi* **53**, 2243 (1970).
260. L. P. Clermont, Canadian Patent 901,759 (1972).
261. E. L. Springer and L. L. Zoch, *Sv. Papperstidn.* **69**, 513 (1966).
262. J. Eliassaf, R. M. Fuoss, and J. E. Lind, Jr., *J. Phys. Chem.* **67**, 1724 (1963).

Author Index

Numbers in parentheses are reference numbers and indicate that an author's work is referred to although his name is not cited in the text. Numbers in italics show the page on which the complete reference is listed.

C

E

F

G

I

J

L

Q

R

V

W

Subject Index

A 6
B 7
C 8
D 9
E 0
F 1
G 2
H 3
I 4
J 5